Phytoremediation
of
CONTAMINATED SOIL and WATER

Phytoremediation

of

CONTAMINATED SOIL and WATER

Edited by

Norman Terry
Gary Bañuelos

CRC Press
Taylor & Francis Group
Boca Raton London New York

CRC Press is an imprint of the
Taylor & Francis Group, an **informa** business

CRC Press
Taylor & Francis Group
6000 Broken Sound Parkway NW, Suite 300
Boca Raton, FL 33487-2742

First issued in paperback 2019

ISBN-13: 978-1-56670-450-2 (hbk)
ISBN-13: 978-0-367-39943-6 (pbk)

Library of Congress Cataloging-in-Publication Data

Phytoremediation of contaminated soil and water / edited by Norman Terry, Gary Bañuelos.
    p. cm.
    Includes bibliographical references and index.
    ISBN 1-56670-450-2 (alk. paper)
    1. Phytoremediation. 2 Soil remediation 3. Water-Purification. I. Terry, Norman.
II Bañuelos, Gary Stephen, 1956-.
    TD192.75.P478 1999
    628.5—dc21
                                       99-30741
                                       CIP

Library of Congress Card Number 99-30741

Visit the Taylor & Francis Web site at
http://www.taylorandfrancis.com

and the CRC Press Web site at
http://www.crcpress.com

Preface

The need to synthesize, critically analyze, and put into perspective the ever-mounting body of new information on phytoremediation in the soil and water environment provided the impetus for the development of this book. It is a compilation of articles provided by speakers at a symposium entitled "Phytoremediation of Trace Elements in Contaminated Soil and Water" that was held in June 1997 as part of the Fourth International Conference on the Biogeochemistry of Trace Elements on the Clark Kerr campus of the University of California, Berkeley. Also included in the book are invited articles on special topics such as the phytoremediation of constructed wetlands and the role of microphytes.

Twenty eminent scientists from around the world spoke at the symposium on topics such as field demonstrations of phytoremediation in trace element cleanup; the role of hyperaccumulator plants in phytoextraction; the genetics, molecular biology, physiology, and ecology of trace element hyperaccumulation and tolerance; phytovolatilization of mercury and selenium in phytoremediation; the role of microbes; and the phytostabilization and immobilization of metals in contaminated soil. We are especially indebted to Dr. Jaco Vangronsveld who helped coordinate the symposium and who was instrumental in developing the list of excellent speakers from Europe. The papers represent the latest research in all of the major aspects of phytoremediation of trace elements in contaminated soil and water.

All of the articles in the book were peer reviewed. We gratefully acknowledge the following reviewers: Husein Ajwa, Robert Brooks, Carolee Bull, Stanley Dudka, Steve Grattan, Satish Gupta, Seongbin Hwang, Elizabeth Pilon-Smits, Mark de Souza, Lin Wu, Jaco Vangronsveld, and Adel Zayed. We also would like to thank the organizers of the conference and especially Drs. I. K. Iskandar and Domy Adriano who had the vision and foresight to develop the idea of having a special symposium on phytoremediation.

A substantial portion of the funds used to support travel and other expenses of symposium participants and to develop this book was provided by the Kearney Foundation of Soil Science. The Foundation's mission in the 1990s has been to research the reactions of toxic pollutants in soil systems. We hope this book will benefit government agencies charged with the cleanup of California's soil and water and for developing policy in this regard. We also acknowledge the generous financial support from other agencies, including the International Lead Zinc Research Organization, Inc., Chevron Research and Technology Company, Phytotech, Inc., and E. I. DuPont DeNemours and Company.

Norman Terry
Gary Bañuelos

Editors

Norman Terry is Professor of Environmental Plant Biology in the Department of Plant and Microbial Biology, and Researcher in the Agricultural Experiment Station at the University of California, Berkeley. Terry received his Ph.D. in Plant Physiology at the University of Nottingham, England, and was awarded a NRC (Canada) Post-doctoral Fellowship to carry out research on phloem translocation (Ottawa, 1966–1968). He joined the Berkeley faculty in 1972 and currently teaches advanced undergraduate courses on plant physiology, biochemistry, and environmental plant biology. During his research career, Terry authored over 120 scientific articles. His early research was on the regulation of photosynthesis *in vivo*, the environmental control of plant growth, mineral nutrition, and salinity.

In 1990, Terry's research interests shifted to phytoremediation. He developed a research program that is a multidisciplinary blend of environmental engineering, microbiology, plant biochemistry, and molecular biology. This approach is unique in phytoremediation research and has facilitated several innovative and creative solutions to environmental problems. He pioneered the use of constructed wetlands for the cleanup of selenium and other toxic elements from oil refinery effluents and agricultural irrigation drainage water. Using cutting edge molecular approaches, Terry developed transgenic plants with superior capacities for the phytoremediation of selenium and heavy metals (e.g., cadmium). And, by using sophisticated high energy x-ray absorption spectroscopy to monitor element speciation changes, he successfully demonstrated that plants have the ability to detoxify metals (e.g., chromium).

Gary S. Bañuelos is a plant/soil scientist at the USDA/ARS' Water Management Research Laboratory in Fresno, CA and an adjunct professor at California State University. Focusing his research activities on the phytoremediation of soil and water contaminated with selenium, boron, and salinity, Dr. Bañuelos is the principal author of over 60 refereed technical articles and a member of the American Chemical Society, American Society of Agronomy, and the International Soil Science Society, among others.

He received his German proficiency degree in 1977 from Middlebury College in Vermont, a B.A. degree in German from Humboldt State University in California (1979) and a German language certification at Goethe Institute in Germany in 1979. In 1984, he received a B.S. degree in crop science and Master's in agriculture from CalPoly Technical University, and in 1987 he was a National Science Foundation Fellow at Hohenheim University in Germany, where he acquired a Ph.D. in plant nutrition/agriculture.

Contributors

J. Scott Angle
University of Maryland
Dept. of Natural Resources and
  Landscape Architecture
College Park, MD 20742

Alan J. M. Baker
Dept. of Animal and Plant Sciences
Environmental Consultancy (ECUS)
University of Sheffield
Sheffield S10 2TN U.K.

Gary S. Bañuelos
USDA–ARS
Water Management Research
  Laboratory
2021 S. Peach Ave.
Fresno, CA 93727

Roland Bernhard
Dept. of Ecology and Ecotoxicology
Faculty of Biology
Vrije Universiteit
De Boelelaan 1087
Amsterdam, The Netherlands

William R. Berti
Environmental Biotechnology Program
DuPont Central Research and
  Development
Glasgow Business Community 301
Newark, DE 19714-6101

Michael J. Blaylock
Phytotech, Inc.
1 Deer Park Drive, #1
Monmouth Junction, NJ 08852

Sally L. Brown
University of Maryland
Dept. of Natural Resources and
  Landscape Architecture
College Park, MD 20742

Rufus L. Chaney
USDA–ARS
Environmental Chemical Laboratory
Beltsville, MD 20705

Mel Chin
Artist
New York, NY

H. Clijsters
Liverpool John Moores University
School of Biological and Earth Sciences
Byrom Street
Liverpool L3 3AF U.K.

P. Corbisier
Environmental Technology
Vlaamse Instelling voor Technologisch
  Onderzoek
VITO, Boeretang 200
B-2400 Mol, Belgium

R. L. Correll
CSIRO Mathematical and Information
  Sciences
PMB2 Glen Osmond
Adelaide 5064
Australia

Scott D. Cunningham
DuPont Company
Centre Road
Wilmington, DE 19805-0708

Keri L. Dandridge
Dept. of Biology
Furman University
Greenville, SC 29613

Mark deSouza
Dept. of Plant and Microbial Biology
University of California
Berkeley, CA 94720

L. Diels
Environmental Technology
Vlaamse Instelling voor Technologisch
  Onderzoek
VITO, Boeretang 200
B-2400 Mol, Belgium

S. J. Dunham
Soil Science Dept.
IACR-Rothamsted, Harpenden
Herts. AL5 2JQ U.K.

R. Edwards
Liverpool John Moores University
School of Pharmacy and Chemistry
Byrom Street
Liverpool L3 3AF U.K.

Teresa W.-M. Fan
Dept. of Land, Air, and Water Resources
University of California
Davis, CA 95616-8627

A. Gilis
Environmental Technology
Vlaamse Instelling voor Technologisch
  Onderzoek
VITO, Boeretang 200
B-2400 Mol, Belgium

Peter Goldsbrough
Dept. of Horticulture and Landscape
  Architecture
Purdue University
West Lafayette, IN 47907-1165

G. Gragson
Dept. of Genetics
Life Sciences Building
University of Georgia
Athens, GA 30602

S. K. Gupta
Institute for Environmental Protection
  and Agriculture
Swiss Federal Research Station for
  Agroecology and Agriculture
Schwarzenburgstrasse 155
CH-3003 Bern, Switzerland

T. Hari
Institute for Environmental Protection
  and Agriculture
Swiss Federal Research Station for
  Agroecology and Agriculture
Schwarzenburgstrasse 155
CH-3003 Bern, Switzerland

T. Herren
Division of Radiation Protection and
  Waste Management
Paul Scherrer Institut
CH-5232 Villigen-PSI
Switzerland

Richard M. Higashi
Crocker Nuclear Laboratory
University of California
Davis, CA 95616-8627

Faye A. Homer
University of Maryland
Dept. of Natural Resources and
  Landscape Architecture
College Park, MD 20742

Alex J. Horne
Ecological Engineering Group
Dept. of Civil and Environmental
  Engineering
University of California
Berkeley, CA 94720-1710

Edward M. Jhee
Dept. of Biology
Furman University
Greenville, SC 29613

M. K. Kandasamy
Dept. of Genetics
Life Sciences Building
University of Georgia
Athens, GA 30602

N. Kato
AgBiotech Center
Rutgers University
Cook College
New Brunswick, NJ 08903-0231

Leon V. Kochian
Plant, Soil, and Nutrition Laboratory
USDA–ARS
Cornell University
Ithaca, NY 14853

U. Krämer
Fakultät für Biologie-W5
Universität Bielefeld
Bielefeld, Germany

R. Krebs
AMT für Umweltschutz SG
Linsebühlstrasse 91
St. Gallen, Switzerland

Mitch M. Lasat
Plant, Soil, and Nutrition Laboratory
USDA–ARS
Cornell University
Ithaca, NY 14853

N. W. Lepp
Liverpool John Moores University
School of Biological and Earth Sciences
Byrom Street
Liverpool L3 3AF U.K.

Yin-Ming Li
USDA–ARS
Environmental Chemical Laboratory
Beltsville, MD 20705

Zhi-Qing Lin
Dept. of Plant and Microbial Biology
University of California
Berkeley, CA 94720

Mercè Llugany
Dept. of Ecology and Ecotoxicology
Faculty of Biology
Vrije Universiteit
De Boelelaan 1087
Amsterdam, The Netherlands

C. Lodewyckx
Environmental Technology
Vlaamse Instelling voor Technologisch
  Onderzoek
VITO, Boeretang 200
B-2400 Mol, Belgium

Mark R. Macnair
Dept. of Biological Sciences
University of Exeter
Prince of Wales Road
Exeter EX4 4PS U.K.

Minnie Malik
University of Maryland
Dept. of Natural Resources and
  Landscape Architecture
College Park, MD 20742

S. P. McGrath
Soil Science Dept.
IACR-Rothamsted, Harpenden
Herts. AL5 2JQ U.K.

R. B. Meagher
Dept. of Genetics
Life Sciences Building
University of Georgia
Athens, GA 30602

M. Mench
INRA Agronomy Unit
Bordeaux-Aquitaine Research Centre
BP 81
F-33883 Villenave d'Ornon cedex
France

M. Mergeay
Environmental Technology
Vlaamse Instelling voor Technologisch
  Onderzoek
VITO, Boeretang 200
B-2400 Mol, Belgium

Elizabeth Pilon-Smits
Dept. of Plant and Microbial Biology
University of California
Berkeley, CA 94720

A. Joseph Pollard
Dept. of Biology
Furman University
Greenville, SC 29613

I. Raskin
AgBiotech Center
Rutgers University
Cook College
New Brunswick, NJ 08903-0231

Roger D. Reeves
Dept. of Chemistry
Massey University
Palmerston North, New Zealand

C. L. Rugh
Dept. of Genetics
Life Sciences Building
University of Georgia
Athens, GA 30602

David E. Salt
Chemistry Department
Northern Arizona University
Flagstaff, AZ 86011-5698

Henk Schat
Dept. of Ecology and Ecotoxicology
Faculty of Biology
Vrije Universiteit
De Boelelaan 1087
Amsterdam, The Netherlands

J. A. C. Smith
Dept. of Plant Sciences
University of Oxford
South Parks Road
Oxford OX1 3RB U.K.

R. D. Smith
De Kalb Genetics Corp.
Mystic, CN 06355

Susanne E. Smith
Dept. of Biological Sciences
University of Exeter
Prince of Wales Road
Exeter EX4 4PS U.K.

N. Spelmans
Limburgs Universitair Centrum
Environmental Biology
Universitaire Campus
B3590 Diepenbeek, Belgium

S. Taghavi
Environmental Technology
Vlaamse Instelling voor Technologisch
  Onderzoek
VITO, Boeretang 200
B-2400 Mol, Belgium

Norman Terry
Dept. of Plant and Microbial Biology
University of California
Berkeley, CA 94720

Gavin H. Tilstone
Grupo de Oceanoloxia
Instituto de Investigacions Marinas-
  CSIC
Eduardo Cabello, 6
36208 Vigo, Spain

D. van der Lelie
Environmental Technology
Flemish Institute of Technological
  Research (VITO)
Boeretang 200
B-2400 Mol, Belgium

J. Vangronsveld
Limburgs Universitair Centrum
Environmental Biology
Universitaire Campus
B3590 Diepenbeek, Belgium

N. J. Wang
Dept. of Genetics
Life Sciences Building
University of Georgia
Athens, GA 30602

K. Wenger
Institute for Environmental Protection
  and Agriculture
Swiss Federal Research Station for
  Agroecology and Agriculture
Schwarzenburgstrasse 155
CH-3003 Bern, Switzerland

Adel Zayed
Dept. of Plant and Microbial Biology
University of California
Berkeley, CA 94720

Table of Contents

1 Field Demonstrations of Phytoremediation of Lead-Contaminated Soils

Michael J. Blaylock

CONTENTS

SUMMARY

Phytoremediation is a new technology that uses specially selected metal-accumulating plants to remediate soil contaminated with heavy metals and radionuclides. Phytoremediation offers an attractive and economical alternative to currently practiced soil removal and burial methods. The integration of specially selected metal-accumulating crop plants (e.g., *Brassica juncea*) with innovative soil amendments allows plants to achieve high biomass and metal accumulation rates from soils.

INTRODUCTION

The use of plants to remove toxic metals from soils (phytoremediation) is being developed as a method for cost-effective and environmentally sound remediation of contaminated soils (Baker et al., 1994; Chaney, 1983; Raskin et al., 1994). Metal (hyper)accumulating plants have been sought that have the ability to accumulate and tolerate unusually high concentrations of heavy metals in their tissue. Accumulators of nickel (Ni) and zinc (Zn), for example, may contain as much as 5% of these metals on a dry-weight basis (Baker et al., 1994, Brown et al., 1995). This process of extracting metals from the soil and accumulating and concentrating metals in the above-ground plant tissues enables plants to be used as part of a soil cleanup technology. For example, plants accumulating metals at the above-mentioned 5% (50,000 mg/kg) dry-weight concentration from a soil with a total metal concentration of 5000 mg/kg results in a 10-fold bioaccumulation factor. The metal-rich plant material can be swathed, collected, and removed from the site using established agricultural practices, without the extensive excavation and loss of topsoil associated with traditional remediation practices. Post-harvest biomass treatments (i.e., composting, compaction, thermal treatments) may also be employed to reduce the volume and/or weight of biomass for disposal. The metal bioaccumulation of the plant shoots above that of the soil concentration coupled with subsequent biomass reduction processes can greatly reduce the amount of contaminated material requiring disposal compared to soil excavation, thereby decreasing the remediation costs.

Successful implementation of phytoremediation in the field depends on a significant quantity of metal being removed from the soil through plant uptake to effectively decrease the soil metal concentration. Several conditions must be met in order for phytoremediation to be effective. The availability of metals in the soil for root uptake is the first critical factor for metal uptake. Soils containing metal contaminants that cannot be solubilized or made available for plant uptake will limit the uptake and therefore the success of phytoremediation.

Metal solubility is dependent on a number of soil characteristics and is strongly influenced by soil pH (Harter, 1983) and complexation with soluble ligands (Norvell, 1984). Chelating agents have been used extensively in the laboratory as extractants to estimate metal availability (Martens and Lindsay, 1990) and also to supply micronutrients in fertilizers. Numerous studies have been conducted to evaluate the effectiveness of soil-applied chelating agents to increase micronutrient availability to crop plants (Wallace, 1983 and references contained therein; Muchovej et al., 1986; Norvell, 1991; Sadiq and Hussain, 1993). The addition of synthetic chelated metals (predominantly polyaminopolycarboxylic acids) to the soil has generally been effective in diminishing micronutrient deficiencies in plants. The effectiveness of the chelate varies depending on soil conditions and the specific micronutrient of interest (Wallace and Wallace, 1983). Although the major portion of the chelate literature addresses amelioration of Fe deficiency, increases in heavy metal uptake have also been demonstrated. Wallace (1977) showed a yield reduction in bush bean (*Phaseolus vulgaris*) coupled with an increase in leaf cadmium concentrations (from 6.7 to 423 μg/g) through the soil application of 100 μg/g of EDTA (ethylenediamine tetraacetic acid) to soils spiked with 100 μg Cd/kg. A much smaller increase in Cd leaf concentrations (from 6.7 to 12.8 µg/g) was observed with a similar treatment of NTA (nitrilotriacetic acid) instead of EDTA. Patel et al. (1977) showed an increase in Pb uptake in bush beans and barley with 100 µg/g additions of DTPA (diethyle-netrinitrilopentaacetic acid) to soil spiked with Pb. The levels of uptake observed (477 µg Pb/g), however, were much less than those required for effective phytore-mediation (>1000 mg/kg).

Blaylock et al. (1997) and Huang et al. (1997) have recently shown the effec-tiveness of applying synthetic chelates to the soil to increase lead solubility and plant uptake as part of the phytoremediation process. In their studies, the application of EDTA and other chelates (DTPA; CDTA, *trans*-1,2-cyclohexylenedinitrilotetraac-etic acid; EGTA, ethylenebis[oxyethylenetrinitrilo] tetraacetic acid; and citric acid) to the soil resulted in enhanced shoot lead concentrations. Concentrations greater than 10,000 mg/kg were achieved with EDTA, DTPA, and CDTA while maintaining biomass production. Key factors involved in the increase of lead uptake were soil pH, chelate concentration, and the total soil lead concentration, as well as water soluble lead concentrations. Plants grown in soils at pH 5 and amended with EDTA accumulated nearly 2000 mg/kg more lead in their shoots than corresponding treat-ments in soil limed to pH 7.5. Shoot lead concentrations also dramatically increased as the total soil lead concentration increased from 150 to 300 mg/kg (Blaylock et al., 1997). Only when the added chelate (EDTA, DTPA, or CDTA) concentration exceeded 1 mmol/kg was substantial lead accumulation (>5000 mg/kg) in the shoots observed. The effectiveness of the chelator can be partially attributed to an increase in lead solubility in the soil coupled with an enhancement of the transport of lead from roots to shoots. EDTA was more effective than DTPA in increasing Pb uptake in the shoots, however, even when both produced equivalent concentrations of water soluble Pb in the soil. Huang et al. (1997) showed that Pb uptake varied with plant species as well as soluble lead concentrations. Lead concentrations in pea (*Pisum sativum* L. cv. Sparkle) were much greater (11,000 mg/kg) than corn (3500 mg/kg) receiving equivalent EDTA applications. In their studies, EDTA was substantially more effective than the other chelates tested at increasing Pb solubility in the soil solution and increasing Pb concentrations in the plant shoots. A correlation value (r^2) of 0.96 was obtained when comparing shoot Pb concentrations in corn to soil solution Pb in soils treated with chelates. From the data of Huang et al. (1997), the soil solution Pb concentration must be greater than 2000 mg/l to achieve substantial shoot Pb concentrations (>5000 mg/kg) in corn.

The plants selected for phytoremediation must also be responsive to agricultural practices and produce sufficient biomass coupled with high rates of metal uptake. The plant must also be adapted to the wide variety of environmental conditions that exist in contaminated soils and waste sites. One crop plant that produces high rates of biomass under field conditions and also has the capacity to accumulate substantial metal concentrations in its shoots is *B. juncea* or Indian mustard (Kumar et al., 1995; Blaylock et al., 1997), which has also been used successfully to decrease the sele-nium content of soils in central California (Bañuelos et al., 1993).

The application of phytoremediation in the field requires the integration of a variety of skills and techniques. The appropriate plant for the field conditions must be combined with agricultural techniques that support the application of soil amendments to enhance plant availability of the metal contaminants in order to achieve a successful remediation program.

Two field demonstrations of phytoremediation were recently conducted at sites in the U.S. to demonstrate the technical feasibility of phytoremediation for remediating lead-contaminated soils. At both sites, total soil lead levels were significantly reduced during a single growing season. This chapter will detail the results of these two studies. A brief description of each site is below.

Bayonne, NJ

The first site is an industrial site in Bayonne, NJ contaminated with various heavy metals, but predominantly high levels of total lead. Due to the shallow water table and potential site flooding, an elevated, plastic-lined lysimeter of approximately 1000 sq. ft in area and 3.5 ft deep was constructed and filled with lead-contaminated soil from the site for the purposes of the field trial. A sump was created at one end of the lysimeter to collect any excess drainage water. The source of metal contamination at this site has been attributed to cable manufacturing operations.

Dorchester, MA

The second site is located in a heavily populated, urban residential area in Dorchester, MA. The site is a backyard to young children who have been treated twice for lead poisoning. A 1081 sq. ft area was selected for the field trial. The source of lead at the site is unknown but is believed to be from paint and aerial deposition. The plot has been used as a home garden for a number of years.

METHODS

Treatability Study

A preliminary site investigation was conducted for each site prior to the field studies to determine the distribution of lead in the soil and to collect bulk surface (0 to 15 cm depth) samples for a laboratory treatability study. The treatability study was conducted to assess the potential of phytoremediation to reduce the lead concentration of the soil. The study determines the forms and concentration of lead in the soil and evaluates plant growth and metal uptake from the soil under greenhouse conditions. The bulk soil samples were sieved to 2 mm and a subsample was submitted to the Rutgers University Soil Testing Laboratory for a standard soil fertility analysis. An additional sample was analyzed for total metals by EPA Method 3050 and also extracted sequentially (Ramos et al., 1994) to assess metal associations with operationally defined soil fractions (i.e., exchangeable, carbonates, oxides, organic matter, and residual). The remaining soil from the treatability sample was fertilized with urea (150 mg N/kg), potassium chloride (83 mg K/kg), and gypsum (70 mg $CaSO_4$/kg). The soil was then placed in 8.75-cm diameter pots (350 g soil/pot) and seeded with *B. juncea*. Phosphate fertilizer was added as a spot placement of triple super phosphate 1 cm below the seeds at planting at the rate of 44 mg P/kg. After seedling emergence, the pots were thinned to two plants per pot.

The plants were grown for 3 weeks in a growth chamber using a 16-h photoperiod and weekly fertilization treatments of 16 and 7 mg/kg N (urea) and K_2O (KCl), respectively. The potassium salt of EDTA (ethylenedinitrilo tetraacetic acid) was applied to the soil surface as a solution to equal 5 mmol EDTA/kg soil 3 weeks after seedling emergence using 4 replications of each treatment. The pots were placed in individual trays to prevent loss of amendments from leaching. The plants were harvested 1 week after the amendment treatment by cutting the stem 1 cm above the soil surface. The plant tissue was dried at 70°C and then wet ashed using nitric and perchloric acids. The resulting solution was analyzed for metal content by inductively coupled plasma spectrometry (ICP; Fisons Accuris, Fisons Instruments, Inc., Beverly, MA).

FIELD PLOTS

Initial Sampling

Based on the results of the treatability study, a field trial was planned and conducted at each site. An initial sampling of the site to obtain baseline soil data was conducted by sampling on a 3 m (10 ft) grid at three depths (0 to 15, 15 to 30, and 30 to 45 cm). The soil samples were collected using a hand-operated, 5 cm diameter, stainless-steel bucket auger. Duplicate samples were collected from 20% of the soil cores. The extracted soil core was mixed in a polyethylene bucket and transferred to a polyethylene bag. Soil samples were collected again at the end of the growing season on the same grid as the initial sampling to determine metal removal efficiency and to monitor changes in Pb concentration in the surface (0 to 15 cm) and subsurface soil (15 to 45 cm).

Site Preparation and Cultivation

The sites were fertilized according to the soil fertility test results and roto-tilled to a depth of 10 to 15 cm before seeding with B. juncea (cv. 426308). Tensiometers were installed at two depths (30 and 45 cm) to monitor soil water content. Irrigation was conducted using overhead impact sprinklers. Soil amendments containing EDTA were applied at a rate of 2 mmol/kg through the irrigation system to enhance metal uptake. The crop of B. juncea was harvested after 6 weeks of growth. Plant samples were collected randomly from 1 m^2 blocks for metal analysis, rinsed with water, and placed in paper bags for drying. The remaining biomass was harvested by mowing and removed from the plot for appropriate disposal. Roots were not collected and were left in the soil to decompose. After harvest, the plot was roto-tilled to a 10-cm depth and replanted within 1 week. A total of three crops were grown and harvested at each site during 1996.

SOIL ANALYSIS

The soil samples were air dried and sieved to 2 mm before analysis. Soil aggregates were crushed to pass through the sieve and the remaining rocks and debris were discarded. The sieved soil samples were extracted for total metals using a modification of EPA SW-846 Method 3050 (U.S. EPA, 1983). The supernatant solution was analyzed for lead. Sequential extraction and fractionation of the soil lead was conducted according to the procedure of Ramos et al. (1994). Lead and total metal content of the soil extracts was determined using ICP by EPA SW-846 Method 6010 (U.S. EPA, 1983). Soil pH was measured in a 1:1 soil:water suspension. Duplicates and spikes were carried through the procedure in combination with National Institutes of Standards and Technology (NIST) Standard Reference Material 2711 to ensure the quality of the data. Contour maps of lead contamination at the site were plotted and areas corresponding to specific levels of metal concentration were calculated using Surfer 6.04 (1996).

Plant Tissue Analysis

Plant tissue samples were dried in a forced-air oven at 60°C, ground to 20 mesh using a stainless steel Wiley Mill, and digested using nitric and perchloric acids. The sample was diluted to 25 mL and analyzed for total metals by ICP using EPA SW-846 Method 6010 (U.S. EPA, 1983). Appropriate duplicates and spikes were carried through the digestion procedure as well as the NIST Peach Leaf Standard (SRM 1547) as part of the Quality Assurance/Quality Control (QA/QC) plan.

RESULTS AND DISCUSSION

Treatability Studies

Bayonne

The soil at the Bayonne site was an alkaline (pH 7.9) sandy loam soil with 2.5% organic matter. Slightly elevated Cu and Zn concentrations were present in the soil, although they did not exceed regulatory limits. Soil characteristics of the bulk sample collected for the treatability studies are presented in Table 1.1. The sequential extraction of the soil sample from the Bayonne site used for the treatability studies showed the soil lead to be predominantly associated with the carbonate fraction (66% of the total lead), with only 211 mg/kg of the 1608 mg/kg total lead associated

---

TABLE 1.1

Soil Characteristics and Total Metal Content of a Surface Soil (0 to 15 cm) Sample Collected at Each Site for the Treatability Study

Site	pH	Texture	Organic Matter %	Cd	Cr	Cu	Ni	Pb	Zn
						(mg/kg)			
Dorchester	6.1	Sandy loam	9.0	5	21	32	13	735	101
Bayonne	7.9	Sandy loam	2.5	8	33	139	19	1438	454

with the residual fraction (Table 1.2). Assuming that most of the lead associated with the exchangeable, carbonate, oxide, and organic fractions can be made plant available through soil amendments, enough available lead existed for plant uptake and removal to reduce the soil concentration to below the 400 mg/kg target level.

TABLE 1.2

Fractionation of Metal Contaminants Based on the Sequential Extraction of a Surface (0 to 15 cm Depth) Soil Sample Collected for the Treatability Study

Fraction	Dorchester, MA	Bayonne, NJ
	(mg/kg)	
Exchangeable	100	34
Carbonates	126	1064
Oxide	75	130
Organic	137	170
Residual	125	211
Sum of Fractions	563	1608

Dorchester

The soil at the Dorchester site is a sandy loam containing 9% organic matter in the surface horizon (0 to 15 cm). Soil characteristics of the bulk sample collected for the treatability studies are presented in Table 1.1. The sequential extraction of the bulk soil sample used for the treatability study showed the soil lead to be fairly evenly distributed between all fractions with the organic fraction containing the highest proportion of the total lead (24%; Table 1.2). Similar to the soil from the Bayonne plot, the lead concentration of the residual fraction (125 mg/kg) was much less than the 400 mg/kg target, indicating a suitable quantity of lead in the available/semi-available fractions (exchangeable, carbonate, oxide, and organic) to allow phytoremediation to be successful.

The greenhouse treatability studies indicated that B. juncea plants were capable of accumulating significant shoot concentrations of lead from these soils. Shoot lead concentrations of 2080 and 8240 mg/kg were achieved from the soils of the Dorchester and Bayonne sites, respectively, through the use of EDTA-containing amendments in the greenhouse experiments. The plant uptake data coupled with the soil chemical fractionation analysis indicating a low proportion of lead in the unavailable residual fraction (Table 1.2), suggested that the soil Pb could be made plant available through additions of chelators and solubilizing agents. Based on this data, the application of phytoremediation in the field as a means to reduce the surface soil lead concentrations to less than 400 mg/kg was selected.

FIELD APPLICATIONS

Bayonne

The excavated soil in the lysimeter at the Bayonne site varied in pH from 7.3 to 8.7. Because surface soil (0 to 15 cm) was used to fill the lysimeter, the Pb contamination was distributed throughout the 3.5-ft deep profile. Initially, the surface (0 to 15 cm) samples ranged in lead concentration from 1000 to 6500 mg/kg with an average of 2055 mg/kg. Average soil Pb concentrations of the subsurface samples were similar (±800 mg/kg) to those of the surface soil samples and ranged from 780 to 2100 at the 15 to 30 cm depth and 280 to 8800 at the 30 to 45 cm depth. After three crops, the lead contamination in the surface soil ranged from 420 to 2300 mg/kg with an average concentration of 960 mg/kg. The average lead concentration in the 15 to 30 cm depth decreased slightly to 992 mg/kg (from 1280 mg/kg, initially) while the 30 to 45 cm depth concentrations remained relatively unchanged.

Dorchester

Initial total lead concentrations in the surface soil at the Dorchester site were lower than at the Bayonne site and ranged from 640 to 1900 mg/kg with an average of 984 mg/kg. The subsurface soil exhibited lower total Pb levels than the surface, averaging 538 mg/kg at the 15 to 30 cm depth and 371 mg/kg at the 30 to 45 cm depth. The Dorchester site exhibited a slightly narrower pH range than the Bayonne site, but was much more acidic with a pH range of 5.1 to 5.9. After three phytoremediation crops, the average concentration in the surface soil decreased from 984 to 644 mg/kg, while the 15 to 30 cm depth samples increased slightly to 671 mg/kg and the 30 to 45 cm depth decreased slightly to 339 mg/kg.

The change in lead concentrations in specific areas of the plot can be evaluated through the surface contour maps created by kriging the data. This allows interpretation of the data based on sample locations and the spatial variability that exists. It also allows one to calculate areas associated with particular Pb concentrations and by comparing the initial and final contour maps to evaluate an increase or reduction in concentration at particular areas. Areas in the plots where the soil exceeded defined Pb concentrations, i.e., 400, 600, 800, or 1000 mg/kg, were calculated based on the initial sampling and then the process repeated after the final sampling. At the Bayonne site, through the process of phytoremediation, the area with lead concentrations exceeding 1000 mg/kg was reduced from 73 to 32% of the plot of the total plot area. Figure 1.1 presents a contour map showing the areas corresponding to specific total soil Pb concentrations before and after one season of phytoremediation (three crops/season). A reduction in area where total soil Pb concentration exceeded the 600, 800, 1200, 1500, and 1700 mg/kg levels was also observed and is quantified in Table 1.3. The greatest reductions were observed in the areas contaminated at the 1000, 1200, and 1500 mg/kg levels.

The implementation of phytoremediation technology at the Dorchester site was also successful in reducing the area of lead-contaminated soil. Figure 1.2 presents

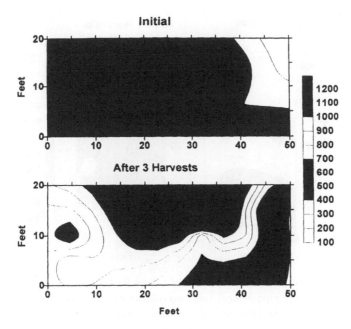

FIGURE 1.1 Contour plot showing the surface soil (0 to 15 cm) lead distribution at the Bayonne site before (top) and after (bottom) three phytoremediation crops. Color contours represent total soil Pb concentrations in mg/kg according to the values on the color scale.

TABLE 1.3

Effect of Phytoremediation on the Area of Surface Soil (0 to 15 cm) Pb Contamination at the Bayonne Site

Soil Pb Concentration (mg/kg)	Initial	After Third Harvest
	(% of Plot Area)	
>600	100	87
>800	80	66
>1000	73	32
>1200	67	20
>1500	49	10
>1700	24	6

Note Values given are the percentage of the plot area that exceed the given total soil Pb concentrations before and after one season of phytoremediation (three harvests)

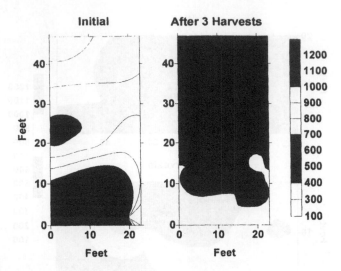

FIGURE 1.2 Contour plot showing initial soil surface lead concentrations (left) and the soil concentration after three phytoremediation crops (right) at the Dorchester site. Color contours represent soil lead concentrations in mg/kg according to the values on the color scale.

a contour map showing the initial soil lead concentration and the soil lead concentration after three phytoremediation crops. At the time of the initial sampling, 68% of the plot was above 800 mg/kg and about 25% of the plot exceeded 1000 mg/kg (Table 1.4). After three crops, none of the treated area exceeded 800 mg/kg.

TABLE 1.4

Effect of Phytoremediation on the Area of Surface Soil (0 to 15 cm) Pb Contamination at the Dorchester Site

Soil Lead (mg/kg)	Initial	After Third Harvest
	(% of Treated Area)	
>500	100	100
>600	100	100
>800	68	0
>1000	25	0

Note: Values given are the percentage of the plot area that exceed the given total soil Pb concentrations before and after one season of phytoremediation (three harvests).

Although none of the area was cleaned below the regulatory limit of 400 mg/kg at the Dorchester and Bayonne sites in the first year, the decrease in the average soil lead concentration shows the potential for phytoremediation to reduce the soil lead concentrations and the associated hazards. An evaluation of the lead concentrations in the surface soil shows that the average concentration at the Bayonne site decreased from 2055 to 960 mg/kg. This is a substantial decrease — greater than one would expect from plant accumulation of Pb in three phytoremediation crops in one growing season. In fact, under ideal conditions based solely on plant uptake, one would generally predict a 50 mg/kg/crop decrease in the total soil Pb concentration. This assumes a perfectly homogeneous soil with Pb uniformly distributed in the <2 mm particle size fraction (a condition that does not exist at contaminated sites). Nevertheless, it is apparent that some processes occur as part of phytoremediation which enhances the reduction observed above that was predicted. Movement of soil lead from the surface to lower depths was not observed eliminating the effect of leaching or movement of the soil Pb to the subsurface. In addition, the results presented indicate the performance for one growing season only. Additional decreases in soil-metal content may be observed in subsequent seasons. Only with additional research can removal rates over longer periods of time be determined.

These results provide an important first step in establishing phytoremediation as a method to reduce soil Pb levels in the field. Phytoremediation, as implemented at these sites, is projected to be comparable in cost to nonpermanent remediation systems, such as capping, while eliminating the liability concerns and requirements for long-term monitoring. In addition, phytoremediation provides an environmentally compatible means of removing the contaminant. Although this technology may not be applicable to all contaminated soils, it is particularly effective for those sites where the average lead contamination is less than 1500 mg/kg. Phytoremediation has the potential to treat many of the urban and industrial sites containing metal concentrations above the required action limits. The substantial cost savings will result in the ability of cities and private industry to remediate many more sites than would otherwise be economically possible.

ACKNOWLEDGMENTS

The authors gratefully acknowledge the assistance of The City of Boston, Department of Health and Hospitals, Division of Public Health, and Office of Environmental Health for their assistance and contributions to this effort.

REFERENCES

Baker, A.J.M., S.P. McGrath, C.M.D. Sidoli, and R.D. Reeves. The possibility of in-situ heavy metal decontamination of polluted soils using crops of metal-accumulating plants. Res. Conserv. Recyc. 11, 41-49, 1994.

Bañuelos, G.S., G. Cardon, B. Mackey, J. Ben-Asher, L. Wu, P. Beuselinck, S. Akohoue, and S. Zambrzuski. Boron and selenium removal in boron-laden soils by four sprinkler irrigated plant species. J. Environ. Qual. 22, 786-792, 1993.

Blaylock, M.J., D.E. Salt, S. Dushenkov, O. Zakharova, C. Gussman, Y. Kapulnik, B.D. Ensley, and I. Raskin. Enhanced accumulation of Pb in Indian mustard by soil-applied chelating agents. Environ. Sci. Technol. 31, 860-865, 1997.

Brown, S.L., R.L. Chaney, J.S. Angle, and A.J.M. Baker. Zinc and cadmium uptake by hyperaccumulator *Thlaspi caerulescens* grown in nutrient solution. *Soil Sci. Soc. Am. J.* 59, 125-133, 1995.

Chaney, R.L., *Land Treatment of Hazardous Wastes.* Noyes Data Corp., Park Ridge, NJ, 50-76, 1983.

Harter, R.D., Effect of soil pH on adsorption of lead, copper, zinc, and nickel. *Soil Sci. Soc. Am. J.* 47, 47-51, 1983.

Huang, J.W., J.J. Chen, W.R. Berti, and S.D. Cunningham. Phytoremediation of lead-contaminated soils: role of synthetic chelates in lead phytoextraction. *Environ. Sci. Technol.* 31, 800-805, 1997.

Kumar, N.P.B.A., S. Dushenkov, H. Motto, and I. Raskin. Phytoextraction: the use of plants to remove heavy metals from soils. *Environ. Sci. Technol.* 29, 1239-1245, 1995.

Martens, D.C. and W.L. Lindsay. Testing soils for copper, iron, manganese, and zinc, in *Soil Testing and Plant Analysis, 3rd ed.,* SSSA Book Series, No. 3, Westerman, R.L., Ed., Soil Science Society of America, Madison, WI, 1990.

Muchovej, R.M.C., V.G. Allen, D.C. Martens, L.W. Zelazny, and D.R. Notter. Aluminum, citric acid, nitrilotriacetic acid, and soil moisture effects on aluminum and iron concentrations in ryegrass. *Agron. J.* 78, 138-145, 1986.

Norvell, W.A. Comparison of chelating agents as extractants for metals in diverse soil materials. *Soil Sci. Soc. Am. J.* 48, 1285-1292, 1984.

Norvell, W.A. Reactions of metal chelates in soils, in *Micronutrients in Agriculture, 2nd ed.,* SSSA Book Series, No. 4, Mortvedt, J.J., F.R. Cox, L.M. Shuman, and R.M. Welch, Eds., Soil Science Society of America, Madison, WI, 1991.

Patel, P.M., A. Wallace, and E.M. Romney. Effect of chelating agents on phytotoxicity of lead and lead transport. *Commun. Soil Sci. Plant Anal.* 8, 733-740, 1977.

Ramos, L., L.M. Hernandez, and M.J. Gonzalez. Sequential fractionation of copper, lead, cadmium and zinc in soils from or near Doana National Park. *J. Environ. Qual.* 23, 50-57, 1994.

Raskin, I., N.P.B.A. Kumar, S. Dushenkov, and D.E. Salt. Bioconcentration of heavy metals by plants. *Curr. Opin. Biotechnol.* 5, 285-290, 1994.

Sadiq, M. and G. Hussain. Effect of chelate fertilizers on metal concentrations and growth of corn in a pot experiment. *J. Plant Nutr.* 16, 699-711, 1993.

Salt, D.E., M.J. Blaylock, N.P.B.A. Kumar, V. Dushenkov, B.D. Ensley, I. Chet, and I. Raskin. Phytoremediation: a novel strategy for the removal of toxic metals from the environment using plants. *Biotechnology* 13, 468-474, 1995.

Surfer 6.04. Surface Mapping System. Golden Software, Inc., Golden, CO. 1996.

U.S. EPA. Test Methods for Evaluating Solid Waste (SW-846). Rev. 0, Sept. 1986. Office of Solid Wastes, Washington, D.C., 1983.

Wallace, A., E.M. Romney, G.V. Alexander, S.M. Soufi, and P.M. Patel. Some interactions in plants among cadmium, other heavy metals, and chelating agents. *Agron. J.* 69, 18-20, 1977.

Wallace, A. (Ed.), The third decade of synthetic chelating agents in plant nutrition 1952-1982, Part 1. *J. Plant Nutr.* 6, 423-526, 1983.

Wallace, G.A. and A. Wallace. Clay fixation of metal chelates as a factor in their usability by soil application to correct micronutrient deficiencies. *J. Plant Nutr.* 6, 439-446, 1983.

2 Phytoremediation by Constructed Wetlands

Alex J. Horne

CONTENTS

1-56670-450-2/00/$0.00+$.50
© 2000 by CRC Press LLC

SUMMARY

Constructed wetlands offer an unlimited potential for the phytoremediation of toxins and pollutants. Their unique advantage is complete low-cost treatment of large volumes of water. High capacity makes wetlands very different from terrestrial phytoremediation or conventional physical–chemical methods that deal with relatively small volumes of contaminated soils or groundwater. No post-treatment such as filtration is needed for wetlands differentiating them from algae-based systems. Another difference between wetlands and terrestrial phytoremediation is that harvesting of pollutant accumulator plants as yet plays only a small role in wetlands, which have a very limited flora. Harvesting large volumes of toxic plants in wetlands considerably increases the cost of treatment. At least for heavy metals and some organics, the anoxic soils that characterize wetlands immobilize pollutants while the oxidized soils of terrestrial phytoremediation mobilize them into plant tissue. Pollutants such as nitrate, some organics, and probably microbial pathogens can be destroyed or detoxified in wetlands. Phosphate, heavy metals, selenium, and organics are usually immobilized and held in nontoxic forms. The greatest drawback of most terrestrial or wetland phytoremediation is the creation of a toxic "attractive nuisance" to wildlife while the pollutant is moved between the source and final sink. A management problem for treatment of wetlands is pollutant release due to seasonal biotic cycles or when the wetland is fully loaded. Natural wetlands are inefficient, but constructed wetlands, designed for specific pollutants, can deliver reliable treatment and even meet strict discharge limits. All the while the wetland provides multiple use benefits ranging from aesthetic enjoyment to enhanced biodiversity. The combination of higher plants, some algae, and bacteria make wetlands an exciting prospect for detoxification and for the control of eutrophication.

Remediation of pollution requires large amounts of energy. As with other phytoremediation, wetlands become competitive with other cleanup methods by employing free solar energy. Wetland phytoremediation differs from other forms in that bacterial transformation rather than plant uptake dominates detoxification. Nonetheless, some combinations of plants increase efficiency. Wetland plants provide the litter layer that provides both microbial habitat and a source of labile organic carbon for bacterial processes. The key to efficient phytoremediation in constructed wetlands is manipulation of the partially decomposed litter layer and sediments whose high horizontal porosity (m/h) compares with cm/week in deeper sediments. Combinations of toxic and anoxic sites and wet and dry cycles aid remediation of recalcitrant toxics. The detoxification mechanisms involved in wetland phytoremediation differ with each class of pollutant. For example, both nitrate and phosphate must be removed to fully reverse eutrophication. Nitrate is best removed as a gas by denitrification, thus emphasizing the role of plants as providers of labile carbon for bacteria. In contrast, phosphate removal in wetlands is primarily by uptake into plant and algal cell material. Here uptake and burial combined with repressing nutrient recycling is most important. With heavy metals such as copper or lead, or metalloids such as selenium, emphasis is on creating conditions for immobilization in the highly reduced sulfite or metallic form. Selenium is unusual in that it can be volatilized as dimethylselenide gas. Less is known about toxic organics or pesticide removal, although recent studies indicate that wetlands efficiently remove some chlorinated compounds present at low levels that are difficult to remove by other means. Finally, removal of bacteria, viruses, and protozoan cysts, currently of great importance in the water industry, would appear to be a major advantage of wetlands.

There are similarities between phytoremediation in wetlands, in soils using seeded crops, and groundwater bioremediation, but wetlands are less easily controlled. Thus, floods and higher trophic level interaction such as insect infestation must be considered if regulatory authorities impose effluent discharge limits.

INTRODUCTION

DEFINITIONS

Phytoremediation can be defined as the clean up of pollutants primarily mediated by photosynthetic plants. *Clean up* is defined as the destruction, inactivation, or immobilization of the pollutant in a harmless form. In this way, both higher plants and algae are included as prime phytoremediation agents, but the use of plants to create a suitable physiochemical environment for pollutant detoxification by bacteria and fungi is also specifically included. Small phytoplankton and attached algae can also be important in wetland phytoremediation (see Chapter 16). Larger wetland algae such as the skunkweed, *Chara,* or its close relative, *Nitella,* that may be 50 cm high, are here considered as part of the true wetlands flora.

Wetlands are shallow water bodies containing higher plants. Technically, jurisdictional wetlands are defined by three common components: shallow water coverage for at least a few weeks per year, permanent or temporarily anoxic soils, and characteristic vegetation (i.e., no roots or roots that can survive anoxia; Lyon, 1993). For the purposes of phytoremediation, however, wetlands are shallow waters with at least a 50% aerial cover of submerged or emergent macrophytes or attached algae. Unfortunately, by common usage, as well as the current European definition, small lakes or ponds surrounded by a thin fringe of aquatic macrophytes are termed wetlands. In practice, lakes and ponds are poor at remediation relative to wetlands. This is primarily because the large plants and a few large algae species that provide reduced carbon and the physical environment for wetland phytoremediation are not present in deeper, open lake waters. In terms of simple primary production, the least productive wetland bog exceeds the most eutrophic green lake or pond.

Wetlands are customarily divided into four groups based on their water regime (and often concomitant productivity) or the general kinds of vegetation plants present (Mitsch and Gosslink, 1993). Marshes are dominated by emergent macrophytes, swamps by trees, acid bogs by *Sphagnum* and other mosses, and alkaline fens by mosses and grasses (Horne and Goldman, 1994). Depending on the water depth and degree of shading, marshes and swamps also typically contain submerged macrophytes, often with abundant periphyton. Wetlands are characterized by anoxic reducing soils and consequently plant roots are very shallow, even absent, forcing pollutant treatment into the upper few centimeters of sediment or the litter layer. Productive seasonal wetlands dry out in summer and are thus distinguished from the less productive permanent wetlands. Tidal wetlands have some energetic advantages over other wetlands since water is pumped through the system at no cost. Finally, the different chemistry and biology of marine and inland saltwater wetlands distinguishes them from the more usual freshwater wetlands. Many of the four classes overlap. For example, the selenium-polluted Kesterson system in central California was an inland, saline, seasonal marsh but it was converted into a freshwater permanent marsh as part of an experimental cleanup (Horne, 1991).

HISTORICAL BACKGROUND OF WETLANDS AND TRADITIONAL REMEDIATION TECHNIQUES

Natural wetlands have long been used for the disposal of wastes. In fact, marshes and bogs were called "wastes" in northern England up until this century. Any treatment occurring in early waste disposal wetlands was incidental and confined to some reduction in the biological oxygen demand (BOD). Currently, the U.S. government encourages the use of simple wetlands for economical treatment of sewage BOD from small communities of less than 5000 people. There are several recent volumes that detail the engineering design required for BOD removal as well as the removal of other pollutants, primarily phosphorus and nitrogen, but also including metals and pesticides (Hammer, 1988, 1996; Marble, 1992; Moshiri, 1993; USEPA, 1993; and a comprehensive survey by Kadlec and Knight, 1996). Given that most wetlands are basically water-saturated anoxic sediments with plants growing on top, they are the least obvious way to remove oxygen-demanding BOD, which is much more efficiently removed with other methods such as algae-based oxidation ponds or small "package" plants using bacteria-based activated sludge. Thus natural or constructed wetlands are best reserved for two purposes: (1) *polishing* of already partially treated (oxidized) industrial or domestic waste or (2) *removal of specific pollutants*, such as nitrogen, phosphorus, copper, lead, selenium, organic compounds, pesticides, viruses, or protozoan cysts from all wastes including agricultural and urban storm runoff.

Traditional remediation of wastes also has a long history (Tchobanoglous and Schroeder, 1985) and in the U.S. has been amplified over the past decade by the need to clean up U.S. EPA Superfund and other lesser-polluted sites (Mineral Policy Center, 1997). If pollution generated by domestic and industrial sewage, agricultural runoff, and storm runoff is added to that from abandoned mines and industrial sites, the range of pollution problems is large. Typical physiochemical remediation methods include addition of bases or metals such as iron that will neutralize and precipitate soluble acid-mine toxic metals such as copper and zinc. Other physiochemical methods are the extraction of polluted groundwater directly or following additions of steam or solvents. Groundwater bioremediation provides additional nutrients and perhaps bacteria to metabolize the toxicant *in situ*. When remediation is not economical, containment by grout walls or other impermeable barriers, including on-site burial, is common. Traditional methods of treating domestic or industrial sewage involve oxygenated activated sludge bacteria, trickling filters, or high rate oxidation ponds. The volumes of agricultural and storm runoff are so large that treatment is rare. Pollutant source control by best management practices (BMPs), usually involving soil conservation but also including wetlands, has been tried but with only moderate success (Meade and Parker, 1985). Finally, a new regulatory tool, total maximum daily load (TMDL) is being implemented to provide the quantitative tool lacking in previous BMP programs.

The most obvious advantage of phytoremediation over traditional techniques is cost. While most traditional remediation methods rely on electricity, pumping, or oxygen additions and often require large concrete or steel vessels, phytoremediation uses free solar energy and requires no sophisticated containment system. Other differences between conventional remediation, terrestrial phytoremediation, and wetlands phytoremediation are shown in Table 2.1.

TABLE 2.1

Similarities and Differences Between Conventional Bioremediation, Phytoremediation, and Wetlands Phytoremediation

Contamination	Conventional Bioremediation	Terrestrial Phytoremediation	Wetlands Phytoremediation
Waste liquid volume	Low	Low	High
Waste solid volume	High	Moderate (roots)	Low
Energy source	Added carbon	*In situ* generation	*In situ* generation
Containment	Tanks, pumps, grout curtains	Not needed on land	Earth berms
Remediation away from site	Yes and no	No	Yes and no
Agricultural runoff	No	No	Yes
Urban storm runoff	No	No	Yes
Domestic wastewater	No	No	Yes
Industrial wastewater	Yes	Yes?	Yes
Acid-mine drainage	No	No	Yes
Heavy metals	NA	Metal accumulation	Metal immobilization
Polluted soils	Yes	Yes	Rarely
Pumped polluted groundwater	Yes	No	Yes
Metals	No	Yes	Yes
Toxic organics	Yes	Potentially	Potentially
Nutrients	No	No	Yes
Pathogens	No	No	Maybe

Note: Conventional bioremediation has concentrated on toxic organics such as solvents and dissolved nonaqueous phase liquids (DNAPL), while terrestrial phytoremediation has focused on heavy metals. Major differences are also due to wetlands normally being used to treat external water inflows while terrestrial phytoremediation and *in situ* bioremediation restore contaminated soils or groundwater on site. The common method of groundwater cleanup "pump and treat," could use any of the three methods.

NA = not applicable

Differences Between Wetlands and Terrestrial Phytoremediation

Terrestrial and wetlands phytoremediation both use plants to provide the main energy source for pollutant mobilization or immobilization. The difference between the two is that growing seeded crops on land or treating groundwater in tanks of fixed plant species is about as reliable as farming. Plants in wetlands are not easily controlled. By their nature, wetlands are more susceptible to floods than other lands. For example, in the 200 ha Prado wetlands tests, very heavy 1993 winter rains caused California's Santa Ana River to change course and deposit 60,000 m³ of sand in the upper end of the wetlands (Figure 2.1). In addition, higher trophic level effects such as insect infestation can wipe out wetlands plants. Even cattails, one of nature's most hardy plants, are subject to at least four species of caterpillar infestation. As a result thousands of acres can turn brown in a few weeks (Snoddy et al., 1989). Duckweed and aquatic grasses, providers of labile organic matter for bacteria, are quite good at removing many pollutants. Unfortunately, as the name suggests, ducks can eat even dense stands of duckweed in just a few days. The toxic effect on the ducks may be serious but has not been explored. On other occasions, winds blow duckweed into piles on the downwind shores that are then useless for pollutant removal. Such uncontrollable potential changes in the ability of wetlands to process

FIGURE 2.1 Aerial view of a full-scale phytoremediation wetland: Prado Wetland, Riverside, CA. This 200-ha (500-acre) wetland removes nitrate from the Santa Ana River which contains more than the 10 mg/l nitrate-N allowed by public health standards. Open water areas alternate with cattail, bulrush, grasses, and duckweed to provide carbon of variable biological lability for bacterial denitrification.

pollutants must be solved by flexible responses such as increasing residence time or constructing excess capacity to meet effluent limits imposed by regulatory authorities.

PHYTOREMEDIATION USING CONSTRUCTED WETLANDS

Natural wetlands are not very efficient at pollution removal. Water often short-circuits through natural wetlands, giving little time for treatment. The annual mass balance for nutrients in natural wetlands often shows seasonal effects but no net loss (Elder, 1985). Pollutants can build up into toxic amounts in seeds and insects resulting in deaths of birds (Ohlendorf et al., 1986). Paradoxically, these reasons for low efficiency in natural wetlands are the reason why constructed wetlands can be so useful. Although many features of large wetlands are uncontrolled, the hydraulic regime, kinds of plants and animals, and drying cycles of a constructed wetland can be modified to maximize treatment. Also, the mass removal of pollutants rises dramatically when the loading of many pollutants is increased. In part, the change is due to moving concentrations to well beyond saturation of the enzyme uptake and cellular transport mechanisms. Additional removal is due to an increase in the gradient in the diffusion barrier between the pollutant stream and its living or dead wetland sink. Unfortunately, the details of how water moves through the leaf litter and fine sediments can only be inferred from laboratory studies with homogeneous materials and the role of aquatic insect larvae in stirring the leaf ooze can only be guessed.

Ideally, constructed wetlands are designed to maximize removal of a specific pollutant or group of pollutants. Such wetlands are now being built. The most important difference between constructed and natural wetlands is the isolation of the water regime from natural patterns. Unlike terrestrial phytoremediation accomplished by planting specific vegetation, few things can be regulated directly in a large wetland that sets its own biotic diversity as well as temperature. For example, cattail-shaded areas of wetlands are 2°C cooler than open water areas. In shallow water, cattails will tend to dominate but pre-planting with bulrush can stave off invasion for decades. Nevertheless, regulation of the water depth and timing in a wetlands can control plant types in a very general sense. For example, many wetlands plants will not grow in water more than 10 cm deep and even cattails and bulrush do not grow well in water over 1.5 m deep. Similarly, drying the wetlands out in summer will kill many larger species allowing the seeds of small annuals to dominate the next year. Thus the initial bed contouring, flooding depth, and hydroperiod of the constructed wetland can control the general kind of plants.

Matching the Wetlands Type to the Pollutant

Wetlands are not simple ecosystems. Phytoremediation in wetlands requires that specific type and management match the pollutant to them. For example, to fully reverse eutrophication and restore a water body to its original condition requires both nitrate and phosphate removal. But wetlands do not carry out each of these removals equally well. Nitrate is easily removed by denitrification, thus emphasizing the role of plants as providers of labile carbon for bacteria. In contrast, phosphate removal is primarily by uptake into plant and algae cell material, so burial and repression of nutrient recycling is most important.

With heavy metals such as copper or lead, or metalloids such as selenium, the emphasis is on creating conditions for immobilization, usually anoxia and the presence of sulfides. Organic removal, other than BOD reduction, is in its early stages of investigation in wetlands but may be more a case of providing physical sorption sites than enhancing bacterial metabolism or plant uptake. It likely that alternation of areas or oxygenated open water (phytoplankton and some submerged macrophytes) with dense anoxic macrophyte stands will give the best results for almost all pollutants. Although large treatment wetlands are difficult to maintain with a required plant mixture, general types of plants can be favored by manipulation of water depth and hydroperiod. Recently, the 200 ha Prado Wetlands in southern California was re-graded to give a variety of water depths. This retrofit has produced a much larger variety of emergent and submergent plants as well as more habitat for attached and planktonic algae. The expected result is a wider variety of organic carbon for bacterial denitrification.

Phytoremediation in wetlands can be used to remove a wide variety of pollutants and toxicants. Some examples of how wetland phytoremediation can solve some of the problems caused for human health and recreation as well as those of the biota in the environment are shown in Table 2.2.

Importance of the Leaf Litter and Fine Sediment Layer

The working hypothesis for the importance of the litter and fine sediments layer is that it is the only site that provides reduced carbon energy, sites for bacterial growth, and any of the other needed but often ill-defined conditions such as protection from predation or provision of anoxia. Therefore, most constructed wetlands differ from terrestrial phytoremediation in that manipulation of the physiochemical environment of the litter layer and fine sediments is more important than any specific plant or algal species. For example, denitrification in both bulrush and cattail marshes increase as leaf litter increases (Bachand and Horne, 1999). Uptake of metal ions from acid-mine wastes takes advantage of the cation uptake sites on the resin-like dead stems of *Sphagnum* which are similar in all species. For immobilization of heavy metals such as copper or lead, the provision of anoxia, no matter what the source of reducing power, is most important.

Even where specific combinations of plants are more efficient than others, it is the provision of leaf litter and dissolved organic carbon that is most important. For example, denitrification in wetlands is greater in pure cattail stands than in pure bulrush stands (Table 2.3). Most researchers have noted that more mature wetlands are better for general pollution clean up and this is primarily due to the time taken to establish the plants, not the kind of plant. At present then, the particular plant species or genetically engineered strains are less important than the manipulation of the total wetland environment to provide specific physiochemical conditions that can detoxify or immobilize the pollutant. Future advances may allow seeding with

TABLE 2.2

Summary of Known Uses of Phytoremediation Wetlands

Pollutant or Toxicant Remediated	Human Problem	Environmental Problem
Biological oxygen demand	Drinking water quality, malodors	Fish kills, slime production
Nitrate	Blue baby disease, lake use[a]	[b]Eutrophication, avian botulism
Particulate-N/P	Lake use	Water clarity
Phosphorus	Lake use	Eutrophication
[c]Heavy metals (Cu, Pb, acid-mine drainage, storm runoff)	Drinking water standards	Toxicity
Metalloid (Se from agriculture, copiers, taillight production)	Toxicity to livestock (blind staggers)	Bird embryo deformities, skeletal deformation in fish
Pesticides	Food chain toxicity, cancers	Nontarget organism deaths
Trace organics (chlorinated organics, estrogen mimics)	Major long-term objection to human water reuse	Subtle toxic effects
Bacterial pathogens	Microbial diseases	None?

Note: Phytoremediation using wetlands ranges more widely than terrestrial phytoremediation in that drinking water supplies, as well as streams and rivers, are targets for clean up. Wetlands used range from acid *Sphagnum* bogs for acid-mine drainage to cattail and duckweed marshes for denitrification and pesticide removal.

[a] Examples of enhanced lake, reservoir, or river use include decreased algae and bacterial growth leading to better water percolation for groundwater recharge and better recreation since the water will be more transparent and blue, not green, in color

[b] Examples of wetlands used for eutrophication control are the 500-acre Prado wetlands (nitrate and phosphate removal), the 60 acres at Irvine Ranch Water District, and the 40,000-acre Everglades Protection Wetland in Florida (the last two are under construction).

[c] Will not work for strongly chelated metals such as nickel

"superplants," but their survival in the highly competitive wetland ecosystem will require further research.

The litter and upper fine sediments layer with their very high horizon porosity is the key to efficient phytoremediation in wetlands. True sediments such as peat and clay are quite compact and rapidly become clogged in wetlands due to settling of small particles such as diatom frustules and release of bacterial mucopolysaccharides. The porosity of peat and clay in wetlands ranges from 10^{-4} to 10^{-8} cm s^{-1} (i.e., cm/week, Mitsch and Gosslink, 1993). In contrast, the fine sediments and leaf litter found in wetlands used for phytoremediation has a high porosity (10^{-1} cm s^{-1} or m/h). This ooze can be so loose that the stirring caused by passing insect larvae and fish feeding reduces clogging. Thus, free water surface wetlands with advective water flows and about 50 cm of water depth are much more efficient than subsurface wetlands where molecular diffusion dominates. Nonetheless, for some purposes subsurface wetlands that are dry on the surface are appropriate. In particular, sub-

TABLE 2.3

Rate of Denitrification in Stands of Pure Bulrush, Cattail, and a
Mixed Growth of Duckweed and Aquatic Grasses in Southern
Californian Marsh

Plant Species	Denitrification Rate mg-N m^{-2} d^{-1}		
	1-Year-Old Marsh	2-Year-Old Marsh	4-Year-Old Marsh
Cattail	570	760	1220
Bulrush	260	320	540
Duckweed/grasses	830	600	550

Note: The kind of plant is apparently less important than the amount of litter it produces, since addition of more leaf litter increased denitrification in all systems.

Source: From Bachand, P. A. M. and A. J. Horne, 1999.

surface wetlands harbor no insect vectors and have obvious advantages where malaria and such diseases are common and vector control authority is weak or absent. In such cases, larger, less efficient subsurface wetlands may be the best choice.

CASE HISTORIES

CLASS 1. NUTRIENT REMOVAL

In terms of sheer mass, nitrate and phosphates are the most common of all pollutants. They are present at quite high concentrations in the huge volumes of water from sewage, agriculture, and urban storm runoff (Bogardi and Kuzelka, 1991). For example, urban storm runoff may contain 50 mg/l of nitrate-N but only a few mg/l of gasoline, 0.1 mg/l of copper and zinc, a similar amount of polycyclic aromatic hydrocarbons, and a few μg/l of pesticides. Treated sewage and agricultural runoff have a similar dominance of nitrogen and phosphorus over metals and anthromorphogenic organics. Excess nutrients cause eutrophication of lakes, rivers, estuaries and coastal oceans (de Jong, 1990). Recent fish kills due to poisonous "red tides" of dinoflagellates in the Carolinas and Virginia or tropical reef losses (Hodgson, 1994) are probably due to excess nitrogen. The often toxic scums of blue green algae in lakes are the characteristic symptoms of eutrophication and have been shown to kill sheep drinking the water (Negri et al., 1995). Wetlands are an excellent site for nitrate removal and can also remove phosphorus.

Case Study #1. A Natural Filter: Removal of Total Nitrogen and Phosphorus from Lake Apopka, FL

Lake Apopka is a large but shallow lake near Orlando, FL. Within living memory it has become polluted with agricultural and other nutrient-laden runoff. The result is that once clear water is now a cloudy mess with large amounts of suspended algae. High-quality fishing has declined and local property values have probably been reduced. Some of the dead algae and sediments are easily stirred up from the bottom by the wind, especially since there are less aquatic plants in the lake due to cloudiness of the water. Typically submerged plants and their roots hold the sediment together and reduce wind-induced turbidity.

Because its large size it is not economically feasible to apply most techniques of lake management to Lake Apopka. In addition, the cloudy water is due as much to dead, resuspended matter as to living algae. Reduction of diffuse sources of nutrients from runoff is a long-term process that may never fully succeed, given the growing population in the watershed. An obvious solution would be to filter out or coagulate the particles, but the huge volumes involved hitherto make this solution impractical. The use of a wetlands as a natural filter would remove suspended matter, but no one had carried out filtration on such a vast scale.

An initial experiment using about 150 ha (over 300 acres) was carried out. Water had only to be pumped to give about one meter of head and the system then worked by gravity flow through a modified existing wetlands. Removal rates of up to 95% were found for total N and total P (Coveney et al., 1994). A full-scale project involving 1500 ha is planned. Removal of soluble nutrients was not high, but soluble material comprised only a small fraction of N and P in this case. This would not be the case for many other wastes such as sewage or agricultural drainage (see below).

Case Study #2. Drinking Water Treatment: Nitrate Removal Followed by Groundwater Recharge in Prado Wetlands, CA

Nitrate may be the most ubiquitous pollutant in modern society (Canter, 1997). Background concentrations in rain and streams were probably less than 0.1 mg/l nitrate-N even 500 years ago. Now concentrations of 0.5 mg/l occur in rain in many places and some streams and groundwater contain over 100 mg/l of nitrate-N. Approximately 3 million people in the U.S. take their drinking water from community service wells where concentrations of nitrate exceed the safe level (U.S. EPA, 1992), and an unknown number of others use single wells that were not surveyed. Above 10 mg/l-NO_3-N, very young children are susceptible to a potentially fatal disease called "blue babies," characterized by poor oxygen transport in the blood. The disease is due to reduction of the ingested nitrate to nitrite in the infant's acid gut. The nitrite then binds with hemoglobin in the bloodstream. Small infants lack the enzymes necessary to reverse the reaction

The Orange County Water District's (OCWD) Prado nitrate removal wetlands began operation in 1992. At over 200 ha (500 acres; Figure 2.1), it is the world's largest engineered wetlands phytoremediation sites with a legally mandated standard of performance. The main water supply for OCWD is the Santa Ana River, and summer flow is dominated by highly treated domestic wastewater containing up to 20 mg/l nitrate-N. For eventual use in drinking water supplies, this source must reliably contain less than 10 mg/l. Additional benefits are gained if nitrate is reduced to only 1 mg/l because nutrient-enhanced algal growth in percolation ponds hampers groundwater infiltration (Horne, 1988). Prado Wetlands reduces nitrate from 10 mg/l to 1 mg/l (as N) in 50 to 200 ha (100 to 500 acres) with a residence time from 2 to 7 days (Reilly et al., 1999).

The essence of this form of phytoremediation is the removal of nitrate by conversion to nitrogen gas by bacterial dentrification (Bachand and Horne, unpublished; Lund and Horne, in press). There is no toxic accumulation, and the wetlands has been designed to allow other uses such as enhancement of biodiversity. Solar-powered phytoremediation in the Prado Wetlands acts by producing organic carbon in emergent plants such as cattails, submergent plants such as pondweeds, and floating plants such as duckweed. Bacteria use the carbon in the anoxic surface sediments. Uptake into plants is probably less than 10%, which is just as well since huge amounts of biomass would have to be harvested to remove the amount attributable to bacterial denitrification. About one million kilograms of vegetation per day would have to be removed from Prado wetlands if plant nitrogen uptake in growth were to equal the measured average summer denitrification rate of 500 $mg/m^2/day$ (Horne, 1995). To remove this amount of fresh vegetation with a volume of about 100,000 m^3 would require a large truck to leave the wetlands every 3.4 minutes, day and night. Therefore, loss of nitrate as nitrogen gas is much more desirable.

Nitrate removal is maximal in the spring–fall period and is only about 30% of maximum in the winter. Temperature or lower carbon supplies may explain the lower rates in winter but further studies are needed to determine which factor is most important. Experimental studies on the kind of plants needed to optimize performance (Philips and Crumpton, 1994; Bachand and Horne, 1999) and changes in carbon signature (Gray et al., 1996) were used to reconstruct the wetlands to an improved design in spring 1997 and at a cost of $4 million.

Case Study #3. National Park Protection: Removal of Phosphorus to Prevent Eutrophication in the Everglades

The Florida Everglades National Park faces the twin threats of a lack of fresh water combined with contamination of the water supplies it does receive from the north. In particular, the U.S. EPA and the State of Florida rather hastily agreed to require a very low total phosphate standard for water entering the park. The new standard approximates to the historical concentrations, but cannot now be met since agricultural drainage, some domestic wastewater, and storm flows from developed land pollute the original water supply.

The solution to this dilemma is wetlands, a $500 million phytoremediation project involving construction of an enormous phosphorus removal wetland of about 17,000 ha (40,000 acres) to intercept and remove total phosphorus (TAP, 1992). The results from this large phytoremediation project will be of great interest for other regions. The land area proposed is much larger than would be needed to remove an equivalent amount of nitrate, since wetlands are not very efficient at P removal (Richardson et al., 1997). In addition, since the P is held in living and dead plants in the wetlands, some will be recycled each year. In the case of phosphorus, the kind of plant in the wetlands may be vital. In still unspoiled areas of the Florida Everglades, natural associations of bladderwort (*Eutricularia*) and blue-green algae form insoluble calcium carbonate-phosphate complexes that permanently sequester phosphate much better than the water hyacinth–cattail–diatom assemblage typical of the eutrophicated regions affected by sugar cane farm runoff (Craft et al., 1995). In addition, bladderwort is out-completed by several other wetlands plants when nutrients rise to stormwater levels.

Class 2. Natural Toxicants: Heavy Metals, Selenium

Many Superfund sites in North America are abandoned metal mines (Eger et al., 1993) or coal mines (Brodie, 1993) that produce acid-mine drainage. When possible, mines are designed to allow water to flow out of the tunnels by gravity to reduce pumping costs. Unfortunately, this results in outflow long after the mine has been abandoned. Given the fractured nature of most natural mineral deposits, water percolates rapidly through the soil once tunnels have been constructed. Rainwater seeping through spoil heaps above ground produces the same result. Another common method of disposing of mine tailings was to dump them into the nearest valley. The stream flow through the tailings combined with abundant oxygen provides bacteria an ideal site to convert solid metal sulfides to free soluble metal.

The resulting leached metals can cause havoc in lakes and streams generally made most evident by massive fish kills following heavy rain. Effects can also be seen year-round and for long distances. For example, the small stream that drains the small Gray Wolf Mine on the California–Nevada border is totally without aquatic insects for many kilometers and contains virtually no algae. In the U.S., Girts and Kleinmann (1986) estimated that acid-mine waste degrades almost 20,000 km of flowing waters. Most of these mines were sulfide ore mines that yielded copper, zinc, cadmium, lead, and mercury. These metals are also common byproducts of silver mines and coal mine acid waste. In mine wastes where pH may be below 2 and is usually below 6, metals are present in the free ion form, usually the most toxic to fish and other wildlife. Fortunately, free metal ions are highly reactive with many sites including dead plant matter and sulfides that are common in the sediments of productive wetlands (Figure 2.2, Table 2.4).

Successes in Metal Removal in Wetlands

In treatment wetlands, the mechanism for removal of metals is primarily immobilization of the sulfide for Cu, Fe, Mn, Zn, Cd. By definition, wetlands are productive habitats with anoxic waterlogged soils. Under these conditions, decay of sulfur-containing proteins and reduction of natural sulfate in the sediments produce sulfide. In this way, most current wetlands phytoremediation differs from terrestrial phytoremediation where the plants are used to extract and concentrate metals from contaminated soils. Using metal radiotracers it can be shown that wetlands convert soluble metals to precipitates within a few hours (Figure 2.2). Sulfides of most metals are very stable under anoxic water-saturated conditions. In addition, the most abundant heavy metal, iron, forms plaques that are stable in reducing conditions. As was elegantly shown by SEM and x-ray microanalysis by Peverly et al. (1995), a variety of metals can be immobilized in the reed rhizosphere if diffusion of oxygen causes

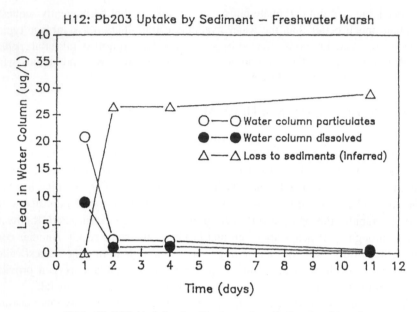

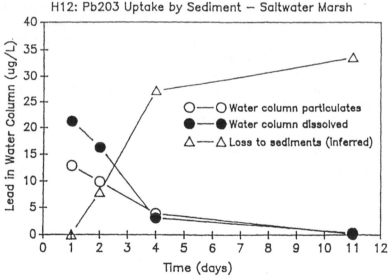

FIGURE 2.2 Metal phytoremediation. Very rapid (hours) and almost complete (>90%) removal of low levels of radio-labeled lead (^{203}Pb) tracer from simulated brackish water wetlands. The lead was incubated for 12 h with secondary effluent to mimic actual sewage treatment practices. In this case, the wetland was required to remove enough metal from a well-treated domestic and industrial wastewater that the effluent from the marsh could be discharged into nearby San Francisco Bay without damage to sensitive estuarine organisms (Adapted from Gregg, J.H. and A.J. Horne. 1993. Environmental Engineering and Health Science Laboratory Report. No. 93-4. December 1993. 159.) This high level of removal (>90%) was not achieved in the actual wetland (net reduction ~36%) due to recycling and input of airborne dust.

TABLE 2.4

Successful Cases of Metals Removal and pH Elevation by Phytoremediation Wetlands

Source/ Metal	Mean Inflow (conc.)	Mean Outflow (conc.)	Percent Removal	Reference
Taconite Tailings Leachate				
Copper	280	13	95	Egar et al., 1993
Zinc	1900	205	90	Egar et al., 1993
Nickel	1940	1075	45	Egar et al., 1993
Acid Coal Mine Runoff				
pH	3.1	6.7	—	Brodie, 1993
Iron (high)	69,000	900	99	Brodie, 1993
Iron (low)	80	1.1	>90	Brodie, 1993
Manganese	9300	1000	80	Brodie, 1993
Manganese	7.7	2.8	65	Brodie, 1993
Acid Metal Mine Runoff				
Lead	12	0.2	98	Tang, 1993
Nickel	52	<5	90	Tang, 1993
Wastewater in Wetland				
Copper	33	17	48	Gregg and Horne, 1993
Lead	11	7	36	Gregg and Horne, 1993
Zinc	50	30	40	Gregg and Horne, 1993
Same Wastewater in Simulated Wetland[a]				
Copper	33	<3	>90	Gregg and Horne, 1993
Lead	11	<1	>90	Gregg and Horne, 1993
Zinc	50	<5	>90	Gregg and Horne, 1993

Note: Most of these examples are for constructed wetlands. Concentrations are in µg/l. Wetlands can remove metals at both high and low concentrations. Note that the percent removal under ideal conditions. e.g., >90% of three metals removed from municipal wastes (bottom of table), is not always realized in actual wetlands where removals of only 36 to 48% were realized (next to bottom). Differences are due to recycling within the wetland, atmospheric fallout, and other inflow sources such as rain or groundwater.

[a]In this experiment, radioisotopic metal tracers were added to the effluent samples to obtain percent removal in simulated wetlands, so final concentrations were inferred.

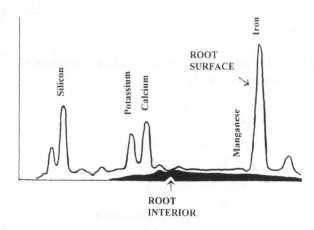

FIGURE 2.3 Metal immobilization on wetland plant roots. X-ray microanalysis spectrum of a cut root of the common reed *Phragmites* exposed to metal-rich leachate. The figure shows the accumulation of four heavy metals and silica on the outer root surface while the interior (lower dark area) showed no metal signal. The iron is probably present as plaque of iron hydoxyoxide, which may absorb other metals and protect from toxicity or uptake. Control roots not exposed to the leachate did not show the metal plaque. (Modified from Peverly et al., 1995. *Ecol. Eng.* 5: 21-35.)

iron and manganese precipitation (Figure 2.3). In this case it was probably the iron hydroxyoxide that acted as a filter or sorption medium for other metals such as copper and zinc (Figure 2.3). Presumably, this suppression of metal uptake does not occur so readily in the oxidized soils required for terrestrial phytoremediation — a major difference between the two techniques.

Some wetlands plants such as the small floating duckweed (Sharma and Gaur, 1995; Bomono et al., 1997), a few emergent macrophytes (Mungar et al., 1997), and potentially even swamp trees can be used to accumulate heavy metals in the same way as in terrestrial phytoremediation. In the future, some wetlands may employ techniques other than sediment immobilization, thus becoming more similar to terrestrial phytoremediation using metal accumulator plants. This will require enhancement of hyperaccumulation and is discussed below.

Failures in Metal Removal with Wetlands

Not all metals are easily removed by wetlands. Highly chelated metals such as soluble nickel, usually present in a strongly chelated form, can pass through wetlands (Table 2.5) while other less strongly chelated metals such as lead or zinc are quickly removed. The high removal of nickel shown for acid-mine wastes in Table 2.4 is possible because the chelating capacity of mine water is very low. Thus, ionic nickel is present and can be precipitated. In contrast, in wastewater there is abundant chelating capacity and so the soluble metal in not precipitated in the wetland (Table 2.5). In addition, the concentrations of almost all metals in mine wastes can be very

TABLE 2.5

Unsuccessful Cases of Metals Removal by Phytoremediation Wetlands

Source/ Metal	Mean Inflow	Mean Outflow	Percent increase	Ref.
Lead (urban stormwater)	2 0	5.5	+180	CH2M-Hill, 1992
Lead (municipal effluent)	11	21	+91	Gregg and Horne, 1993
Nickel (municipal effluent)	2 8	3.5	+27	CH2M-Hill, 1991
Nickel (municipal effluent)	17	25	+47	Gregg and Horne, 1993
Iron (municipal effluent)	240	770	+218	CH2M-Hill, 1992

Note: Release of metals due to saturation of absorption sites or seasonal biotic effects is probably more common than is shown in the published literature In nature, nickel is so strongly chelated that it may be unaffected by wetlands and the inflow and outflow concentrations are probably about the same in these examples. Lead is removed by wetlands (see Figure 2.2) but may increase in treatment wetlands due to external atmospheric loading of the metal from gasoline additives or dust More failures are probable but such results are rarely publicized Metal concentrations in μg/l

high, so even 90% removal of nickel (Table 2.4) still leaves a concentration well above that desirable in natural waters.

Wetlands cannot remove metals forever, unlike nitrate removal, which is theoretically infinite. A wetland will eventually come to equilibrium with any substance when binding and release are equal. Thus, in a mature marsh, metal release may provide a constant baseline that will influence the apparent removal rate. Drying out the wetland should create oxidizing conditions. In turn, this should re-oxidize some metals or metalloids into soluble forms that may become extremely dangerous when the wetland is reflooded. Excavating the contaminated metal concentrate is economically feasible in smaller constructed wetlands

If the inflow concentration is low, as can happen if clean stormwater dilutes the metal, the wetland may actually appear to be a source of the metal for a time. This is quite common (Table 2.5) and may be due to loading of metals as particles from wet and dry atmospheric fallout, especially in urban wetlands.

Hyperaccumulation

The plant accumulation of heavy metals and other trace elements is the basis of much terrestrial phytoremediation and could be used in wetlands (Mungur et al., 1997). However, there are many more terrestrial than aquatic plants, and suitable metal-accumulator wetlands plants are not always available. Despite these drawbacks, phytoremediation using 26 genera of aquatic plants (Guntenspergen et al., 1989) including water hyacinth, duckweed (Bonomo et al., 1997), *Typha*, and *Phrag-*

mites (Tang, 1993; Mungur et al., 1997), and even grasses and sedges (Eger et al., 1993), has been proposed for small-scale cleanups. Storage of excess metals may occur in the leaves and stems of plants as well as in the sediments. The amount and location of the metal varies with plant species (Table 2.6), but is often greatest in roots and seeds. Unfortunately, seeds are a prime food for wildlife, especially migrating birds. Therefore, some means of excluding them from the treatment site is needed.

The problem of economically harvesting and disposal of large volumes of contaminated vegetation remains. Scaling up from the small-scale laboratory studies mentioned above to full-sized wetlands would incur the same problems of harvesting vast quantities of biomass (quantified for nitrogen in Case Study #2). However, the same problem exists for terrestrial phytoremediation, although harvesting on dry land is easier than in 1 m of water.

TABLE 2.6

Storage Sites for Heavy Metals in Wetland Plants Exposed to Heavy Metals

Plant Type	Leaf (ppm))				Root (ppm			
	Cu	Mn	Zn	Se	Cu	Mn	Zn	Se
Cattail (*Typha*)								
exposed	240	2400	82	25	155	770	72	81
Control	5	250	36	<1	16	200	36	<1
Tule (*Scirpus*)								
exposed	90	1300	37	1.6	95	600	110	9.5
Control	11	700	54	<1	38	450	82	<1

Note: Experimental studies in 1-m^2 mesocosms in series, with residence times from 2 to 8 days using contaminated waste from the Ma An Shan, one of China's eight large, open iron mines. Production of acidic wastewater from this site is 1000 t/y. Metal removal from the water ranged from 50% (Mn) to 98% (Cu). (Modified from Tang, S.-Y. 1993. *Ecol. Eng.* 2: 253-260.) Selenium storage from a 40,000-m^2 macrocosm. (From Horne, A.J. and J.C. Roth, 1989. University of California Berkeley. Environmental Engineering Health Science Laboratory Report No. 89-4.) Removal of Se from water was over 90%.

Case Study #4. Duck Deaths: Phytoremediation of Selenium in Kesterson Marsh, CA

Selenium (Se) is a metalloid with properties of both heavy metals and nonmetals such as sulfur, its close neighbor in the periodic table. It is a vital cofactor in antioxidant removal in mammals, but becomes toxic when more concentrated. Once thought to be a rare agricultural problem in dry inland regions (*San Francisco Examiner*, 1987), Se from agriculture, oil and coal industries, photocopying, and auto manufacturing has become a major concern in several areas of North America (*Science*, 1986; Frankenburger and Benson, 1994). Unfortunately, Se is unique in that the range between sufficiency and toxicity is much closer than for metals such as copper or cadmium (Combs and Combs, 1986). The range of needs and toxicity for each species overlaps to such an extent that the U.S. EPA primary maximum contaminant level in drinking water concentration for humans (50 (μg/l) exceeds the threshold of chronic toxicity for freshwater aquatic biota by an order of magnitude. Bioconcentration of Se into the food chain by algae resulted in numerous grotesque deaths of birds in the past 2 decades. Mammals are also affected. Thus, the removal of excess Se in surface waters used by wildlife is imperative.

In dry regions it has long been known that Se will be flushed out of the shallow vadose zone by heavy rains and irrigated agriculture. Agriculture in semi-arid climates, such as the Central Valley of California, requires irrigation and subsurface drainage to prevent salt accumulation in the surface soils. The percolated tile drain effluent is collected and disposed of in various ways. Some currently irrigated land in California and elsewhere contains elevated Se levels. The origin of the Se in the sedimentary rocks of the Coast Range hills is marine algae, deposited as sediment (Barnes, 1985). Similar selenium toxicity problems occur in the dry parts of the world, including Colorado, the Dakotas, and parts of China. In wetter regions such as the Carolinas, coal-fired power plants use fuels high in Se, and lakes and marshes near the fly-ash leachate piles now contain deformed fish.

In 1978, a large 500-ha (1200-acre) wetland was created to store and evaporate agricultural drainwater without knowing that it was Se-contaminated. Following algal bioaccumulation, Se passed up the food chain resulting in severe bird embryo problems by 1983 (Ohlendorf et al., 1986), although lower trophic levels were apparently unaffected (Horne and Roth, 1989). Deliveries of soluble Se ceased in 1995 and the problem was how to remediate the site. Several methods were proposed including digging out the site (about 500,000 m^3 of contaminated soil) and various kinds of flooding. In addition, enhancement of volatilization of dimethyselinide by fungi supplied with additional carbon sources was tested experimentally (Frankenberger and Benson, 1994). Research on plant roots by Terry and coworkers (Chapter 4) suggests another method that could have been applied.

Several cleanup techniques were tested at Kesterson in the 1980s. They included permanent flooding, seasonal flooding, shallow and deep soil excavation, and enhanced fungal volatilization. The most successful technique tested at that time was permanent flooding. This method immobilized Se using a mixture of the giant algae *Chara*, its associated aufwuchs (i.e., the entire attached microbial community of bacteria, algae, fungi, protozoans, and rotifers), and some contribution from emergent macrophytes such as cattails. The main purpose of the phytoremediation was to take advantage of the fact that Se, like many heavy metals, becomes immobilized and biologically unavailable under anoxic conditions. Once new Se inflow had ceased, soluble Se concentration fell from 400 to 3 μg/l in a matter of weeks. Selenium in the food web fell more slowly (Horne, 1991; 1994) taking 2 to 3 years for the base of the food chain to reach below the standard of 5 mg/kg dry weight.

POND 5E

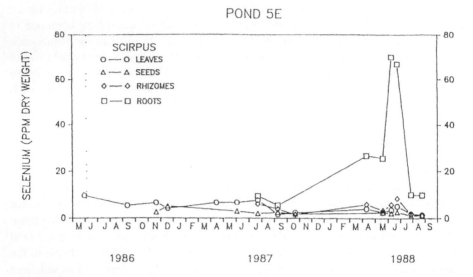

FIGURE 2.4 Phytoremediation of selenium in Kesterson Reservoir-marsh. Alkal: bulrush growing in selenium-rich wetland soil showed relatively low concentrations in above-ground parts. The accumulation of Se in or on the roots (see also Figure 2.3) was one sink for Se. (From Horne, A.J. and J.C. Roth, 1989. Environmental Engineering and Health Science Laboratory Report No. 89-4.)

As an example, the decline of Se in the above-ground parts (but not roots) of alkali bulrush (*Scirpus*) in Kesterson marsh is shown in Figure 2.4. Once the wetlands had been made safe for wildlife, parts could have been dried out, Se removed by volatilization, and then returned to productive use in supporting wild birds. This sophisticated, university research-based approach did not meet with overwhelming support from regulators or environmental groups, and the wetlands was filled in. However, the problem remains in many other areas, and a wetlands solution may be used in the future to treat the Se that has accumulated in farm soils since 1985.

CLASS 3. NATURAL AND SYNTHESIZED ORGANIC COMPOUNDS: DISSOLVED ORGANIC CARBON, PESTICIDES, SOLVENTS, CHLORINATED ORGANICS IN WASTEWATER

It is this class of compounds that is the target of most conventional groundwater physiochemical remediation as well as microbial bioremediation efforts. Perhaps for that reason wetlands organics phytoremediation has developed slowly. For example, no studies of the solvent TCE (trichloroethylene) and wetlands were found, despite this compound being perhaps the most common organic groundwater pollutant. The situation can best be summed up in a quotation by Kadlec and Knight (1996). "What we know thus far shows great promise for wetland technologies for organics control, especially in those situations where these passive systems can be fit into the land-scape and accumulated organics will not pose a threat to wetland biota."

Pesticides

One of the most common selective broad leaf herbicides used for weed control on roadsides and farm crops is the triazine ring compound Atrazine®.* Up to 30,000 tons is applied annually in the U.S., and its sister compound, Simazine®,* has been extensively used for weed control in lakes and canals. Atrazine is moderately persistent, decomposing by less than 0.1% per day (Grover, 1988).

Atrazine control by wetlands has been studied in the laboratory and in the Des Plains experimental wetlands near Chicago, and removal rates were determined (Alord and Kadlec, 1995; Kadlec and Knight, 1996). The superiority of the cattail over mineral sediments was shown indicating that the half-life of Atrazine was only 5 days on cattail peat while it was 40 to 90 days on mineral soil. The latter is similar to published literature values for terrestrial soils. However, the amount initially sorbed depends on the hydraulic residence time of the water. Recent studies have shown that the pesticides are immobilized rather than degraded in wetlands soils (Crumpton, personal communication). Importantly, the pesticide–humic acid bond is very strong, thus release is unlikely to be rapid if the wetlands soil conditions are changed. The half-life of humic substances is measured in tens or hundreds of years.

Case Study #5. Macromolecular Halogen Removal: DOC and Modification of the Organic Signature in Prado Wetlands, CA

An exciting finding relative to wetlands phytoremediation is that wetlands can modify the entire carbon signature of water entering them. In particular, a recent study (Gray et al., 1996) showed that a potentially harmful macromolecular halogenated organic fraction was eliminated (Figure 2.5). Since the water at this time largely consisted of secondarily treated, chlorinated and dechlorinated wastewater, it is likely that the halogenated material originally consisted of large organic moieties that were not easily destroyed by the rapid conventional dechlorination with sulfur dioxide. A likely explanation of the mechanism of detoxification is reductive dechlorination since wetlands sediments are always anoxic when flooded. Other changes in nitrogenous and other carbon fragments are shown in Table 2.7.

Wetlands generally release dissolved organic carbon (DOC) which is not welcome in drinking water supplies since the most common methods of disinfection can produce harmful byproducts when DOC levels are above 2 to 4 mg/l. Recent research at Prado Wetlands indicates that DOC increases are not necessarily inevitable in constructed wetlands. In the 200 ha marsh, DOC increased from 5 to 9 mg/l (~0.6 mg/l DOC/day). Prado Wetlands is a heterogeneous mixture of open water and stands of at least a dozen plant types. Small experimental macrocosms within the marsh planted with dense macrophytes showed only about half of this DOC increase (Table 2.8). The balance between open water and closed canopies of vegetation may control DOC release, which is a balance between release and uptake. Labile DOC is needed to power bacterial activity in treatment wetlands, so it is not an unmixed blessing.

---

* Registered trademark of Polysciences Corp

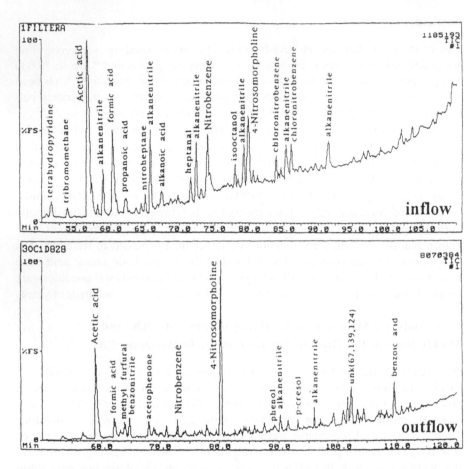

FIGURE 2.5 Removal of halogenated organics by wetlands. The carbon fraction signature shown here is derived from low-temperature flash pyrolysis followed by GC-MS analysis. Residence time in this experimental 1000-m² bulrush wetland macrocosm was about 1 week. The inflow peaks of tribromomethane peak (far left), chloronitrile and chloronitrobenzene (right section) are removed. (From Gray et al., 1996. Report to Orange County Water District, Santa Ana River Water Quality and Health Study, April 1996.)

CLASS 4. PATHOGENS, BACTERIA, VIRUSES, AND PROTOZOAN CYSTS

As with organics removal in wetlands, treatment of microbial pathogens is in its infancy. Human pathogens will die because they are separated from their hosts or are eaten by the myriad filter-feeding microbes in wetlands. On the other hand, some pathogens from wild animals, including *Giardia* from beaver or elk, will increase in some wetlands. An excellent discussion of this topic is given in Kadlec and Knight (1996).

There is as yet no dependable mechanism for understanding pathogen removal. In particular, public health authorities prefer "multiple levels of protection" in any waste treatment process. A long detention time prior to water reuse is one such protection level to be added after conventional treatment. Wetlands, with their dense

TABLE 2.7

Changes in Organic Fingerprint in the Experimental Marsh

Process/Dominant Vegetation	Influent	1-Year-Old Cell A	4-Year-Old Cell C	Change
Halogenated				
Bulrush		7.6	5 1	-2.6
Cattail	7 9	NA	6 1	-1.8
Mean: emergent plants	7.9	7 6	5.6	-2.3
Duckweed		5.0	3.0	-4.9
Unknown-N				
Bulrush		20.6	25	+4.9
Cattail	20.9	22.1	24	+3.9
Mean: emergent plants	20.9	21.3	24.5	+3.6
Duckweed		37.3	20	-0.9
Known-N				
Bulrush		25	25	-3
Cattail	28	21	24	-4
Mean: emergent plants	28	23	24.5	-3.5
Duckweed		25	20	-8
All-N				
Bulrush		45	57	+8
Cattail	49	43	51	+2
Mean: emergent plants	49	44	54	+5
Duckweed		62	56	+7
Aliphatics				
Bulrush		35	27	+1
Cattail	26	31	30	+4
Mean: emergent plants	26	33	29	+3
Duckweed		21	26	0
Aromatics				
Bulrush		12	11	-6
Cattail	17	10	13	-4
Mean: emergent plants	17	11	12	-5
Duckweed		13	15	-2

Note: Changes in the percentages of identifiable compounds in the organic fingerprint of DOC as water passed through the experimental macrocosms in August 1994. Note that the fingerprint fragment changes are frequently consistent, especially between the two emergent macrophytes, cattails and bulrush. Changes in the fingerprint appear to occur much more rapidly in the duckweed macrocosm, probably due to the more rapid depletion of oxygen which is almost zero in cell A compared with 1.1 to 2.7 mg/l in the emergent plant stands. Marsh age appears to speed up the transformation of the carbon fingerprint for the halogenated, unknown-N, and all-N fractions. Aromatics are rapidly reduced to a constant background regardless of additional amounts of carbonaceous detritus. (Derived from data from experiments by Alex Horne and Kim Gray, Northwestern University, Chicago)

TABLE 2.8

Small Changes in Dissolved Organic Carbon (DOC, mg/l) in Well-Treated
Secondary Effluent as It Passed Through the 0.1-ha University of
California, Berkeley Experimental Wetland

Process/Dominant Vegetation	Influent	1-Year Old Cell A	2-Year Old Cell B	4-Year Old Cell C	Change
Bulrush	3.9	4.1	4.3	4.6	+0.7
Cattail	3.9	4 3	4 5	4.8	+0.9
Duckweed	4.1	4.3	4 3	4 7	+0 6
Mean (sd)	3 97 (0.12)	4 23 (0 12)	4.37 (0.12)	4.7 (0.10)	+0 73 (0.15)

Note: The small DOC increases (<1 mg/l in 2 days) relative to the entire marsh (4 mg/l in 7 days).
DOC increased steadily regardless of marsh age (i.e., amount of carbonaceous detritus) or dominant
macrophyte plant.

array of filtering organisms, may be more valuable than groundwater, the current
detention medium of choice. For example, over 500,000 filter-feeding rotifers and
small nematodes were found in a liter of water in submerged wetland plants (Horne
and Roth, 1989). This compares with orders of magnitude less in other aquatic
environments, e.g., a few dozen per liter in reservoir water and an unknown but
presumably very small number in deep groundwater where food is scarce.

An exciting prospect is the removal of *Cryptospiridium* in wetlands. *Cryptospiridium* cysts are about 5 μm in diameter, which is similar to the small algae and
bacterial clumps that are the food of many attached and planktonic rotifers. The
cysts of this protozoan are very resistant to conventional chlorination ingestion of
even a few cysts can cause death in humans with weakened immune systems
(Goldstein et al., 1996). Obviously, this matter needs urgent experimental investigation.

ATTRACTIVE NUISANCES: POTENTIAL DANGERS IN FULL-SCALE IMPLEMENTATION OF PHYTOREMEDIATION

By design, terrestrial phytoremediation often acts by bringing the pollutant into the
biological cycle where it can be removed by harvesting the accumulator plants for
disposal or reuse. Sometimes the pollutant will be passed on into the air or immobilized in the soils but usually it accumulates in leaves and especially in seeds. In
either case the plants may become attractive nuisances loaded with toxicants in a
readily available form for birds and insects. In some cases the plants pose a threat
after death when they lose carbon and contribute to the detritus pool used by
invertebrates.

Most wetlands plants are not accumulators so the attractive nuisance aspect is muted. However, there is much research with more edible wetlands plants such as duckweed that concentrates metals and is an attractive food for waterfowl such as coot. It is virtually impossible to cover wetlands with nets or otherwise keep out wild birds and insects. Perhaps duckweed-type metal removal is best reserved for small-scale applications inside metal plating factories rather than large outdoor systems. Because an international bird migration treaty was broken, the attractive nuisance problem caused Kesterson Marsh to be an international toxic problem for the U.S. vis-à-vis Mexico and Canada. It would be wise for both terrestrial and wetlands phytoremediation to design treatments to avoid repetition of such an event.

REFERENCES

Alord, H. H. and R. H. Kadlec. 1995. The interaction of Atrazine with wetland sorbents. *Ecol. Eng.* 5: 469-480.

Bachand, P. A. M. and A. J. Horne. 1999. Denitrification in constructed free-water surface wetlands: II. Vegetation community effects. *Ecol. Eng.*

Barnes, I. 1985. Sources of selenium (in the Central Valley of California), in *Selenium and Agricultural Drainage.* Proc. 2nd Symp. Berkeley, CA, 41-51.

Bogardi, I. and Kuzelka. 1991. *Nitrate Contamination.* Springer-Verlag Berlin, 520.

Bomono, L., G. Pastorelli, and N. Zambon. 1997. Advantages and limitations of duckweed-based wastewater treatment systems. *Water Sci. Technol.* 35: 239-246.

Brix, H. 1997. Do macrophytes play a role in constructed treatment wetlands? *Water Sci. Technol.* 35: 11-17.

Brodie, G. A. 1993. Stages, aerobic constructed wetlands to treat acid drainage: case history of Fabius Impoundment #1 and overview of the Tennessee Valley Authority's program, in G. A. Moshiri (Ed.) *Constructed Wetlands for Water Quality Improvement.* Lewis Publishers, Boca Raton, FL, 157-165.

Canter, L.W. 1997. *Nitrates in Groundwater.* Lewis Publishers, Boca Raton, FL, 263.

CH2M-Hill, 1991; Grand Strand Water and Sewer Authority Central Wastewater Treatment Plant Wetlands Discharge. Fifth Annual Report, CH2M-Hill Eng., Gainesville, FL.

CH2M-Hill, 1992. Carolina Bay Natural Land Treatment Program. Final Report. CH2M-Hill Eng., Gainesville, FL.

Coveney, M. F., D. L. Stites, E. P. Lowe, and L. E. Battoe. 1994. Nutrient removal in the Lake Apopka marsh flow-way demonstration project. *Lake Reserv. Manage.* 9: 66.

Combs, G. F. and S. B. Combs. 1986. *The Role of Selenium in Nutrition.* Academic Press, New York, 532.

Craft, C. B., J. Vymazal, and C. J. Richardson. 1995. Response of everglades plant communities to nitrogen and phosphorus additions. *Wetlands.* 15: 258-271.

de Jong, J. 1990. Management of the River Rhine. *Water Environ. Technol.* April 44-51.

Eger, P., G. Melchert, D. Antonson, and J. Wagner. 1993. The use of wetland treatment to remove trace metals from mine drainage, in G. A. Moshiri (Ed.) *Constructed Wetlands for Water Quality Improvement.* Lewis Publishers, Boca Raton, FL, 171-178.

Elder, J. F. 1985. Nitrogen and phosphorus speculation and flux in a large Florida river-wetland system. *Water Resour. Res.* 21: 724-732.

Frankenberger, W. T., Jr. and S. Benson (Eds.). 1994. *Selenium in the Environment.* Marcel Dekker, New York.

Goldstein, S. T., D. Juranek, O. Ravenholt, A. W. Hightower, D. G. Martin, J. L. Masnik, S. D. Griffiths, R. R. Riech, and B. L. Herwaldt. 1996. Cryptosporidiosis: an outbreak associated with drinking water despite state-of-the-art water treatment. *Ann. Intern. Med.* 124: 459-468.

Grover, R. 1988. *Environmental Chemistry of Herbicides*, Vol. 1. CRC Press, Boca Raton, FL, 207.

Gray, K. A., S. McAuliffe, R. Bornick, A. Simpson, A. J. Horne, and P. A. M. Bachand. 1996. Evaluation of organic quality in Prado Wetlands and Santa Ana River by Pyrolysis GC-MS. Report to Orange County Water District, Santa Ana River Water Quality and Health Study, April 1996, 100.

Gregg, J. H. and A. J. Horne. 1993. Short-term distribution and fate of trace metals in a constructed wetland receiving treated municipal wastewater: a microcosm study using radiotracers. University of California, Berkeley. Environmental Engineering and Health Science Laboratory Report, No. 93-4, December 1993, 159.

Girts, M. A. and R. L. P. Kleinmann. 1986. Constructed wetlands for treatment of acid mine drainage: a preliminary review, in *National Symposium on Mining: Hydrology, Sedimentology, and Reclamation*. University of Kentucky Press, Lexington, 165-171.

Guntenspergen, G. R., F. Sterns, and J. A. Kadlec. 1989. Wetland Vegetation, in D.A. Hammer (Ed.) *Constructed Wetlands for Wastewater Treatment: Municipal, Industrial and Agricultural*. Lewis Publishers, Boca Raton, FL, 73-88.

Hammer, D. A. 1988. *Constructed Wetlands for Wastewater Treatment: Municipal, Industrial and Agricultural*. Lewis Publishers, Boca Raton, FL, 831.

Hammer, D. A. 1996. *Creating Freshwater Wetlands*. 2nd ed., Lewis Publishers, Boca Raton, FL, 406.

Hodgson, G. 1994. Coral reef catastrophe. *Science*, 266: 2930-1931.

Horne, A. J. 1988. Potential Causes and Cures for the Historical Reduction in Percolation Pond Flow Rates in the Orange County Water District: Review of Data to December 1988 in Kraemer and Anaheim Basins. Report to Orange County Water District, 10.

Horne, A. J. 1991. Selenium detoxification in wetlands by permanent flooding. I. Effects on a macroalga, an epiphytic herbivore, and an invertebrate predator in the long-term mesocosm experiment at Kesterson Reservoir. *Water Air Soil Pollut.* 57-58: 43-52.

Horne, A. J. 1994. Kinetics of selenium uptake and loss and seasonal cycling or selenium by the aquatic microbial community in the Kesterson Wetlands, in W. T. Frankenburger, Jr. and S. Benson (Eds.), *Selenium in the Environment*. Marcel Dekker, New York, 9-1 to 9-4.

Horne, A. J. 1995. Nitrogen removal from waste treatment pond or activated sludge plant effluents with free-surface wetlands. *Water Sci. Technol.* 31: 341-351.

Horne, A. J. and J. C. Roth. 1989. Selenium detoxification studies at Kesterson Reservoir wetlands: depuration and biological population dynamics measured using an experimental mesocosm and pond 5 under permanently flooded conditions. University of California, Berkeley. Environmental Engineering Health Science Laboratory Report. No. 89-4. 107 and Appendix.

Horne, A. J. and C. R. Goldman. 1994. *Limnology* (2nd ed.). McGraw-Hill, New York.

Kadlec, R. H. and R. L. Knight. 1996. *Treatment Wetlands*. Lewis Publishers, Boca Raton, FL, 893.

Lund, L. J. and A. J. Horne. 1999. Role of denitrification in nitrate losses in a large constructed wetland at Prado, California estimated using stable nitrogen isotope ratios. (In press, *Ecol. Eng.*).

Lyon, J. G. 1993. *Wetland Identification and Delineation*. Lewis Publishers, Boca Raton, FL, 157.

Meade, R. H. and R. S. Parker. 1985. Sediments in rivers of the United States. National Water Summary, 1984. Water Supply Paper 2275. U.S. Geological Survey, Reston, VA.

Marble, A. D. 1992. *A Guide to Wetland Functional Design*. Lewis Publishers, Boca Raton, FL, 221.

Mitsch, W. J. and J. G. Gosslink. 1993. *Wetlands*. Van Nostrand Reinhold, 722.

Mineral Policy Center. 1997. Golden Dreams, Poisoned Streams, Mineral Policy Center, Washington, D.C. 209.

Moshiri, G. A. 1993. *Constructed Wetlands for Water Quality Improvement*. Lewis Publishers, Boca Raton, FL, 632.

Mungur, A. S., R. E. Shutes, D. M. Revitt, and M. A. House. 1997. An assessment of metal removal by a laboratory scale wetland. *Water Sci. Technol.* 35: 123-133.

Negri, A. P., G. P. Jones, and M. Hindmarsh. 1995. Sheep mortality associated with paralytic shellfish poisons from the cyanobacterium *Anabaena circinalis*. *Toxicon.* 33: 1321-1329.

Ohlendorf, H. M., D. J. Hoffman, M. K. Saiki, and T. W. Aldrich. 1986. Embryonic mortalities and abnormalities of aquatic birds: apparent impacts by selenium from irrigation drainwater. *Sci. Total Environ.* 52: 49-63.

Peverly, J. H., J. M. Surface, and T. Wang. 1995. Growth and trace metal absorption by *Phragmites australis* in wetlands constructed for landfill leachate treatment. *Ecol. Eng.* 5: 21-35.

Philips, R. G. and W. G. Crumpton. 1994. Factors affecting nitrogen loss in experimental wetlands with different hydrologic loads. *Ecol. Eng.* 3: 399-408.

Reilly, J. F., A. J. Horne, and C. D. Miller. 1999. Nitrogen removal in large-scale free-surface constructed wetlands used for pre-treatment to artificial recharge of groundwater. (In press, *Ecol. Eng.*).

Richardson, C. J., S. Qian, C. B. Craft, and R. G. Qualls. 1997. Predictive models for phosphorus retention in wetlands. *Wetlands Ecol. Manage.* 4: 159-175.

Sharma, S. S. and J. P. Gaur. 1995. Potential of Lemna polyrhiza for removal of heavy metals. *Ecol. Eng.* 4: 37-44.

Tang, S-Y. 1993. Experimental study of a constructed wetland for treatment of acidic wastewater from an iron mine in China. *Ecol. Eng.* 2: 253-260.

TAP. 1992. Review of the Everglades Protection Project conceptual design of stormwater treatment areas. Review to Technical Advisory Panel (TAP) by Nolte & Assoc., Sacramento, CA, 57.

Tchobanoglous, G. and E. D. Schroeder. 1985. *Water Quality*. Addison-Wesley, Reading, MA, 768.

San Francisco Examiner. 1987. State calls for costly U.S. Kesterson clean-up. March 19.

Science. 1986, Selenium Threat in the West. 231: 111.

Snoddy, E. L., G. A. Brodie, D. A. Hammer, and D. A. Tomljanovich. 1989. Control of the armyworm, *Simyra henrici*, on cattail plantings in acid drainage treatment wetlands at Widows Creek steam-electric plant, in D. A. Hammer (Ed.) *Constructed Wetlands for Wastewater Treatment: Municipal, Industrial and Agricultural*. Lewis Publishers, Boca Raton, FL, 808-811.

U.S. EPA. 1992. Another Look: National Survey of Pesticides in Drinking Water Wells – Phase II Report. EPA 579/09-91-020. Office of Water, Washington, D.C.

U.S. EPA. 1993. Created and natural wetlands for controlling non-point source pollution. U.S. Environmental Protection Agency. R. K. Olson, (Ed.), C. K. Smoley/CRC Press, Boca Raton, FL, 216.

3 FACTORS INFLUENCING FIELD PHYTOREMEDIATION OF SELENIUM-LADEN SOILS

Gary S. Bañuelos

CONTENTS

INTRODUCTION

In California's San Joaquin Valley and in numerous other irrigated agricultural areas in the western U.S., irrigation effluent may accumulate in confined shallow aquifers, eventually rising to levels that adversely affect crops (Ayars et al., 1994). To sustain long-term agricultural productivity in these regions, subsurface drainage systems for the removal of this effluent must be installed (Mercer and Morgan, 1991; Ayars, 1996). On the western side of San Joaquin Valley, there are several thousand hectares of irrigated land (possessing subsurface drains) with high water tables resulting from over irrigation. Because the drainage system was never completed, the saline effluent produced in this region was eventually routed and discharged into Kesterson National Wildlife Refuge. The wetlands receiving the drainage water, in the course of being used as wildlife habitat, were also operated as evaporation ponds to reduce the volume of agricultural wastewater. Deleterious effects on birds and fish were documented on biological systems inhabiting or frequenting Kesterson Reservoir (Ohlendorf and Hothem, 1995). Selenium was identified as the element of primary concern. Studies showed exposure to high Se diets resulted in high tissue Se concentrations in waterfowl (Presser and Ohlendorf, 1987; Tanji et al., 1986; Ohlendorf and Hothem, 1995; Sylvester et al., 1991; U.S. Department of the Interior, 1986). These findings resulted in closing of the Kesterson Reservoir for receiving drainage effluent from agricultural lands.

Growers on the west side of the San Joaquin Valley have tried alternative practices to reduce their production of Se-laden effluent and thus sustain their agricultural land. After more than a decade of extensive research on Se remediation in California, several strategies to reduce loads of Se from entering the effluent have been proposed by the Salinity Drainage Task Force Committee in California (UC Salinity/Drainage Program, 1993). Some of these include improvement of irrigation and drainage management practices (Grattan, 1994; Mercer and Morgan, 1991), microbial volatilization (Frankenberger and Karlson, 1994), and vegetation management (phytoremediation) with perennial and annual crops (Bañuelos and Meek, 1990; Wu et al., 1988; Parker et al., 1991; Parker and Page, 1994).

Phytoremediation is a plant-based technology that is being considered for managing Se in central California soils (Bañuelos and Meek, 1990; Terry and Zayed, 1994, 1998; Parker and Page, 1994; Wu and Huang, 1991) and for removing other toxic trace elements in soils (Baker et al., 1994; Chaney et al., 1994; McGrath et al., 1993; Salt et al., 1995; Cunningham and Lee, 1995; Blaylock et al., 1997). The phytoremediation technology for Se implies the use of plants in conjunction with microbial activity to extract, accumulate, and volatilize Se. Any one or a combination of these plant responses may lead to lower concentrations of soluble Se in the soil and thus lower amounts entering the effluent. Most greenhouse studies on phytoremediation of Se have shown that Se added as soluble selenate can be extracted and/or volatilized from soil, translocated to shoot tissue, and removed as Se-laden plant material. Studies are needed, however, that demonstrate the effectiveness of phytoremediation for reducing the amount of naturally occurring Se entering effluent (Martens and Suarez, 1997), since options for disposing of Se-laden effluent are still unclear.

Field research on Se phytoremediation is still in the nascent stage (Bañuelos et al., 1993); however, field studies are crucial to develop sound phytoremediation strategies for remediating soils and sediments (Schnoor et al., 1995). Growing crops to manage soluble Se by field phytoremediation requires the application of a wide range of knowledge about the chemistry and transformation of Se in soil, Se uptake and its toxicity in plants and animals, and sustainable agronomical practices necessary for long-term crop production. The successful implementation of phytoremediation requires growing selected crops in Se-containing soils as part of a crop rotation and simultaneously reducing amounts of soil Se primarily by plant uptake. Factors to consider for phytoremediation under field conditions in central California include: (1) soil salinity and high concentrations of toxic elements, (2) presence of competitive ions affecting Se uptake, (3) adverse climatic conditions, (4) water management strategies that produce less effluent, (5) unwanted consumption of high Se plants by wildlife and insects, and (6) acceptance of phytoremediation as a remediation technology by the public and growers in regions known to have Se.

The objective of this chapter is to summarize results from recent field studies conducted by the USDA–ARS, Fresno, and the University of California on managing levels of naturally occurring Se in west side soils of central California.

GENERAL OBSERVATIONS ON REPORTED FIELD STUDIES

Field sites with moderate levels of naturally occurring Se near Los Baños, CA were selected for evaluating crops used in phytoremediation. The soils at these sites contained very little plant-available Se (soluble forms of Se). Because of regulatory restrictions placed upon growers in water districts of the west side of the San Joaquin Valley, the load of soluble Se leaving these soils via drainage effluent is closely controlled. Thus, Se concentrations in drainage water produced from all west side soils must be constantly monitored.

Cropping, irrigation, cultivation of the soil, and organic matter (Neal and Sposito, 1991) contribute to the solubilization and movement of immobile or complexed soil Se. For this reason, some growers are committed to planting crops considered for phytoremediation as a preventative measure for reducing the amount of soluble Se entering their effluent. Information regarding growth performance and uptake of naturally occurring Se under field conditions, e.g., high salinity, boron, and sulfate, for these crops is limited, however. Moreover, there is no information available on managing Se in the soil with selected phytoremediation crops.

INFLUENTIAL FACTORS IN FIELD PHYTOREMEDIATION

Crop Rotation

Crop selection is an important factor for successful field phytoremediation of Se. Phytoremediation strategies should initially consider rotations among phytoremediation crops under field conditions. This practice will likely contribute to a constant production of biomass and to a reduction of plant disease (e.g., Fusarium, Rhizoctonia root rot, Alternaria block spot), insect population, and weed buildup. Different crops used in rotation may extract Se from different zones of the soil profile and deposit it at more accessible depths for eventual uptake by subsequent crops used in phytoremediation. For long-term maintenance of Se-containing soils in Se-sensitive areas of the west side of central California, selected crops should be tried in rotation with other agronomic crops, e.g., cotton, wheat, tomatoes, etc., typically grown in these saline soils (Shennan et al., 1995).

Bañuelos and colleagues (1997) evaluated a rotation of selected crops as a preventative measure for reducing amounts of naturally occurring Se entering effluent from soils located near Kesterson Reservoir. The following crops that can reduce soil Se levels were evaluated near Los Baños, CA on 15 10 x 10 m plots: Indian mustard (*Brassica juncea* Czern L.), tall fescue (*Festuca arundinacae*), birdsfoot trefoil (*Lotus corniculatus*), kenaf (*Hibiscus cannibinus*), and bare plots (without plants). The four different phytorotations used from 1992 to 1995 consisted of the following: (I) bare plots; (II) Indian mustard, Indian mustard, tall fescue, tall fescue;

(III) birdsfoot trefoil, birdsfoot trefoil/tall fescue mixture, birdsfoot trefoil/tall fescue mixture, tall fescue; and (IV) kenaf, kenaf, tall fescue, tall fescue.

Table 3.1 presents the dry matter of the above-ground biomass production for the tested crops, including two annual clippings for tall fescue, birdsfoot trefoil, tall fescue, and birdsfoot trefoil mixture on an area (m^2) basis for each year. Kenaf produced the greatest amount of biomass among the tested species. Tissue Se concentrations were all under 1 mg Se kg^{-1} DM, except for Indian mustard which exceeded 2 mg Se kg^{-1} DM (Table 3.1). Plots from crop rotation II, which had Indian mustard for the first 2 years, had the greatest reduction of total soil Se compared to all plots after 4 years (Table 3.1). Total soil Se concentrations between 0 to 60 cm were lower in all cropped plots than the bare plots after 4 years. Overall, the cropped plots were more effective in lowering total soil Se in years 1992 and 1993 than in 1994 and 1995. The percentage changes between preplant and post-harvest soil Se concentrations (lost Se) after 4 years for each crop rotation is as follows: I – 17%, II – 60%, III – 34%, and IV – 41%.

In another multiyear field study conducted near Los Baños, Bañuelos and colleagues (1995) planted tall fescue on six 17 x 17 m plots and left six plots bare. Table 3.2 shows that, despite the low tissue Se concentrations, the cropped plots had 25% lower soil Se concentrations after 4 years from 0 to 45 cm vs. 11% in bare plots and 25% lower in cropped plots from 45 to 90 cm vs. 3% in bare plots, respectively. Tall fescue is a perennial grass with an extensive root system and a high transpiration rate. The species appears to be moderately effective at reducing soil Se concentrations near the soil surface, as well as in the subsurface profiles. Moreover, tall fescue is salt tolerant and thus a likely candidate for use on soils with relatively high levels of salinity. Although such perennial crops as tall fescue may take up less Se, their compatibility with conventional fodder crop equipment (e.g., mechanical swather and baler) make them ideal candidates as low-maintenance crops used for phytoremediation. Selenium inventory of lost Se under field conditions will be discussed later in this chapter.

Water Management

Water requirements are not known for crops used in phytoremediation, yet water management strategies are essential for growing Se-accumulating plants in irrigated regions of the west side of central California regions to produce the greatest amount of biomass with the minimum application of water. Efficient irrigation reduces percolation losses and the production of Se-laden effluent. Bañuelos et al. (1993, 1995) used data collected weekly by neutron probe to a depth of 1.5 m and data collected from the California Irrigation Management Information System (CIMIS) weather station (Howell et al., 1983) to estimate crop water use and schedule irrigation on crops used in phytoremediation. Responses of such crops as canola and kenaf cultivars to different regimens of irrigation are presently being evaluated in a multiyear field study conducted in central California (Tables 3.3 and 3.4). Based on the preliminary data, production of biomass increased with the amount of water applied up to reference evapotranspiration (Et$_r$). The greater the yields, the more Se

TABLE 3.1
Changes in Naturally Occurring Se Concentrations from 0 to 60 cm Depth and Mean Tissue Concentrations of Se in Crops Grown in Different Rotations for Phytoremediation

Year	Plant Species[a]	Rotation #	Total Soil Se Concentration at:		Change[b] (%)	Shoot Se (mg kg^{-1} DM)	DM Yield (g m^{-2})
			Preplant (mg kg^{-1} soil)	Post-Harvest			
1992	K° (bare plot)	I	1.32(.08)[c]	1.27(.06)	4	NA	NA
	Indian mustard	II	1.20(.06)	0.90(.09)	25	2.15(.06)	1328
	Birdsfoot trefoil	III	1.18(.12)	1.06(.09)	10	0.58(.01)	435
	Kenaf	IV	1.41(.09)	1.29(.10)	9	0.70(.02)	3125
1993	K° (bare plot)	I	1.22(.10)	1.16(.08)	5	NA	NA
	Indian mustard	II	0.85(.09)	0.69(.06)	19	1.70(.06)	1212
	Birdsfoot trefoil/ tall fescue	III	1.09(.06)	0.92(.05)	16	0.61(.02)	721
	Kenaf	IV	1.26(.13)	1.09(.10)	13	0.59(.03)	3450
1994	K° (bare plot)	I	1.18(.09)	1.13(.09)	4	NA	NA
	Tall fescue	II	0.66(.07)	0.58(.07)	12	0.41(.01)	400
	Birdsfoot trefoil/ tall fescue	III	0.86(.07)	0.77(.09)	10	0.56(.02)	902
	Tall fescue	IV	1.13(.09)	1.01(.08)	11	0.52(.03)	350
1995	K° (bare plot)	I	1.15(.10)	1.10(.12)	4	NA	NA
	Tall fescue	II	0.56(.06)	0.51(.08)	9	0.39(.01)	705
	Birdsfoot trefoil/ tall fescue	III	0.81(.07)	0.78(.07)	4	0.36(.01)	1121
	Tall fescue	IV	0.95(.08)	0.83(.06)	13	0.42(.01)	802

Note: NA = Not applicable.
[a] Indian mustard and kenaf were planted and harvested and then replanted the following year. Tall fescue and birdsfoot trefoil were only planted once in their respective plots.
[b] Percent change of total Se concentrations between preplant and post-harvest soil sampling for each respective year.
[c] Values represent the mean from six replicates followed by the standard error in parenthesis.

TABLE 3.2
Annual Changes in Naturally Occurring Se Concentrations from 0 to 90 cm Depth Planted to Tall Fescue or Allowed to Remain as Bare Plot[a,b]

Year	Treatment	Soil Depth (cm)	Total Se Concentrations Preplant (mg kg⁻¹ soil)	Total Se Concentrations Post-Harvest (mg kg⁻¹ soil)	Change (%)	Shoot Se (mg kg⁻¹ DM)	DM Yield (kg m⁻²)
1992	Bare Plot	0–45	1.65(.12)	1.61(.14)	2		NA
		45–90	1.59(.09)	1.65(.13)	4		
	Planted	0–45	1.47(.10)	1.32(.09)	10	1.02	4.4
		45–90	1.63(.04)	1.49(.09)	9		
1993	Bare Plot	0–45	1.71(.14)	1.63(.09)	5		NA
		45–90	1.72(.12)	1.67(.13)	3		
	Planted	0–45	1.37(.10)	1.24(.08)	9	1.25	4.5
		45–90	1.43(.11)	1.41(.09)	1		
1994	Bare Plot	0–45	1.58(.13)	1.55(.12)	3		NA
		45–90	1.60(.10)	1.63(.09)	2		
	Planted	0–45	1.31(.14)	1.20(.13)	8	2.10	5.4
		45–90	1.30(.10)	1.34(.09)	3		
1995	Bare Plot	0–45	1.57(.10)	1.53(.10)	3		NA
		45–90	1.58(.13)	1.62(.12)	2		
	Planted	0–45	1.19(.08)	1.15(.09)	3	1.95	6.3
		45–90	1.25(.12)	1.17(.09)	6		
1996	Bare Plot	0–45	1.49(.13)	1.47(.11)	5		NA
		45–90	1.49(.14)	1.55(.15)	4		
	Planted	0–45	1.20(.12)	1.09(.11)	9	2.41	6.9
		45–90	1.34(.10)	1.23(.09)	8		

Note: NA = not applicable.
[a] Values are means from 24 samples collected each year followed by standard error in parenthesis.
[b] Values are means from three clippings collected each year.

that can be extracted from the soil. If a minimum amount of water will grow a successful crop, then the acceptance and use of phytoremediation by growers in water districts where water supplies are stringently regulated might be greater.

Other studies have shown the effect of field irrigation practices on biomass and Se accumulation using various plant species on a two-acre field plot by Lawrence Berkeley Laboratory and University of California (Wu, 1994). Generally, water applied efficiently contributed to an increase in biomass. Wu reported that irrigation increased biomass plant tissue Se concentrations in *Brassica hyssopifola* Kuntze (summer weed) and in *Melilotus indica* (winter weed). Irrigation scheduling influences growth of the rooting system. Thus, encouraging deeper root development with planned water deficits may permit some plant species to access bioavailable Se in the deeper subsoil horizons. Information is needed to evaluate the influence of soil water levels on promoting volatilization of Se by the plant or soil microbes, as well as whether sequential drying and rewetting cycles promote microbial activity (Frankenberger and Karlson, 1994).

Water Reuse

Se-laden drainage effluent that contains dissolved salts and other constituents must be managed to minimize its detrimental effects on the ecology of the water or land where it is discharged (van Schilfgaarde, 1990; Shennan et al., 1995). Field studies were conducted in central California by Watson and colleagues (Watson and O'Leary, 1993; Watson et al., 1994; Bañuelos et al., 1993) with Se-laden effluent used as a source of irrigation water on *Atriplex* (saltbush) spp. and *B. juncea* (Indian mustard). The *Atriplex* species were grown on 30 x 124 m plots, while the *B. juncea* was planted three successive times on 5 x 15 m plots and replicated 4 times, respectively. The agricultural effluent used as the source of irrigation water had an electrical conductivity (EC) of 19 dS m^{-1} and a Se concentration ranging from 150 to 200 μg l^{-1}. Mustard plants accumulated Se up to 3.1 mg Se kg^{-1} DM, whereas the *Atriplex* species did not exceed 0.6 mg Se kg^{-1} DM (Table 3.5). If water reuse is to be considered as a disposal option for Se-laden effluent, long-term feasibility of reusing Se-laden drainage water is dependent not only on crop selection (e.g., Indian mustard, saltbush), but also on well planned strategies related to managing chemical and physical changes of the soil (Shennan et al 1995; Ayars et al., 1994; Grattan, 1994). Management strategies should include: (1) maintaining a favorable salt balance — the mass of salts leaving the area must be greater than or equal to that entering the area; (2) maintaining good soil physical conditions; (3) considering the mobility and retention of specific elements within the soil that can be toxic to the plant (e.g., B) or to biological consumers (e.g., Cd, Se); and (4) finding economically viable salt-tolerant crops that accumulate Se. *Brassica* species may be a suitable candidate to receive Se-laden effluent. However, unless plants take up Se faster than evapotranspiration, the net effect of irrigating land with Se-enriched water may increase the soil Se level (Parker and Page, 1994), unless leaching or volatilization by plants and microorganisms is occurring.

TABLE 3.3

Parameters Used for Approximating Water Application Rate for Kenaf with Subsurface Irrigation Under Field Conditions

Treatment (% Et$_c$)	Total Applied Water[a] (mm)	Net Soil Water Depletion (mm)	Calculated Crop Et$_c$ (mm)	Potential Et$_r$ (mm)	Actual Et$_c$/Et$_r$
25	95	125	220	418	0.23
50	190	141	331	418	0.45
100	379	175	554	418	0 91
125	473	49	522	418	1 13
150	593	78	671	418	1 47

[a] Effective precipitation was 0 mm during crop year and assumes no deep percolation losses

Predators (Insects and Wildlife)

Identification of insects frequenting crops and soils used in phytoremediation is important in agriculture-producing regions like central California, especially with flowering plant species, i.e., Indian mustard, birdsfoot trefoil; Tables 3.6 and 3.7. Flowering plants tend to attract greater numbers of potential predators which could be harmful or beneficial to other near-growing agricultural crops. In addition, the transfer of Se from soil to crop, from crop to insect, from insect to insect, and from insect to animal, is a biological Se cycle (bioaccumulation) that should be monitored in long-term field phytoremediation of Se-laden soils. There has been contradictory evidence as to whether biomagnification of Se exists in the food chain (Kay, 1984; Lemly, 1985). Research is needed to examine the dynamics of Se bioaccumulation in insects frequenting field sites. Factors affecting bioaccumulation depend upon the availability of soil Se, plant species, the feeding behavior of the food chain consumers, and mobility of insects. A herbivore may have a choice in the quality of its diet. For example, grasshoppers (*Dissosteria pictipennis* Brunner) may reject plants that accumulated higher tissue Se concentrations, whereas mantises (*Litaneutria minor* Scudder) have little choice if they rely on grasshoppers (Table 3.6; Wu et al., 1995). Unpublished data by Bañuelos and colleagues have shown that aphids (*Aphididae* spp.) prefer feeding upon non-Se-containing *Brassica* species compared to Se-containing plants. Plants that have accumulated high concentrations of Se may inadvertently discourage the infestation by some insect species. High tissue Se concentrations by *Brassica* plants and subsequent bioaccumulation by insects and animals are concerns because of the deleterious effects Se may exert on birds and mammals that eat the insects. Ohlendorf and Santolo (1994) have illustrated in great detail exposure pathways and projected Se concentrations in biota at Kesterson Reservoir. Strategies that are used in fruit production (e.g., metal reflectors and noise guns) may be useful in frightening off certain animals from field-grown *Brassica* planted for phytoremediation. In field experiments, Bañuelos and colleagues (1997;

TABLE 3.4

Biomass of Canola and Cultivars of Kenaf Exposed to Different Water Application Rates Given in Table 3.3 with Subsurface Irrigation Under Field Conditions[a]

Treatment % Et_c	7-N		C-531		Kenaf cultivars: C-533 ($Mg\ ha^{-1}$)		EU-41		TA-2		Canola ($Mg\ ha^{-1}$)	
	Leaves	Stem	Leaves	Stem	Leaves	Stem	Leaves	Stem	Leaves	Stem	Leaves	Stem
25	9.4	12.7	11.4	17.4	10.1	14.8	8.4	13.0	10.5	17.7	0.86	3.9
50	10.7	13.2	12.0	24.0	10.9	22.3	10.9	23.9	17.9	21.6	1.83	5.3
100	12.9	23.5	15.5	36.6	16.2	28.9	12.9	26.6	13.9	28.3	2.36	8.3
125	12.2	2.25	15.6	30.1	15.2	26.7	11.3	21.3	12.3	26.5	4.33	7.5
150	13.7	26.2	16.0	28.8	12.2	27.6	11.2	25.4	19.2	28.2	3.10	9.3

[a] Based on a plant population of 160,000 plants ha^{-1}

TABLE 3.5

Shoot Tissue Concentrations of Se and Other Selected Elements in *Brassica juncea* and *Atriplex* Species Irrigated with Se-Laden Saline Effluent

Plant Species	Tissue Concentrations of						
	Se	B	Fe	Mn	Zn	Cu	S
				(mg kg^{-1} DM)			
Brassica juncea[a]	3.1(.10)	275(.10)	250(.04)	177(.08)	55(.03)	2(.01)	20400(.11)
Atriplex canescens[b]	0.5(00)	126(.07)	411(.09)	60(.02)	36(.04)	6(.01)	10400(.12)
A. undulata	0.6(00)	131(.05)	348(.09)	68(.01)	37(.02)	7(.01)	7500(.10)
A. deserticola	0.6(00)	121(.03)	388(.07)	53(.02)	35(.03)	6(.01)	10900(.09)
A. nummularia	0.6(00)	142(.04)	391(.05)	44(.01)	30(.01)	5(.01)	9260(1.1)
A. polycarpa	0.6(00)	135(.03)	401(.08)	70(.02)	39(.02)	6(.01)	9960(.07)

[a] Values for *B. juncea* are the means from three plantings followed by the coefficients of variation.
[b] Values from the *Atriplex* species are the means from three harvests followed by the coefficients of variation. (From Watson et al., *J. Environ. Qual.* 48: 157-162, 1994.)

TABLE 3.6

Tissue Selenium Concentrations Found in Grasshoppers and Mantis from Different Sites at Kesterson Reservoir

| | Soil Se | | Range of Se in | |
| | Total | Extractable | Grasshopper[a] | Mantis[b] |
Site #	(mg kg⁻¹ soil)	(mg l⁻¹)	(mg kg⁻¹ insect)	
1	0.6 ± .03[c]	0.05 ± .02	1.2–9 8	9–22
2	53 7 ± 17.5	0 72 ± 63	9 1–27 5	31–52
3	0.1 ± 09	0.02 ± .01	3.1–7.0	10.2–18 0
4	4 2 ± 2 3	0.16 ± .04	1 0–4 6	5 5–10 3

[a] *Dissosteria pictipennis* Brunner.

[b] *Litaneutria minor* Scudder

[c] Selected values are the means followed by the standard deviation (From Wu et al., *Environ. Toxicol. Chem.* 14: 733-742,1995.)

data not shown) have observed that grown *Brassica* seedlings are vulnerable to vertebrae herbivores. Ground squirrels that forage throughout the winter were especially fond of eating *Brassica* seedlings.

Selenium Inventory

A complete soil and plant Se inventory was not attempted in our field studies because of the inability to accurately account for (1) the variability of soil Se; (2) the transformation of naturally occurring Se into forms of Se not measured (e.g., volatile Se); (3) the movement of Se below sampling depth (90 cm); and (4) the inability to completely recover the root system of the specific crop. An estimated amount of soil Se removed in only the above-ground portion of a crop as a function of shoot tissue Se concentrations (on a DM basis) did not exceed 10% of the lost soil Se reported in the field studies presented in Tables 3.1 and 3.2 (calculation used 4 x 10⁶ kg soil/ ha/30 cm soil). This small percentage implies that a substantial amount of Se was also lost or removed through other processes, e.g., volatilization, leaching. Limited attempts at constructing Se inventories with vegetation management have been made by Parker and Page (1994). However, because of the above factors, an accurate mass-balance is difficult to construct. Caution should be taken when extrapolating greenhouse data to field scenarios because measurements and cultural practices in the field are not as precise as those in pot experiments and, more importantly, plant tissue concentrations of Se are always lower under field conditions. This assumption is illustrated in the following scenario with Se-laden soil collected from Kesterson Reservoir. Bañuelos and colleagues (1998) collected Se-laden soil from different depths ranging from 0 to 90 cm. Concentrations of total Se concentrations were as high as 112 mg kg⁻¹ soil (at the surface) and as low as 10 mg kg⁻¹ soil (at the deepest depth). Tissue concentrations in canola (*B. napus*) grown in the collected

TABLE 3.7

General Survey of the Predominate Insects Habitating Different Crop Species Used for Phytoremediation of Selenium Soils in Central California

Crop Species	Family	Insects[a] — Genus sp. or Common Name
Birdsfoot trefoil	Miridae	*Lygus* sp.
	Thripidae	*Frankliniella* sp. (western flower thrips)
	Cicadellidae	*Empoasca* sp (leafhopper)
		Aceratagallia sp
		Colladonus montanus
Indian mustard	Miridae	*Lygus* sp
	Thripidea	*Frankliniella* sp. (western flower thrips)
	Chrysomelidae	Subfamily: Alticinae (flea beetle)
	Rhopalidae	*Liorhyssus hyalinus* (hyaline grass bug); adults and nymphs were found.
	Pentatomidae	Stink bugs
	Cicadellidae	*Empoasca* sp. (leafhoppers)
Tall fescue	Thripidae	*Frankliniella* sp. (western flower thrips)
	Aeolothripidae	Banded thrips
	Chrysomelidae	Subfamily: Alticinae (flea beetles)
	Rhopalidae	*Liorhyssus hyalinus* (hyaline grass bug)
	Cicadellidae	*Amblysellus grex* (hopper)
		Euscelis obsoletus
		Other
	Delphacidae	Plant hoppers
	Coccincellidae	Ladybird beetles and other
Kenaf	Miridae	*Lygus* sp.
	Cocincellidae	*Hypersapis* sp.
	Cicadellidae	*Empoasca* sp. (leafhoppers)
		Aceratagallia sp.
	Lepidoptera	Salt marsh caterpillar
Cotton	Miridae	*Lygus* sp.
	Cicadellidae	*Empoasca* sp. (leafhoppers)
	Chrysomelidae	Subfamily: Alticinae (flea beetle)
Fallow (weeds)	Thripidae	*Frankliniella* sp. (western flower thrips)
	Lygaeidae	*Nysius raphanus* (false chinch bug)

[a] Insects are not ranked in order of predominance. Predominance varies with time of year. Sampling took place 15 times during the growing season using an insect sweep net. An in-depth identification of insects is presently being pursued.

soil were as high as 200 mg kg⁻¹ DM. When a mass-balance was attempted after harvest and absolute amounts of lost soil Se were calculated (based upon differences between preplant and post-harvest soil Se concentrations), approximately 54 mg of

Se from the potted soil was missing. Just 23% of this missing Se was accounted for by accumulation in plant tissues, while 77% went unaccounted. Under field conditions, total soil Se concentrations were on the average of 26 mg kg^{-1} soil from 0 to 90 cm depth. Canola planted in this soil at Kesterson Reservoir accumulated approximately 50 mg Se kg^{-1} DM in the plant tissue. When a mass-balance was attempted based on 30-cm depth, approximately 312 mg Se m^{-2} was missing from the soil (based upon differences between preplant and post-harvest soil Se accounted for in plant tissue). Canola accumulated less than 10% of the Se missing from the top 30 cm of soil.

The efficacy of phytoremediation is generally greater under controlled greenhouse conditions than under field conditions. In the above greenhouse and field studies conducted with Kesterson soil, it is clear that a percentage of lost soil Se (difference measured between preplant and post-harvest Se concentrations) was not recovered in plant tissue or in soil at post-harvest soil sampling. This "unaccounted for Se" may have been lost through biological volatilization of Se. The phenomenon of Se volatilization is suggested as an alternative explanation for the lower Se concentrations in soils, especially under field conditions. Certain fungi can convert Se species, selenite and selenate, to dimethylselenide and/or dimethyldiselenide, which are volatile compounds of Se (Frankenberger and Karlson, 1990). Although direct measurements of field volatilization were not made in the previously described studies, we acknowledge the ability of microorganisms and plants (Terry et al., 1992; Biggar and Jayaweera, 1993; Wu and Huang, 1991; Doran, 1982) to volatilize Se. Under field conditions, Frankenberger and Karlson (1994) and Karlson and Frankenberger (1989) found that volatilization of Se can be enhanced in the field with the addition of a carbon source and by having sufficient aeration and moisture under high temperatures. Irrigation practices (wetting and drying) may enhance the organically bound release of organic-bound Se to the methylating organisms in other field studies. Biggar and Jayaweera (1993) found that soil planted to barley volatilized 20 times more Se than soil alone, while Wu and Huang (1991) measured rates of 180 μg Se m^{-2} d^{-1} from *Distichlis* spp. (salt grass). Although it is difficult to distinguish between plant and microbial volatilization, the presence of plants increases the total rate of Se volatilization (Biggar and Jayaweera, 1993). Terry and his research group at the University of California, Berkeley are determining the physiological, biochemical, and microbial mechanisms involved in the volatilization process of Se. Volatilization of Se by plants will be discussed in Chapter 4.

Utilization of Plants in Phytoremediation

Selenium, while not required by plants, is an essential trace element for adequate nutrition and health for mammals (Se deficiencies are probably a greater problem than Se toxicities in animals, especially on the east side of central California). Generally, diets containing 0.1 to 0.3 mg kg^{-1} Se will provide adequate Se for animal feed (Mayland, 1994). Animal producers wishing to ensure adequate supplies of Se to their livestock have a variety of techniques at their disposal, which include giving Se by injection or by mouth as a feed supplement. Alternatively, harvested plant material, e.g., *Brassica*, used in phytoremediation can be carefully blended with animal forage and fed to animals (pending approval by regulating agencies) in Se-deficient areas. The quality of *Brassica* herbage is more comparable to a concentrate than to a traditional forage because of the relatively low fiber and high protein content (Wiedenfoeft and Barton, 1994). Therefore, plant species used for phytore-mediation in Se-laden soils may not only minimize the Se load eventually entering agricultural effluent by plant uptake and volatilization, but also the crop may become a product of economic importance for the grower as part of an agronomic crop rotation scheme.

Another means of improving the Se status of animals is to add plant material with high concentrations of Se (e.g., *Brassica*) to soils as a source of organic Se fertilizer for such forage plants as alfalfa, birdsfoot trefoil, or tall fescue. These plant species will eventually absorb some of the organically added Se (Bañuelos et al., 1993; Ajwa et al., 1998). Table 3.8 shows the accumulation of Se and selected nutritional values in potential forage crops grown in soil with incorporated Se-laden plant material used previously for phytoremediation. Ajwa and colleagues (1998) have evaluated carefully the fate of Se as a mineralized product excreted from animals and followed its subsequent uptake by plants. They found that canola and tall fescue accumulated less than 10 mg Se kg^{-1} DM after three plantings in the same soil amended with seleniferous organic materials. It is important to be aware of the absorption and eventual fate of Se fed to animals. Different crops may store Se as different seleno-amino acids (Abegaz, 1997) and thus affect its rate of absorption after consumption by different animals.

Other disposal possibilities for Se-laden plant material include its use in the production of paper products or utilization as a fuel for biogeneration power plants. The latter option must be considered if concentrations of toxic trace metals (i.e., Cd and As) found in the crop exceed concentrations deemed environmentally safe for

---

TABLE 3.8

Forage Quality of Different Plant Species Grown Under Field Conditions in Soil Enriched with Se-Laden Indian Mustard

Planted Species[a]	Crude Protein (%)	Digestible Dry Matter (%)	Se Content (mg kg^{-1} DM)
Tall fescue	19	92	1.3
Alfalfa	22	93	4.7
Birdsfoot trefoil	19	92	3.1
Canola	25	92	51 0

[a] Harvested Indian mustard, previously irrigated with selenate-containing water, was incorporated into soil to achieve a preplant total soil Se concentration of 0.7 mg kg^{-1} soil.

animal consumption (National Research Council, 1980). Table 3.9 shows concentrations of selected trace elements in four crops used in the phytoremediation of naturally occurring Se under field conditions. Successful long-term phytoremediation strategies must consider the trace element accumulation in harvested plant material. If the harvested plant material is to be considered for animal forage, factors such as age and physiological status of the animal (e.g., growth and lactation), nutritional status, levels of various dietary components, duration and rate of exposure, and biological availability of the compound influence the level at which a mineral element causes an adverse effect in an animal or on the environment (Oldfield, 1998).

Oil extracted from the seed of *Brassica* species, e.g., canola, grown for Se phytoremediation under field conditions is presently being evaluated by Bañuelos et al. (1999, unpublished). Like other vegetable oils, *Brassica* oil is composed predominantly of fatty acid containing triacyclglycerols with lesser amounts of phospholipids and glycolipids (Uppström, 1995). The canola oil is characterized by a low level of monounsaturated fatty acids and contains significant amounts of oleric and inolenic acids, which have important nutritional properties explained elsewhere (McDonald, 1995). The meal remaining after oil extraction is high in protein as a result of accumulation during seed development of napin and cruciferin storage proteins and oleasin, a structural protein associated with the oil bodies. This potential product may increase the growers' interest in considering a *Brassica* species in their crop rotation for managing soil Se, since canola oil, a product of *Brassica* seeds, is already an established and viable oil product throughout the world.

CONCLUSION

Field phytoremediation requires an integrated approach, which must consider initial selection of crops used for phytoremediation, crop rotation, irrigation and drainage water management strategies, chemical and mineralization transformations within the soil, pest management, harvest techniques, biomass use, economic feasibility, and social acceptance. Field studies should be conducted to effectively evaluate phytoremediation as a technology to reduce and manage Se levels in Se-containing soils and thus reduce the amount of soluble Se entering drainage effluent. More importantly, field phytoremediation requires time and must be considered as a long-term commitment.

TABLE 3.9

Mean Concentrations of Selected Trace Elements in Four Crops Used for the Phytoremediation of Se-Laden Soils Under Field Conditions for 2 Years

Species	Organ	Elemental Concentrations:							
		Zn	Cd	Mn	Fe	Al	Cu	Mo	B
					(mg kg^{-1} DM)				
Kenaf	Shoot	42(1)[a]	22(.4)[b]	222(5)	354(12)	420(15)	11(.4)	1(.1)	737(41)
	Root	29(.8)	21(.6)	35(1)	1268(52)[b]	1335(70)[b]	6(.2)	1(.1)	138(6)
Indian mustard	Shoot	26(.7)	15(.4)[b]	48(2)	162(12)	177(13)	5(.2)	4(.1)	273(10)
	Root	20(.8)	12(.5)	30(3)	773(39)	1045(50)[b]	4(.2)	.3(.1)	120(7)
Tall fescue	Shoot	24(.6)	<1	46(3)	84(6)	77(4)	1(.1)	<1	99(3)
	Root	28(.5)	<1	22(1)	497(12)	941(44)	4(.2)	<1	112(9)
Birdsfoot trefoil	Shoot	20(.4)	<1	37(2)	97(4)	80(3)	6(.1)	<1	117(9)
	Root	20(.5)	<1	36(2)	477(16)	980(57)[b]	5(.1)	<1	112(8)

[a] Values are the means followed by the standard error mean in parenthesis for 2 years.

[b] Concentrations could be considered potentially toxic for long-term direct animal consumption. (Based upon values reported and established by the National Research Council. 1980.)

REFERENCES

Abegaz, M. Investigation of the chromatographic retention behavior of organic and inorganic selenium compounds and determination of total selenium and of selenium compounds in environmental and biological samples with selenium-specific detection. Ph.D. dissertation, Universität Graz, Austria, 1997.

Ayars, J.E. Managing irrigation and drainage systems in arid areas in the presence of shallow groundwater: Case studies. *Irrig. Drainage Sys.* 10: 227-244, 1996.

Ayars, J.E., R.B. Hutmacher, R.A. Schoneman, S.S. Vail, and T. Pflaum. Long term use of saline water for irrigation. *Irrig. Sci.* 14: 27-34, 1994.

Ajwa, H.A., G.S. Bañuelos, and H.F. Mayland. Selenium uptake by plants grown in soils treated with inorganic and organic materials. *J. Environ. Qual.* 27: 1218-1227, 1998.

Baker, A.J.M., S.P. McGrath, C.M.D. Sidoli, and R.D. Reeves. The possibility of *in situ* heavy metal decontamination of polluted soils using crops of metal-accumulating crops. *Res. Cons. Recyc.* 11: 41-49, 1994.

Bañuelos, G.S., H.A. Ajwa, L. Wu, and S. Zambrzuski. Selenium accumulation by *Brassica napus* grown in Se-laden soil from different depths of Kesterson Reservoir. *J. Contaminated Soil.* 7: 481-496, 1998.

Bañuelos, G.S., H.A. Ajwa, B. Mackey, L. Wu, C. Cook, S. Akohoue, S. Zambrzuski. Evaluation of different plant species used for phytoremediation of high soil selenium. *J. Environ. Qual.* 26: 639-646, 1997.

Bañuelos, G.S., B. Mackey, L. Wu, S. Zambrzuski, and S. Akohoue. Bioextraction of soil boron by tall fescue. *Ecotoxicol. Environ. Safety* 31: 110-116, 1995.

Bañuelos, G.S., R. Mead, and G. Hoffman. Mineral composition of wild mustard grown under adverse saline conditions. *Agri. Ecosys. Environ.* 43: 119-126, 1993.

Bañuelos, G.S. and D.W. Meek. Accumulation of selenium in plants grown on selenium-treated soil. *J. Environ. Qual.* 19: 772-777, 1990.

Biggar, J.W. and G.R. Jayaweera. Measurement of selenium volatilization in the field. *Soil Sci.* 155: 31-35, 1993.

Blaylock, M.J., D.E. Salt, S. Dushenkov, O. Zakharova, C. Gussman, Y. Kopulnik, B.D. Ensley, and I. Raskin. Enhanced accumulation of Pb in Indian mustard by soil-applied chelating agents. *Environ. Sci. Technol.* 31: 860-865, 1997.

Chaney, R.L., S.L. Brown, F.A. Homer, Y.M. Li, A.J.M. Baker, and J.S. Angle. Hyperaccumulators: how do they differ in uptake, translocation, and storage of toxic elements. *Abstract for Dept. of Energy Workgroup on Bioremediation/Phytoremediation,* July 24-26, Santa Rosa, CA, 1994.

Cunningham, S.D. and C.R. Lee. *Phytoremediation: Plant-Based Remediation of Contaminated Soils and Sediments.* SSSA Special Pub. No. 43: 145-156, 1995.

Doran, J.W. Microorganisms and the biological cycling of selenium. *Adv. Microbial Ecol.* 6: 1-31, 1982.

Frankenberger Jr., W.T. and U. Karlson. Environmental factors affecting microbial production of dimethylselenide in a selenium-contaminated sediment. *Soil Sci. Soc. Am. J.* 53: 1435-1442, 1990.

Frankenberger Jr., W.T. and U. Karlson. Microbial volatilization of selenium from soils and sediments, in *Selenium in the Environment.* W.T. Frankenberger, Jr. and S. Benson, Eds., Marcel Dekker, Inc., New York, 369-389, 1994.

Grattan, S.R. Irrigation with saline water, in *Water Use in Agriculture.* K.K. Tanji and B. Yaron, Eds., *Adv. Series in Agric. Sci.* Springer-Verlag, Berlin, 22: 179-198, 1994.

Howell, T.A., C.J. Phene, D.W. Meek, and R.J. Miller. Evaporation from screened class A pans in a semi-arid climate. *Agri. Meteor.* 29: 111-124, 1983.

Karlson, U. and W.T. Frankenberger, Jr. Accelerated rates of selenium volatilization from California soils. *Soil Sci. Soc. Am.* 53: 749-753, 1989.

Kay, S.H. Potential for Biomagnification of Contaminants Within Marine and Freshwater Food Webs. Tech. Rpt. D-84-7, U.S. Army Corps of Engineers, Washington, D.C., 1984.

Lemly, A.D. Ecological basis for regulating aquatic emissions from the power industry: the case with selenium. *Regul. Toxicol. Pharmacol.* 5: 405-486, 1985.

Martens, D.A. and D.L. Suarez. Selenium speciation of soil sediment determined with sequential extractions and hydride generation atomic absorption spectrophotometry. *Environ. Sci. Technol.* 31: 133-139, 1997.

Mayland, H.F. Selenium in Plant and Animal Nutrition, in *Selenium in the Environment.* W.T. Frankenberger, Jr. and S. Benson, Eds., Marcel Dekker, Inc., New York, 29-46, 1994.

McDonald, B.E. Oil properties of importance in human nutrition, in *Brassica Oilseeds: Production and Utilization.* D. Kimber and D.I. McGregor, Eds., Cab International Oxon, U.K., 217-242, 1995.

McGrath, S.P., C.M.D. Sidoli, A.J.M. Baker, and R.D. Reeves. The potential for the use of metal-accumulating plants for the *in situ* decontamination of metal-polluted soils, in *Integrated Soil and Sediment Research: A Basis for Proper Protection.* J.P. Eysakers and T. Hamers, Eds., Kluwer Academy Publishers, Dordrecht, The Netherlands, 673-676, 1993.

Mercer, L.J. and W.D. Morgan. Irrigation, drainage, and agricultural development in the San Joaquin Valley, in *The Economics and Management of Water and Drainage in Agriculture.* Kluwer Press, Boston, 1991.

National Research Council. *Mineral Tolerances of Domestic Animals: Washington, D.C. Subcommittee on Animal Nutrition,* National Academy of Sciences, Washington, D.C., 577, 1980.

Neal, R.H. and G. Sposito. Selenium mobility in irrigated soil columns as affected by organic carbon amendments. *J. Environ. Qual.* 20: 808-814, 1991.

Ohlendorf, H.M. and R.L. Hothem. Agricultural drainwater effects on wildlife in central California, in *Handbook of Ecotoxicology.* D.J. Hoffman, B.A. Rattner, G.A. Burton, Jr. and J. Cairns, Eds., Lewis Publishers, Boca Raton, FL, 577-595, 1995.

Ohlendorf, H.M. and G.M. Santolo. Kesterson Reservoir — past, present, and future: an ecological risk assessment, in *Selenium in the Environment.* W.T. Frankenberger, Jr. and S. Benson, Eds., Marcel Dekker, New York, 69-117, 1994.

Oldfield, J.E. Environmental implications of uses of selennium with animals, in *Environmental Chemistry of Selenium.* W.T. Frankenberger, Jr. and R.A. Engberg, Eds., Marcel Dekker, New York, 129-142, 1998.

Parker, D.R. and A.L. Page. Vegetation management strategies for remediation of selenium contaminated soils, in *Selenium in the Environment.* W.T. Frankenberger, Jr. and S. Benson, Eds., Marcel Dekker, New York, 327-342, 1994.

Parker, D.R., A.L. Page, and D.N. Thomas. Salinity and boron tolerances of candidate plants for the removal of selenium from soils. *J. Environ. Qual.* 20: 157-164, 1991.

Presser, T.S., and H.M. Ohlendorf. Biogeochemical cycling of selenium in the San Joaquin Valley, Calif, USA. *Environ. Manage.* 11: 805-821, 1987.

Salt, D.E., M.J. Blaylock, D.B.A. Kumar, V. Dushenkov, B.D. Ensley, I. Chet, and I. Raskin. Phytoremediation: a novel strategy for the removal of toxic metals from the environment using plants. *Environ. Sci. Technol.* 13: 468-474, 1995.

Schnoor, J.L., L.A. Licht, S.C. McCutcheon, N.L. Wolfe, L.H. Carreira. Phytoremediation of organic and nutrient contaminants. *Environ. Sci. Technol.* 29: 7, 1995.

Shennan, C., S.R. Grattan, D.M. May, C.J. Hillhouse, D.P. Schachtman, M. Wander, B. Roberts, S. Tafoya, R.G. Burau, C. McNeish, and L. Zelinski. Feasibility of cyclic reuse of saline drainage in a tomato cotton rotation. *J. Environ. Qual.* 24: 476-486, 1995.

Sylvester, M.A., J.P. Deason, H.R. Feltz, and R.A. Engberg. Preliminary results of the Dept. of the Interior's irrigation drainage studies. *Proc. Planning Now for Irrigation and Drainage. Lincoln, NE., July 18-21, 1988.* American Society of Civil Engineering, New York, 665-677, 1991.

Tanji, K., A. Läuchli, and J. Meyer. Selenium in the San Joaquin Valley. *Calif. Agric.* 28: 6-39, 1986.

Terry, N., C. Carlson, T.K. Raab, and A.M. Zayed. Rates of selenium volatilization among crop species. *J. Environ. Qual.* 21: 341-344, 1992.

Terry, N. and A.M. Zayed. Selenium volatilization by plants, in *Selenium in the Environment.* W.T. Frankenberger, Jr. and S. Benson, Eds., Marcel Dekker, New York, 343-369, 1994.

Terry, N. and A.M. Zayed. Phytoremediation of selenium, in *Environmental Chemistry of Selenium.* W.T. Frankenberger, Jr. and R.A. Engberg, Eds., Marcel Dekker, New York, 633-656, 1998.

UC Salinity/Drainage Program. *Annual Report. WRC Prosser Trust, Centers for Water and Wildland Resources.* Division of Agriculture and Natural Resources, University of California, 1998.

Uppström, B. Seed chemistry, in *Brassica Oilseeds: Utilization.* D. Kimber and D.I. McGregor, Eds., Cab International, Oxon, U.K., 291-300, 1995.

U.S. Department of the Interior. Contaminant Issues of Concern on National Wildlife Refuges. U.S. Fish and Wildlife Service. Washington D.C., 1986.

van Schilfgaarde, J. Irrigated agriculture: is it sustainable? in *Agricultural Salinity Assessment and Management.* K.K. Tanji, Ed., ASCE Manuals and Reports on Engineering Practices 71. Am. Soc. Chem. Eng. New York, 584-594, 1990.

Watson, C.M. and J.W. O'Leary. Performance of *Atriplex* species in the San Joaquin Valley, California, under irrigation and with mechanical harvests. *Agric. Ecosys. Environ.* 43: 255-266, 1993.

Watson, F., G.S. Bañuelos, and J. O'Leary. Trace element composition of *Atriplex* species. *J. Environ. Qual.* 48: 157-162, 1994.

Wiedenfoeft, M.H. and B.A. Barton. Management and environmental effects on *Brassica* forage quality. *Agron. J.* 86: 227-232, 1994.

Wu, L. Selenium accumulation and colonization of plants in soils with elevated selenium and salinity, in *Selenium in the Environment.* W.T. Frankenberger, Jr. and S. Benson, Eds., Marcel Dekker, New York, 279-326, 1994.

Wu, L., J. Chen, K.K. Tanji, and G.S. Bañuelos. Distribution and biomagnification of selenium in a restored upland grassland contaminated by selenium from agricultural drain water. *Environ. Toxicol. Chem.* 14: 733-742, 1995.

Wu, L. and Z.H. Huang. Selenium tolerance, salt tolerance, and selenium accumulation in tall fescue lines. *Ecotoxicol. Environ. Safety* 21: 47-56, 1991.

Wu, L., Z.H. Huang, and R.G. Burau. Selenium accumulation and selenium-salt co-tolerance in five grass species. *Crop Sci.* 28: 517-522, 1988.

4 Remediation of Selenium-Polluted Soils and Waters by Phytovolatilization

Adel Zayed, Elizabeth Pilon-Smits, Mark deSouza, Zhi-Qing Lin, and Norman Terry

CONTENTS

INTRODUCTION

Selenium (Se) is a naturally occurring element that has a wide distribution in almost all parent materials on Earth. It is a Group VI metalloid which is found directly below sulfur in the Periodic Table. Selenium and sulfur share, therefore, very similar chemical properties. Selenium has an important role in biochemical systems. At low

1-56670-450-2/00/$0.00+$.50
© 2000 by CRC Press LLC

levels of nutritional supply, it is essential to the health of animals and humans. At high levels, it is poisonous. The concentration range from trace element requirement to lethality is quite narrow (Wilber, 1983). While the minimum nutritional level for animals is about 0.05 to 0.1 mg Se kg^{-1} dry forage feed, exposure to levels of 2 to 5 mg Se kg^{-1} dry forage will lead to toxic problems in animals (Wu et al., 1996; Wilber, 1983).

Selenium toxicity is encountered in arid and semi-arid regions of the world with alkaline, seleniferous soils derived from marine sediments. Soluble, oxidized forms of Se (e.g., selenate) may be leached from seleniferous soils into the shallow water table. In the western U.S., the leaching process is accelerated by excessive irrigation. The collection of the Se-contaminated irrigation leachate in the Kesterson Reservoir located in the San Joaquin Valley, CA, and the subsequent bioaccumulation of Se in the avian food chain led to a well-publicized environmental disaster: high levels of Se resulted in mortality, developmental defects, and reproductive failure in migratory aquatic birds (Ohlendorf et al., 1986) and fish (Saiki and Lowe, 1987). Selenium pollution of waters, soils, and sediments may also occur as a result of industrial activities. Oil refineries release substantial amounts of Se into coastal waters. Electric utilities generate Se-contaminated aqueous discharges from the storage of coal and other byproducts (e.g., coal pile runoff, coal pile seepage, coal ash landfill discharges, etc.).

Once present in soils and waters at high concentrations, Se is very complicated and highly expensive to remove with conventional physical and chemical techniques. These techniques (e.g., chemical or electrochemical treatments) may also produce large amounts of highly polluted sludges. In some instances, the sludges produced by these treatments may have to be managed as a hazardous waste that must be shipped to a toxic landfill. An inexpensive and environmentally friendly alternative is phytoremediation. Plants are highly effective in removing Se from contaminated sites. With their copious root systems, plants may scavenge large areas and volumes of soils, removing Se as selenate or selenite. Once absorbed by plant roots, Se is translocated to the shoot where it may be harvested and removed from the site. In addition to their uptake and immobilization of Se in their tissues, plants have the capacity to remove Se from contaminated substrates and metabolize it into a nontoxic volatile gas, e.g., dimethyl selenide (DMSe). Volatilization of Se by plants, i.e., phytovolatilization (Terry et al., 1995), is a highly attractive technique for the phytoremediation of Se pollution because it removes Se completely from the ecosystem while at the same time minimizing entry of toxic Se into the food chain. This is because Se is volatilized by roots and not appreciably accumulated in aboveground vegetation in forms available to wildlife. Recent research showed that Se volatilization by plant roots may require rhizosphere microorganisms. Microorganisms are also known to volatilize Se from soils and waters directly into the atmosphere (Frankenberger and Karlson, 1988; Doran, 1982). This process of microbial transformation of Se may also be enhanced by the presence of plants.

Plants used for phytoremediation of Se may generate other useful byproducts depending on their structural, chemical, and energetical characteristics (Nyberg, 1991), e.g., fibers for the production of paper and building materials. Plants may also be used as a source of energy for heat production by the combustion of the dried plant material or by fermentation to methane or ethanol. In addition, a large variety of chemical compounds (e.g., oil, sugars, fatty acids, proteins, pharmacological substances, vitamins, and detergents) are naturally produced by plants and may be useful byproducts of the phytoremediation process (Nyberg, 1991).

In this chapter we discuss the use of plants and their associated microorganisms to remove Se from polluted soils and water. Our emphasis is on the role of biological volatilization in the Se removal process. We also present recent advances in the molecular biology of Se uptake and volatilization by plants.

SE UPTAKE, TRANSPORT, AND METABOLISM

Selenium can exist in a variety of oxidation states: selenide (Se^{2-}), elemental or "colloidal" Se (Se^0), selenite (Se^{4+}), selenate (Se^{6+}), and several organic Se compounds (Rosenfeld and Beath, 1964). Selenium also exists in volatile forms, including DMSe, dimethyl diselenide (DMDSe), dimethyl selenone ($DMSeO_2$), and probably dimethyl selenenylsulfide (DMSeS) and methaneselenol (Reamer and Zoller, 1980; Fan et al., 1997). Plants take up Se mostly as selenate, selenite, and/or organic Se. The uptake of selenate and organic Se is driven metabolically, whereas the uptake of selenite may have a passive component (Abrams et al., 1990; Arvy, 1993). According to Williams and Mayland (1992), organic forms of Se may be more readily available to plant uptake than some inorganic forms. They demonstrated, for both Se-accumulator and nonaccumulator plant species, that the order of Se bioavailability for uptake was selenomethionine > selenocysteine = sodium selenite.

Once selenate and selenite are present in the root, they show differential translocation to the plant shoot (Arvy, 1993; Asher et al., 1977). The Se concentration of the xylem exudate in selenate-supplied tomato plants was 6 to 18 times higher than in the external solution and selenate moved in the xylem without chemical modification (Asher et al., 1977). In selenite-supplied plants, xylem Se concentrations were lower than the Se concentration in root medium and selenite was converted to selenate and another unidentified organic form of Se prior to translocation in the xylem (Asher et al., 1977). In our research, we found that when broccoli plants are supplied with 20 μm Se as selenate, selenite, or selenomethionine, Se translocation from root to shoot was highly dependent on the form of Se supplied (Zayed et al., 1998). Selenium accumulation in leaves was greatest when Se was supplied as selenate followed by selenomethionine and then selenite. In roots, selenomethionine was accumulated the most, followed by selenite and selenate. It seems, therefore, that selenate is more mobile inside plant tissues than selenite and selenomethionine.

Selenate is the highly bioavailable form of soluble Se that is most commonly found in soils and subsurface drainage waters. Selenate is believed to be taken up and assimilated by the enzymes of the sulfate assimilation pathway. Metabolism of Se through this pathway leads to the formation of Se analogs of the S-containing amino acids (e.g., selenocysteine, selenocystathionine, selenohomocysteine, and selenomethionine). There is evidence that Se analogs of S compounds compete for various enzymes in the S-assimilation pathway. Ng and Anderson (1978) demonstrated that selenide inhibits the synthesis of cysteine and sulfide inhibits the synthesis of selenocysteine. Dawson and Anderson (1988) showed that the enzyme cystathionine γ-synthase (which catalyzes the synthesis of cystathionine from cysteine) has a greater affinity for selenocysteine than for cysteine. McClusky et al. (1986) found that the enzyme cystathionine β-lyase (which catalyzes the production of homocysteine from cystathionine) metabolizes selenocystathionine with a similar affinity to cystathionine. Figure 4.1 summarizes the possible steps in the assimilation pathway for selenate in Se nonaccumulator plants. Details of this pathway are given elsewhere (Terry and Zayed, 1994).

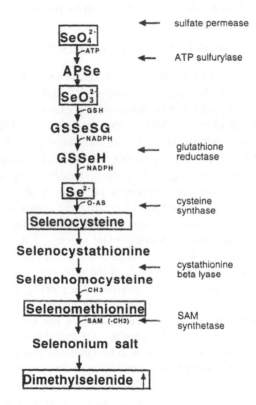

FIGURE 4.1 Proposed selenium volatilization pathway in nonaccumulator plants, and the target enzymes that are overexpressed in our studies. (Modified from Zayed, A. and N. Terry, *J. Plant Physiol.* 140, 646-652, 1992.)

BIOLOGICAL VOLATILIZATION OF SE

The idea that biological volatilization of Se could be used to clean up Se from contaminated soils was first proposed by Frankenberger and Karlson (1988) who suggested the use of soil microorganisms, especially soil fungi, for this purpose. Later, it was demonstrated that plants too possess the capacity to remove substantial amounts of soil Se by volatilization (Terry et al., 1992; Duckart et al., 1992; Biggar and Jayaweera, 1993). Biological volatilization of Se is an attractive method for the bioremediation of Se pollution because it removes the element completely from the local ecosystem, thereby minimizing entry into the food chain (Frankenberger and

Karlson, 1988; Terry et al., 1992; Terry and Zayed, 1994). In the following section we review recent information on the role of soil microorganisms and higher plants in the removal of Se from contaminated sites by biological volatilization.

PHYTOVOLATILIZATION OF SE

The first indication that volatile Se can be released from plant tissues was reported by Beath et al. (1935), who noticed that volatile Se compounds were released from *Astragalus bisulcatus* plant materials during storage. Later, Lewis et al. (1966) demonstrated that volatile Se compounds were released from intact higher plants when grown in a Se-rich root medium. This process occurs in Se- and non-accumulator species (Evans et al., 1968; Lewis, 1971); Se nonaccumulators typically release DMSe and do not produce DMDSe, while Se-accumulators mainly volatilize DMDSe (Evans et al., 1968; Lewis, 1971). The importance of plants in the removal of Se through volatilization has been established by several researchers who showed that the addition of plants to soil increased the rate of Se volatilization above that found for soil alone (Zieve and Peterson, 1984a; Biggar and Jayaweera, 1993; Duckart et al., 1992). Greenhouse experiments conducted by Zayed and Terry (unpublished) showed that the addition of barley to Se-contaminated Kesterson soil increased the volatilization rate from 392 μg m^{-2} day^{-1} for soil only to 1340 μg m^{-2} day^{-1} for soil plus barley.

Rates of Se Volatilization Among Different Plant Species

Plant species differ substantially in their ability to take up and volatilize Se (Terry et al., 1992). Rice, broccoli, cabbage, cauliflower, Indian mustard, and Chinese mustard volatilized Se at rates exceeding 1500 μg Se day^{-1} kg^{-1} plant dry weight when supplied with 20 μm Se as sodium selenate (Figure 4.2). Sugarbeet, bean, lettuce, and onion exhibited very low rates of Se volatilization, below 250 μg Se day^{-1} kg^{-1} plant dry weight. Other plant species tested, including carrot, barley, alfalfa, tomato, cucumber, cotton, eggplant, and maize, showed intermediate values of 300 to 750 μg Se day^{-1} kg^{-1} plant dry weight (Figure 4.2). Plant species from the Brassicaceae family were particularly effective Se volatilizers (of the top six species, only rice was not a member of this family). There was evidence that the ability to volatilize Se was associated with the ability to accumulate Se in plant tissues — the rate of Se volatilization by different plant species was strongly correlated with the plant tissue Se concentration (Terry et al., 1992). Duckart et al. (1992) also obtained large differences among species. Of the five different plant species tested in their study, *A. bisulcatus* and broccoli showed the highest rates of Se volatilization, both on a leaf area and soil dry weight bases, followed by tomato, tall fescue, and alfalfa, respectively.

Chemical Form of Se and the Rate of Volatilization

Several researchers have pointed out that Se volatilization by plants proceeds more rapidly if Se is supplied in more reduced forms, e.g., selenite or selenomethionine as compared to selenate. Early research by Lewis et al. (1974) showed that cabbage

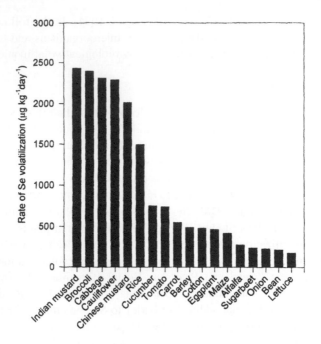

FIGURE 4.2 Rates of selenium volatilization of various plant species. Plants were grown hydroponically in half-strength Hoagland solution and supplied with 20 μm Se as sodium selenate.

leaves from plants supplied with selenite released 10 to 16 times more volatile Se than those taken from plants supplied with selenate. Similarly, roots of selenite-grown plants released 11 times more volatile Se during oven drying than did roots of selenate-grown plants (Asher et al., 1967). When soils from the low-Se region in China were amended with selenomethionine, the rate of Se volatilization was enhanced by 14-fold, while the addition of less reduced forms of Se resulted only in a 4-fold increase (Zijian et al., 1991). In a recent study, Zhang and Moore (1997) found that the concentration of dissolved organic Se is a more important factor affecting Se volatilization from plants and sediments than dissolved inorganic Se. They also illustrated that the decomposition of plants greatly enhances Se volatilization due to the increase in the dissolved organic Se. Decomposing plants also provide soil microorganisms with methyl groups required for Se methylation and serve as a carbon source to stimulate microbial growth.

In our research, we examined the influence of the chemical form of Se in the root medium on the rate of Se volatilization by plants. We supplied plants of broccoli, Indian mustard, sugarbeet, and rice with 20 μm Se as Na_2SeO_4, Na_2SeO_3, or L-selenomethionine. We found that the highest rates of Se volatilization were attained when selenomethionine was supplied, followed by SeO_3, then SeO_4 (Zayed et al., 1998). The rate of Se volatilization by plants significantly correlated with Se accumulation in roots, but not in shoots. X-ray absorption spectroscopy speciation analysis showed that most of the Se taken up by SeO_4-supplied plants remained unchanged, whereas plants supplied with SeO_3 or selenomethionine contained only selenomethionine-like species. These data indicate that reduction from SeO_4 to SeO_3 appears to be a rate-limiting step in the production of volatile Se compounds by plants (Zayed et al., 1998).

Se Volatilization as Affected by Competitive Ions

Plants are thought to take up selenate by the same carrier in root cell membranes as sulfate (Leggett and Epstein, 1956). Consequently, sulfate ions are antagonistic to the uptake of selenate ions (Mikkelsen and Wan, 1990). Zayed and Terry (1992) showed that, when broccoli plants were supplied with 20 μm Se as sodium selenate, increasing the sulfate supply in the culture solution from 0.25 to 1 mM had no effect on the concentration of Se in plant tissue; but, with further increase in sulfate from 1 to 5 or 10 mM, tissue Se concentrations were substantially decreased. Broccoli plants also volatilized selenate less efficiently as the concentration of the competing sulfate ion was increased. Zayed and Terry (1994) showed that with increase in the sulfate concentration in nutrient solution from 0.25 to 10 mM, the rate of Se volatilization by broccoli plants decreased from 97 to 14 μg Se day^{-1} m^{-2} leaf area. The decrease in Se volatilization rate was best correlated with a decrease in the ratio of Se:S in plant tissues (Figure 4.3). This suggests that sulfur compounds out competed their Se analogs for the active sites of enzymes responsible for the conversion of inorganic Se to volatile forms. Sulfate levels also affect Se translocation in plants. Singh et al. (1980) showed that, in the absence of S, Se tends to accumulate in the roots of Indian mustard plants and in the presence of S more Se is transported to

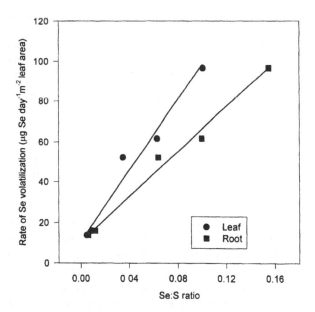

FIGURE 4.3 Relationship between rates of selenium volatilization and the ratio of total Se to total S in broccoli plant leaf and root tissues. Plants were grown hydroponically in half-strength Hoagland solution and supplied with 20 μm Se as sodium selenate and various levels of sulfate ranging from 0.25 to 10 mM.

the shoot. The incorporation of Se into plant proteins is also affected by S supply. Zayed and Terry (1994) demonstrated that with each increase in sulfate supply from 0 to 10 mM the amount of Se incorporated into proteins in shoot and root progressively decreased (Table 4.1).

TABLE 4.1

Influence of Sulfate Supply on Se Accumulation and Incorporation into Protein in Shoot and Root Tissues of Broccoli Plants

Sulfate Supply	Total Se	Protein-Se	
(mM)	(mg kg^{-1} dry wt)	(mg kg^{-1} dry wt)	(mg kg^{-1} protein)
Shoot			
0.00	1421.2	273.4	0.955
0.25	875.6	127.5	0.429
0 50	976.1	104.6	0.375
1.00	551 6	40.7	0.138
5.00	107.9	11.7	0.035
10 0	62.7	7.8	0.025
Root			
0 00	465.7	86.6	0.757
0.25	331 7	68.0	0.470
0.50	280.5	35.9	0.265
1.00	93.1	24.5	0.210
5.00	91.9	7.5	0.043
10.0	32 7	4 7	0 038

Note: Plants were grown hydroponically and supplied with 20 μm Se as sodium selenate

The competitive effect of S on Se may be of less importance when Se is present in soil or water in other forms than selenate. Zayed et al. (1998) found that increasing the sulfate level from 0.25 to 10 mM inhibited SeO$_3$ and selenomethionine uptake by 33% and 15 to 25%, respectively, as compared to an inhibition of 90% of SeO$_4$ uptake. Similar results were observed with regard to sulfate effects on volatilization. Some plant species may also exhibit a differential response to sulfate supply (e.g., Se accumulators). Calculation of the Se/S discrimination coefficient (the ratio of the plant Se/S ratio to the solution Se/S ratio) indicated that *A. bisulcatus* (mean coefficient = 5.43), rice (mean coefficient = 3.66), and Indian mustard (mean coefficient = 1.7) were able to take up Se preferentially in the presence of a high sulfate supply (Bell et al., 1992). Other species (e.g., alfalfa, wheat, ryegrass, barley, broccoli) had discrimination coefficient values lower than 1, and selenate uptake was profoundly inhibited by the increases in sulfate supply.

MICROBIAL SE VOLATILIZATION

Numerous microorganisms were shown to be capable of producing volatile Se compounds from soils and sediments, including fungi, bacteria, yeast, and microalgae (Challenger et al., 1954; Abu-Erreish et al., 1968; Doran and Alexander, 1977; Reamer and Zoller, 1980; Karlson and Frankenberger, 1989; Fan et al., 1997). The predominant form of Se volatilized by soil microorganisms is DMSe, but other forms of volatile Se are also produced (e.g., DMDSe, dimethyl selenone) depending on the form and concentration of soil Se, the microbial species and soil type (Reamer and Zoller, 1980; Doran, 1982; Karlson and Frankenberger, 1989). Frankenberger and his coworkers (Karlson and Frankenberger, 1989, 1990) optimized the dissipation of soil Se through microbial volatilization as a bioremediation technique to detoxify Se-contaminated soils. They isolated three species of fungi which, when added to soils or sediments, substantially accelerated the rate of soil Se volatilization. They also identified several soil amendments that can greatly increase the rate of microbial Se volatilization.

Microbiology research on Se volatilization in our laboratory has two major foci: (1) the role of microorganisms in facilitating Se volatilization by plants and (2) Se volatilization by rhizosphere microorganisms independent of the plant.

Role of Microorganisms in Se Volatilization by Plants

Selenium volatilization by plants is strongly influenced by the microbial population in the root zone. This conclusion is supported by two types of investigation carried out in our laboratory: (1) antibiotic studies, where plants were grown with different species of Se, with and without antibiotics, and (2) sterile plant studies, where plants were germinated and grown under aseptic conditions, amended with Se, and Se uptake and volatilization characteristics compared to those for nonsterile plants (Tables 4.2 and 4.3). In the first study, broccoli plants were grown hydroponically with selenate, plus and minus penicillin-G and chlortetracycline (Zayed and Terry, 1994). These antibiotics inhibited 95% of the Se volatilization activity from selenate-amended detopped roots, supporting the view that bacteria in the rhizosphere are essential for plant Se volatilization. In a subsequent experiment where sterile and nonsterile plants were supplied with selenate or selenite, the rate of Se volatilization in sterile plants was only 10% that of the nonsterile plants; however, when selenomethionine rather than selenate or selenite was supplied, the rate of Se volatilization was similar from sterile and nonsterile plants. These data suggest that bacteria in the rhizosphere of broccoli are involved in Se volatilization from selenate and selenite but not from selenomethionine.

Using a different antibiotic, ampicillin, recent antibiotic experiments with broccoli and Indian mustard also support the view that bacteria are necessary to achieve maximum rates of Se volatilization from selenate-supplied plants. Selenium volatilization from selenate was inhibited by 35 to 50% in ampicillin-amended plants, compared to unamended plants (de Souza et al., 1999). When the tissues of the plants used in the antibiotic or sterile plant experiments were analyzed for Se, it was observed that about 70% of the selenate uptake into the roots of plants was

TABLE 4.2

Role of Rhizosphere Bacteria in Se Volatilization by Plants

Plant	Se Species	Experiment	Bacterial Contribution (%) to Plant Se Volatilization
Broccoli	Selenate	Antibiotic	95
	Selenite	Penicillin	ND
	Se-methionine	+ Chlortetracylcine	ND
Broccoli	Selenate	Sterile vs.	>90
	Selenite	nonsterile	78
	Se-methionine		0
Broccoli	Selenate	Antibiotic	50
	Selenite	Ampicillin	0
	Se-methionine		0
Brassica juncea	Selenate	Antibiotic	35
	Selenite	Ampicillin	4
	Se-methionine		0

TABLE 4.3

Role of Rhizosphere Bacteria in Se Uptake by Plant Roots

Plant	Se Species	Experiment	Bacterial Contribution (%) to Plant Se Uptake
Broccoli	Selenate	Sterile vs.	88
	Selenite	nonsterile	2
	Se-methionine		30
Broccoli	Selenate	Antibiotic	70
	Selenite	Ampicillin	18
	Se-methionine		0
Brassica juncea	Selenate	Antibiotic	55
	Selenite	Ampicillin	17
	Se-methionine		0

inhibited by ampicillin or sterile conditions. Thus, it appears that bacteria in the rhizosphere facilitate the uptake and volatilization of selenate supplied to plants. However, with selenite uptake, antibiotic-amended or sterile plants showed similar levels of tissue Se compared to antibiotic-unamended or nonsterile plants. This suggests that bacteria in the rhizosphere are not involved in selenite uptake. Selenite uptake is most likely passive, and the role of bacteria in selenite volatilization may lie in the transformation of selenite to organic Se inside the plant.

Because the role of bacteria in enhancing Se volatilization (of broccoli and Indian mustard plants supplied with selenate) appears to involve increased selenate uptake, the rhizosphere microorganisms that were best at enhancing selenate uptake were identified by inoculating sterile plants with different strains of rhizosphere bacteria. This research yielded two strains of bacteria which enhanced the ability of sterile plants to increase their selenate uptake three- and fourfold, respectively, in two different plant species. The combination of plants with bacteria gave much higher rates of selenate uptake than sterile plants alone or bacteria alone. Research is underway to determine if the newly discovered plant–microbe associations exhibiting high selenate uptake also enhance Se volatilization.

Se Volatilization by Microorganisms Independent of Plants

Our research has shown that rates of Se volatilization from selenite-contaminated rhizosphere sediment samples taken from a constructed wetland were higher than from nonvegetated soil samples (Azaizeh et al., 1997). In lab cultures, the addition of a carbon and energy source (e.g., glucose and peptone), and aeration of the sediment cultures, enhanced Se volatilization. These results for constructed wetlands samples were similar to those found in Se-contaminated pond water, where casein amendments stimulated deselenification and Se biomethylation (Thompson-Eagle and Frankenberger, 1990), presumably by increasing bacterial numbers (Thompson-Eagle and Frankenberger, 1991). In sediments from the Kesterson reservoir, fungi were reported to be the dominant Se volatilizing microorganisms (Karlson and Frankenberger, 1989), but in pond water bacteria were thought to play the dominant role (Thompson-Eagle and Frankenberger, 1991). Many species of fungi (Thompson-Eagle et al.,1989; Brady et al., 1996) and bacteria (Chau et al., 1976; Doran and Alexander, 1977; Rael and Frankenberger, 1996; de Souza and Terry, unpublished) which show high rates of Se volatilization have been isolated.

In our constructed wetlands study (Azaizeh et al., 1997), bacteria were the predominant volatilizers of Se rather than fungi. Similarly, rhizosphere bacteria, and not fungi, were the dominant microorganisms that enhanced selenate volatilization from broccoli (Zayed and Terry, 1994). In order to identify superior volatilizing bacteria, we have screened the rhizospheres of different plant species from uplands and wetlands environments contaminated with Se. We have isolated many species of bacteria (from broccoli, Indian mustard, bulrush, and cattail) that can volatilize selenium at high rates from selenate, selenite, and selenomethionine. The highest rates of Se volatilization were measured from selenomethionine and the lowest rates from selenate. These bacteria are tolerant of high levels of Se, e.g., up to 2 mM Se in the growth medium. We plan to test the best strains of Se-volatilizing bacteria under field conditions by inoculation into Se-contaminated wetlands mesocosms. This approach has been carried out successfully by Zieve and Peterson (1981), who inoculated a yeast, *Candida humicola*, into Se-contaminated sediment resulting in a doubling of the rate of Se volatilization.

Se Remediation by Volatilization

The preceding discussion concludes that both bacteria and plants are needed for optimum phytoremediation of selenium by phytovolatilization. Although bacteria can volatilize Se independently of the plant, they are limited in their remediation potential because they can clean up only limited volumes of contaminated environments and because they require an energy source for optimum activity. Plants, on the other hand, can remediate large polluted areas by virtue of their ability to generate a large biomass and a plentiful supply of carbon to support rhizosphere microorganisms.

Se Volatilization by Wetlands

Wetland plants remove pollutants through uptake and immobilization in plant tissues. In the case of Se, an additional, and potentially very important, pathway of removal is through biological volatilization. This involves the biomethylation of Se by plants or plant/microbe associations to produce volatile forms of Se. Field and laboratory measurements show that substantial amounts of Se can be removed from wastewater by volatilization and tissue accumulation (Cooke and Bruland,1987; Velinsky and Cutter, 1991; Allen,1991). Recently, Zhang and Moore (1997) examined the role of Se volatilization in the removal of Se by a wetland system. They used wetland microcosms spiked with different forms of Se. Their results indicate that natural Se volatilization is an important process in removing Se from wetland systems. Sediment and plants were the major producers of volatile Se from the system. High temperature, higher air flow, alternate flooding–drying cycles, and the decomposition of wetland plants greatly increased the removal rates of Se through volatilization.

The effectiveness of natural Se volatilization in the cleanup of Se-polluted industrial wastewater using constructed wetlands is exemplified by the Chevron Oil Water Enhancement Wetland at Point Richmond, CA. Chevron Oil Company created this 36-ha constructed wetlands in 1988 to clean up polluted wastewater from their refinery. The Chevron wetlands was found to remove a substantial portion of the Se in the 10 million liters of oil refinery wastewater entering the wetland each day (Duda, 1992). In collaboration with Chevron, Terry and his colleagues monitored Se movement in the wetland over a period of 4 months and found that the wetland removed an average of 148.4 g Se day^{-1}; this represents 89.6% of the Se entering the wetlands (Hansen et al., 1998). Most of this removal occurred in the first pass (about 12 ha in area). Measurements of Se volatilization over a small area (75 x 75 m) of the wetland indicated rates from 10 to 350 µg Se m^{-2} day^{-1}. Biological volatilization of Se may have accounted for 10 to 30% of the Se removed. However, since there was a very large variation in rates with respect to site location and time of measurements, much more research is needed before an accurate estimate of volatilization as a pathway for Se removal can be provided for this wetland. This study clearly illustrates that constructed wetlands are effective in removing Se with a significant, but as yet not well-determined, portion being removed by biological volatilization. Based on the success of the constructed wetlands in removing Se from industrial wastewaters, another wetlands study was initiated in 1996 in Corcoran,

CA to treat selenate-contaminated agricultural drainage wastewater. To determine which species would be the most efficient in removing Se, 10 quarter-acre experimental wetland cells were constructed and planted with a variety of different plant species. This study (led by N. Terry, and supported by the University of California and Tulare Lake Drainage District) is still underway and preliminary results indicate that substantial reductions in Se concentrations of the inlet water are occurring as the water passes through the wetlands cells.

FATE OF THE VOLATILE SE IN THE ATMOSPHERE

Some concerns have been expressed that biomethylation of Se may lead to the production of toxic forms of gaseous Se, or that Se may be redeposited in other areas to become toxic there. Dimethyl selenide is reported to be 500 to 700 times less toxic than selenate and selenite (LD_{50} of DMSe = 1600 to 2200 mg Se kg^{-1} rat; Wilber, 1980; Ganther et al., 1966; McConnell and Portman, 1952). Using a volatilization rate of 250 μg m^{-2} h^{-1} (an emission rate which would be promising for remediation), the highest 24-h average exposure of Se under stagnant conditions was computed to be 837 ng m^{-3} (Frankenberger and Karlson, 1988). An acceptable intake level documented by the EPA in guidance for Superfund sites is considered to be 3500 ng m^{-3}, which is substantially higher than that calculated for Se bioremediation. Furthermore, studies on the tropospheric transformation of DMSe on the west side of San Joaquin Valley, CA indicate that during the short lifetime of gaseous DMSe (9.6 days), the gas will be dispersed and diluted by air currents directly away from the contaminated areas, with deposition possibly occurring in the Se-deficient areas (Atkinson et al., 1990). At the highest annual deposition flux (4.5 g Se ha^{-1}) mixed in the upper 10 cm of soil, the soil Se content would be increased by about 0.005 ppm (Frankenberger and Karlson, 1988).

MEASUREMENT OF SE VOLATILIZATION IN THE FIELD

Laboratory and greenhouse volatilization studies have provided valuable information on mechanisms of Se volatilization from the water–soil–plant system, and showed the potential significance and importance of Se volatilization for the remediation of Se-contaminated soils and waters. Measurements of Se volatilization in these studies were conducted using different types of sampling chambers under well-controlled environmental conditions (Zieve and Peterson, 1984b; Terry et al., 1992; Zayed and Terry, 1992, 1994; Zhang and Moore, 1997). Recently, there have been major efforts to determine the rate of biological Se volatilization under field conditions. Different types of sampling chamber enclosures have been developed and demonstrated as a simple and convenient method for collecting volatile Se over soil or soil–plant ground surfaces in the field (Karlson and Frankenberger, 1988, 1990; Biggar and Jayaweera, 1993; Hansen et al., 1998). Most of these chambers were designed as open-bottom cubic compartments made of rigid transparent material such as Plexiglas. These chambers were equipped with different trapping systems to trap the volatile Se compounds from air. These trapping systems include the use of concentrated HNO_3 (Zieve and Peterson, 1984b), alkaline-peroxide solution (Weres et al., 1989; Terry et al., 1992), activated carbon columns (Karlson and Frankenberger, 1988; Zhang and Moore, 1997), and activated carbon filters (Biggar and Jayaweera, 1993; Hansen et al., 1998). Due to simplicity, convenience, and effectiveness of activated carbon filters in trapping and extracting volatile Se, they have frequently been used for trapping volatile Se in most recent studies (Hansen et al., 1998; Lin et al., 1999). When activated carbon trapping systems were used for trapping volatile Se, methanol (Karlson and Frankenberger, 1990; Zhang and Moore, 1997) or alkaline-peroxide solution (Jayaweera and Biggar, 1992; Hansen et al., 1998) were used for extraction of Se from surfaces of the activated carbon materials.

Unlike studies of Se volatilization under controlled environments, measurements of Se volatilization in the field are generally faced with various extreme environmental conditions (in particular, the heat buildup in the sampling chamber due to the greenhouse effect). In an effort to overcome such deficiencies, Biggar and Jayaweera (1993) developed a diffusive Plexiglas sampling chamber system equipped with a cooling radiator connected to an evaporative cooler. Such a cooling system was shown to effectively reduce heat buildup without affecting light conditions in the chambers. However, it has not been demonstrated that this system would be effective when chambers with a larger internal space are required for Se volatilization measurements from tall plants.

In a recent study of the role of Se volatilization in the removal of Se by constructed wetlands, Hansen et al. (1998) developed an open-flow Plexiglas sampling chamber system for measurements of volatile Se under field conditions. The open-flow chamber system allowed for a minimal heat and humidity buildup inside the chamber and a maximum trapping efficiency of the volatile Se produced by plants and microbes growing in soil and waters. With an open-flow sampling chamber, ambient air is continuously scrubbed of volatile Se before entering the chamber through charcoal filters at the inlet port. Selenium-free air mass is then mixed with volatile Se produced inside the chamber and drawn under vacuum into the outlet port passing through another set of charcoal filters where volatile Se is trapped and collected. With further improvement on the sampling chamber system used by Hansen et al. (1998), Lin et al. (1999) reported a modified open-flow sampling chamber system with detailed information on the design, construction, and calibration of the trapping system. Trapping efficiency and capacity of activated carbon filters to trap volatile Se were examined with consideration of airflow velocity, moisture condensation, and sampling residence time (Lin et al., 1999). To minimize the effect of rising temperature inside sampling chambers on the volatile Se measurement in the field, Lin et al. (1999) set up sun canopies at a height of 1.5 m above the ground. These canopies effectively shaded the open-flow chambers from direct solar radiation. The monitoring results showed that the difference in air temperature between the inside and outside of the chambers varied within only 1.5 ± 0.1°C during a 24-h sampling period in summer (Lin et al., 1999). The level of photosynthetically active radiation (PAR) at 0.5 m above ground under a sun canopy was comparable to the PAR level found below plant canopies under field conditions.

Micrometeorological flux measurement is another attractive technique for the estimation of the rate of Se volatilization in the field. It provides an average integrated flux of volatile Se over a large area by measuring the difference in Se concentrations in ambient air at two different heights above ground (Haygarth et al., 1995). However, this flux measurement technique is difficult to use in practice as it is costly in instrumentation and often associated with some uncertainties (Haygarth and Jones, 1997).

MOLECULAR BIOLOGY OF SE UPTAKE AND VOLATILIZATION BY PLANTS

GENETIC ENGINEERING OF SUPERIOR SE PHYTOREMEDIATORS

Genetic engineering provides a powerful tool to broaden our insight into the biochemistry of Se metabolism in plants. We can increase the activity of one or more enzymes by inserting extra copies of their specific genes. Analysis of the capacity of these transgenic plants to accumulate, tolerate, and volatilize Se can pinpoint specific enzymes that are rate-limiting for these processes. Those transgenic plants displaying superior rates of volatilization, accumulation, or tolerance for Se may be used for the phytoremediation of Se-contaminated soils or waters. Higher Se accumulation will mean improved extraction of Se from the substrate, improved volatilization will enhance the proportion of Se that is detoxified and removed completely from the site, and improved Se tolerance will result in better growth and biomass production on heavily contaminated substrates, and thus in increased Se removal.

Ideally, the plant species to be used for Se phytoremediation should have the ability to accumulate and volatilize large amounts of Se, grow rapidly and produce a large biomass, tolerate salinity and other toxic conditions, and provide a safe source of forage for Se-deficient livestock. Bearing this in mind, we have chosen Indian mustard for our studies. This plant is one of the best phytoremediators of Se known thus far, with fast growth, large biomass, and high levels of Se volatilization and accumulation (Bañuelos et al., 1996). Genetic transformation of Indian mustard is fairly easy (Barfield and Pua, 1991). The advantage of using this species is that transgenics with superior Se-metabolizing properties will be among the best Se phytoremediators available. Thus, by genetically engineering Indian mustard we hope to gain knowledge about the metabolic processes controlling Se metabolism in plants, and at the same time obtain plants that are superior Se phytoremediators.

For the genetic engineering of Indian mustard, we are using two different strategies. In one approach we try to speed up existing processes involved in Se metabolism by overexpressing possible rate-limiting enzymes. In the second approach we are attempting to introduce an additional metabolic pathway in Indian mustard, which now only exists in Se-hyperaccumulating plants, e.g., *A. bisulcatus*. For our first approach we have selected six target enzymes based on the Se volatilization pathway proposed earlier by Zayed and Terry (1992) (see Figure 4.1). Selenium and sulfur assimilation in plants is probably mediated by the same enzymes (Anderson, 1993; Lauchli, 1993), a hypothesis which is based on the chemical similarity of sulfur and selenium, as well as on biochemical and physiological competition studies with S and Se homologs.

The six enzymes we have targeted for overexpression are as follows: (1) sulfate permease, the root membrane protein which takes up selenate; (2) ATP sulfurylase, which activates selenate in the reduction to selenite (Burnell, 1981); (3) glutathione reductase, an enzyme involved in the further reduction of selenite to selenide (Ng and Anderson, 1979); (4) cysteine synthase, which incorporates selenide into selenocysteine (Ng and Anderson, 1978); (5) cystathionine-β-lyase, which, along with cystathionine-γ-synthase, mediates the conversion of selenocysteine to selenomethionine via the intermediate selenocystathionine (Burnell, 1981; Dawson and Anderson, 1988); and (6) SAM-synthetase, which generates the methyl donor, S-adenosylmethionine (SAM), for the methylation of selenomethionine (Lewis et al., 1974). Below, we summarize the results obtained thus far regarding each of the above specified enzymes.

ATP Sulfurylase

Transgenic Indian mustard plants that overexpress ATP sulfurylase (APS) have been created using a gene construct generously provided by Tom Leustek (Rutgers University, NJ). This construct contains the *Arabidopsis thaliana aps 1* gene (Leustek et al., 1994), with its own chloroplast transit sequence, fused to the constitutive 35S promoter. The APS activity was four times higher in the shoots of the transgenic APS plants, but not in roots. Compared to wildtype plants, the APS transgenics showed increased tolerance to selenate, both at the seedling stage and as mature plants. APS seedlings had significantly longer roots (+50%) and larger fresh weights (+50%) than wildtype seedlings when grown for 7 days on agar medium with 400 μm selenate. Four-week-old plants showed less stress symptoms and higher fresh weights when treated with 50 μm selenate. These APS plants contained higher Se concentrations in their shoots (2- to 3-fold), as well as their roots (1.5-fold), compared to wildtype. We performed x-ray absorption spectroscopy (XAS) to analyze which chemical form of Se was present inside wildtype and APS plants. Selenate-supplied APS plants were shown to accumulate mainly selenomethionine, whereas wildtype plants accumulated selenate in both roots and shoot. Thus, the APS plants had increased selenate reduction. Detopped APS roots, however, accumulated selenate, indicating that the shoot has an important role in selenate reduction. As expected, the APS plants did not accumulate more Se when supplied with selenite instead of selenate. Se volatilization studies are now being performed to investigate whether the APS plants are also better volatilizers of Se.

These studies confirm that APS is responsible for selenate reduction, confirming the general hypothesis, which was based on *in vitro* studies. Furthermore, APS appears to be rate-limiting for selenate reduction and for Se accumulation. We conclude that overexpression of APS is a very promising approach for improving Se phytoremediation efficiency. If transgenic plants can be used which accumulate threefold more Se per unit surface area, the cost and duration of phytoremediation will be reduced threefold. The same gene encoding APS may also be overexpressed in other plant species, such as trees or aquatic plants, and these may be used under different environmental conditions.

Glutathione Reductase

We have obtained transgenic plants which overexpress glutathione reductase (GR) either in the cytosol or in the chloroplast, using constructs containing the *E. coli gr* gene, coupled to the 35S promoter, with or without a chloroplast transit sequence (Foyer et al., 1995). In the transgenic Indian mustard plants obtained, the *E. coli* GR enzyme was shown to be present and active in all tissues. The total GR activities were about two- to threefold higher than in wildtype plants for the cytGR construct, and up to 50-fold higher for the cpGR construct. In spite of the increased GR activity, Se accumulation and volatilization was not significantly different from wildtype plants when supplied with selenate or selenite. This indicates that the GR enzyme is either not rate-limiting for Se accumulation and volatilization under the conditions used, or that the GR enzyme is not involved in Se assimilation. The hypothesis that GR enzyme is not rate-limiting agrees with the fact that selenite-supplied wildtype plants accumulate an organic form of Se, probably selenomethionine, indicating that the conversion of selenite to selenomethionine is a very rapid process.

Cysteine Synthase

We have genetically engineered Indian mustard plants which overexpress cysteine synthase (CS), using a construct that contains the spinach CS gene, fused to a chloroplast transit sequence, under the control of the 35S promoter (Saito et al., 1994). The CS enzyme activity in leaves of transgenic CS plants was about twofold higher than in untransformed plants. The transgenic CS plants did not show any significantly different rates of Se volatilization or Se accumulation when supplied with selenate or selenite. Thus, we can conclude that the CS enzyme is not rate-limiting for Se accumulation or volatilization under our assay conditions. Again, this finding is in agreement with our XAS data, which show that selenite-supplied wildtype plants accumulate selenomethionine and, therefore, that the step mediated by CS is a very rapid one in wildtype plants, and not likely to be rate-limiting.

SAM Synthetase (SAMS)

We have obtained transgenic SAMS plants (i.e., plants overexpressing SAM synthetase) using a construct that contains the *A. thaliana* SAMS gene under control of the 35S promoter (Boerjan et al., 1994). The transgenic SAMS plants had significantly lower rates of Se volatilization when supplied with selenate; Se accumulation was not consistently different from wildtype. When supplied with selenomethionine, there were no significant differences in Se volatilization or Se accumulation compared to wildtype. From these preliminary data, it does not appear that SAMS is a pivotal enzyme for Se volatilization.

Generally, the data obtained to date on the transformation of Indian mustard plants indicate that the uptake of selenate by sulfate permease and the reduction of selenate by APS are the key steps in Se assimilation by plants that control the flow of the pathway. It has been shown that overexpression of these two enzymes results in enhanced Se accumulation and, in the case of APS, increased Se reduction. We are still measuring the Se volatilization properties of these transgenics, as previous measurements have not given consistent results. In any case, since selenomethionine accumulates in Indian mustard plants supplied with selenite, as well as in APS plants supplied with selenate, one of the final steps in the Se volatilization pathway, between selenomethionine and DMSe appears to be another bottleneck. Therefore, future plans must focus on developing transgenic plants that overexpress the enzymes involved in the conversion of selenomethionine to DMSe.

CONCLUSION

Although the concept of phytoremediation has been in existence for many years, only in the last decade has interest in this technology accelerated to the point of commercial application. It is estimated that phytoremediation of heavy metals will have an annual market worth $400 million or more in North America and Europe by the turn of the century (Salt et al., 1995). However, before this technology becomes available for use on a large scale, it will be necessary to integrate research in many disciplines, including molecular biology, plant physiology, and biochemistry and microbiology at the molecular, cell, and plant levels, as well as ecology, agronomy, hydrology, soil science, agricultural engineering, and entomology at the field level. In turn, these studies must be considered in relation to economics, environmental impact assessment, and government regulatory policies. In our laboratory, we have assembled a multidisciplinary team of plant physiologists/biochemists, ecologists, molecular biologists, and microbiologists to establish phytovolatilization of Se as a low-cost, low-impact, and environmentally sound technology.

Research on Se phytovolatilization has successfully identified superior volatilizing plant species, characterized the optimum environmental conditions for maximum rates of Se volatilization, and begun to unravel, using modern molecular biology techniques, the mechanisms by which Se is evolved by plants. In addition, recent findings on the microbial involvement in plant uptake and volatilization of Se has provided another tool to accelerate the phytovolatilization process through the creation of superior plant/microbe associations. Furthermore, the possibility of using constructed wetlands for the removal of Se from agricultural and industrial wastewaters is particularly exciting. After the success of the Chevron wetlands in removing Se from oil refinery effluent, the use of constructed wetlands for the removal of Se from irrigation drainage water is currently underway. If this experiment is successful, it will revolutionize the treatment of Se-contaminated agricultural drainage water and dramatically increase the acreage of wetlands throughout the western U.S.

REFERENCES

Abrams, M.M., C. Shennan, J. Zasoski, and R.G. Burau. Selenomethionine uptake by wheat seedlings. *Agron. J.* 82, 1127-1130, 1990.
Abu-Erreish, G.M., E.T. Whitehead, and O.E. Olson. Evolution of volatile selenium from soils. *Soil Sci.* 106, 415-420, 1968.

Allen, K.N. Seasonal variation of selenium in outdoor experimental stream-wetland systems. *J. Environ. Qual.* 20, 865-868, 1991.

Anderson, J.W. Selenium interactions in sulfur metabolism. In: *Sulfur Nutrition and Assimilation in Higher Plants — Regulatory, Agricultural and Environmental Aspects.* De Kok, L.J., Ed., SPB Academic Publishing. 1993.

Arvy, M.P. Selenate and selenite uptake and translocation in bean plants (*Phaseolus vulgaris*). *J. Exp. Bot.* 44, 1083-1087, 1993.

Asher, C.J., C.S. Evans, and C.M. Johnson. Collection and partial characterization of volatile selenium compounds from *Medicago sativa* L. *Aust. J. Biol. Sci.* 20, 737-748, 1967.

Asher, C.J, G.W. Butler, and P.J. Peterson. Selenium transport in root systems of tomato. *J. Exp. Bot.* 23, 279-291, 1977.

Atkinson, R., S.M. Aschmann, D. Hasegawa, E.T. Thompson-Eagle, and W.T. Frankenberger, Jr. Kinetics of the atmospherically important reactions of dimethyl selenide. *Environ. Sci. Technol.* 24, 1326-1332, 1990.

Azaizeh, H., S. Gowthaman, and N. Terry. Microbial selenium volatilization in rhizosphere and bulk soils from a constructed wetland. *J. Environ. Qual.* 26, 666-672, 1997.

Bañuelos, G.S., A. Zayed, N. Terry, B. Mackey, L. Wu, S. Akohoue, and S. Zambrzuski. Accumulation of selenium by different plant species grown under increasing sodium and chloride salinity. *Plant Soil* 183, 49-59, 1996.

Barfield, D.G. and E.-C. Pua. Gene transfer in plants of *Brassica juncea* using *Agrobacterium tumefaciens*-mediated transformation. *Plant Cell Rep.* 10, 308-314, 1991.

Beath, O.A., H.F. Eppson, and C.S. Gilbert. Selenium and other toxic minerals in soils and vegetation. *Wyo. Agric. Stand. Bull.* 206, 1-9, 1935.

Bell, P.F., D.R. Parker, and A.L. Page. Contrasting selenate-sulfate interactions in selenium-accumulating and nonaccumulating plant species. *Soil Sci. Soc. Am. J.* 56, 1818-1824, 1992.

Biggar, J.W. and G.R. Jayaweera. Measurement of Selenium volatilization in the field. *Soil Sci.* 155, 31-35, 1993.

Boerjan, W., G. Bauw, M. van Montagu, and D. Inze. Distinct phenotypes generated by overexpression and suppression of S-adenosyl-methionine synthetase reveal developmental patterns of gene silencing in tobacco. *Plant Cell* 6, 1401-1414, 1994.

Brady, J.M., J.M. Tobin, and G.M. Gadd. Volatilization of selenite in aqueous medium by a *Penicillium* species. *Mycol. Res.* 100, 955-961, 1996.

Brown, T.A. and A. Shrift. Selenium: toxicity and tolerance in higher plants. *Biol. Rev.* 57, 59-84, 1982.

Burnell, J.N. Selenium metabolism in *Neptunia amplexicaulis*. *Plant Physiol.* 67, 316-324, 1981.

Challenger, F., D.B. Lisle, and P.B. Dransfield. Studies on biological methylation. Part XIV. The formation of trimethylarsine and dimethyl selenide in mould cultures from methyl sources containing [14]C. *J. Chem. Soc.* 1760-1771, 1954.

Chau, Y.K., P.T. Wong, B.A. Silverberg, P.L. Luxon, and G.A. Bengert. Methylation of selenium in the aquatic environment. *Science* 192, 1130-1131, 1976.

Cherest, H. J.-C. Davidian, D. Thomas, V. Benes, W. Ansorge, and Y. Surdin-Kerjan. Molecular characterization of two high affinity sulfate transporters in *Saccharomyces cerevisiae*. *Genetics* 145, 627-635, 1997.

Cooke, T.C. and K.W. Bruland. Aquatic chemistry of selenium: evidence of biomethylation. *Environ. Sci. Technol.* 21, 1214-1219, 1987.

Dawson, J.C. and J.W. Anderson. Incorporation of cysteine and selenocysteine into cystathionine and selenocystathionine by crude extracts of spinach. *Phytochemistry* 27, 3453-3460, 1988.

de Sousa, M.P., D. Chu, M. Zhao, A.M. Zayed, S.E. Ruzin, D. Schichnes, and N. Terry. Rhizosphere bacteria enhance selenium accumulation and volatilization by Indian mustard. *Plant Physiol.* 119, 565-573, 1999.

Doran, J.W. Microoganisms and the biological cycling of selenium. *Adv. Microb. Ecol.* 6, 1-32, 1982.

Doran, J.W. and M. Alexander. Microbial formation of volatile Se compounds in soil. *Soil Sci. Soc. Am. J.* 40, 687-690, 1977.

Duckart, E.C., L.J. Waldron, and H.E. Donner. Selenium uptake and volatilization from plants growing in soil. *Soil Sci.* 153, 94-99, 1992.

Duda, P.J. *Chevron's Richmond Refinery Water Enhancement Wetland.* Technical Report prepared for the Regional Water Quality Control Board, Oakland, CA, December, 1992.

Evans, C.S., C.J. Asher, and C.M. Johnson. Isolation of dimethyl diselenide and other volatile selenium compounds from *Astragalus racemosus* (Pursh.). *Aust. J. Biol. Sci.* 21, 13-20, 1968.

Fan, T.W-M., A.N. Lane, and R.M. Higashi. Selenium biotransformations by a euryhaline microalga isolated from a saline evaporation pond. *Environ. Sci. Technol.* 31, 569-576, 1997.

Foyer, C.H., N. Souriau, S. Perret, M. Lelandais, K.J. Kunert, C. Pruvost, and L. Jouanin. Overexpression of glutathione reductase but not glutathione synthetase leads to increases in antioxidant capacity and resistance to photoinhibition in poplar trees. *Plant Physiol.* 109, 1047-1057, 1995.

Frankenberger Jr., W.T. and U. Karlson. Dissipation of soil selenium by microbial volatilization at Kesterson Reservoir. Final report submitted to U.S. Department of the Interior. *Project 7-F.C-20-05240, November. U.S. Dept. of the Interior, Bureau of Reclamation,* Sacramento, CA., 1988.

Ganther, H.E., O.A. Levander, and C.A. Saumann. Dietary control of selenium volatilization in the rat. *J. Nutr.* 88, 55-60, 1966.

Hansen, D., P.J. Duda, A. Zayed, and N. Terry. Selenium removal by constructed wetlands: role of biological volatilization. *Environ. Sci. Technol.* 32, 591-597, 1998.

Haygarth, P.M. and K.C. Jones. Balance for selenium in grassland soil: net accumulation or net loss? In: *Contaminated Soil, Proc. 3rd International Conference on the Biogeochemistry of Trace Elements.* Prost, R., Ed. C1. May 15-19, 1995, Paris, France. INRA. 1997.

Haygarth, P.M., D.S. Fowler-Sturup, B.M. Davison, and K.C. Jones. Determination of gaseous and particulate selenium over a rural grassland in the U.K. *Atm. Environ.* 28, 3655-3663, 1995.

Jayaweera, G.R. and J.W. Biggar. Extraction procedure for volatile selenium from activated carbon. *Soil Sci.* 153(4), 288-292, 1992.

Karlson, U. and W.T. Frankenberger, Jr. Determination of gaseous ^{75}Se evolved from soil. *Soil Sci. Soc. Am. J.* 52, 673-681, 1988.

Karlson, U. and W.T. Frankenberger, Jr. Accelerated rates of selenium volatilization from California soils. *Soil Sci. Soc. Am. J.* 53, 749-753, 1989.

Karlson, U. and W.T. Frankenberger, Jr. Volatilization of selenium from agricultural evaporation pond sediments. *Sci. Total Environ.* 92, 41-54, 1990.

Läuchli, A. Selenium in plants: uptake, functions, and environmental toxicity. *Bot. Acta* 106, 455-468, 1993.

Leggett, J.E. and E. Epstein. Kinetics of sulphate absorption by barley roots. *Plant Physiol.* 31, 222-226, 1956.

Leustek, T., M. Murillo, and M. Cervantes. Cloning of a cDNA encoding ATP sulfurylase from *Arabidopsis thaliana* by functional expression in *Saccharomyces cerevisiae*. *Plant Physiol.* 105, 897-902, 1994.

Lewis, B.G. Volatile selenium in higher plants: the production of dimethyl selenide in cabbage leaves by the enzymatic cleavage of methylselenomethionine selenonium salt. Ph.D. thesis, University of California, Berkeley, CA, 1971.

Lewis, B.G., C.M. Johnson, and C.C. Delwiche. Release of volatile selenium compounds by plants: collection procedures and preliminary observations. *J. Agric. Food Chem.* 14, 638-640, 1966.

Lewis, B.G., C.M. Johnson, and T.C. Broyer. Volatile selenium in higher plants. The production of dimethyl selenide in cabbage leaves by enzymatic cleavage of Se-methyl selenomethionine selenonium salt. *Plant Soil* 40, 107-118, 1974.

Lin, Z.Q., D. Hansen, A.M. Zayed, and N. Terry. Biological selenium volatiliation of selenium: method of measurement under field conditions. *J. Environ. Qual.* 28, 309-315, 1999.

McCluskey, T.J., A.R. Scarf, and J.W. Anderson. Enzyme calatysed α,β-elimination of selenocystathionine and selenocystine and their sulfur isologues by plant extracts. *Phytochemistry* 25, 2063-2068, 1986.

McConnell, K.P. and O.W. Portman. Toxicity of dimethyl selenide in the rat and mouse. *Proc. Soc. Exp. Biol. Med.* 79, 230-231, 1952.

Mikkelsen, R.L. and H.F. Wan. The effect of selenium on sulfur uptake by barley and rice. *Plant Soil* 121, 151-153, 1990.

Neuhierl, B. and A. Bock. On the mechanism of selenium tolerance in selenium-accumulating plants. Purification and characterization of a specific selenocysteine methyltransferase from cultured cells of *Astragalus bisulcatus*. *Eur. J. Biochem.* 239, 235-238, 1996.

Ng, B.H. and J.W. Anderson. Synthesis of selenocysteine by cysteine synthases from selenium accumulator and non-accumulator plants. *Phytochemistry* 17, 2074, 1978.

Ng, B.H. and J.W. Anderson. Light-dependent incorporation of selenite and sulphite into selenocysteine and cysteine by isolated pea chloroplasts. *Phytochemistry* 18, 573-580, 1979.

Nyberg, S. Multiple use of plants: studies on selenium incorporation in some agricultural species for the production of organic selenium compounds. *Plant Foods Hum. Nutr.* 41, 69-88, 1991.

Ohlendorf, H.M., D.J. Hoffman, M.K. Saiki, and T.W. Aldrich. Embryonic mortality and abnormalities of aquatic birds: apparent impacts of selenium from irrigation drain water. *Sci. Total Environ.* 52, 49-63, 1986.

Rael, R.M. and W.T. Frankenberger. Influence of pH, salinity, and selenium on the growth of *Aeromonas veronii* in evaporation agricultural drainage water. *Wat. Res.* 30, 422-430, 1996.

Ravanel, S., M.-L. Ruffet, and R. Douce. Cloning of an *Arabidopsis thaliana* cDNA encoding cystathionine β-lyase by functional complementation in *Escherichia coli*. *Plant Mol. Biol.* 29, 875-882, 1995.

Reamer, D.C. and W.H. Zoller. Selenium biomethylation products from soil and sewage sludge. *Science* 208, 500-502, 1980.

Rosenfeld, I. and O.A. Beath. *Selenium, Geobotany, Biochemistry, Toxicity, and Nutrition*. Academic Press, New York, 1964.

Saiki, M.K. and T.P. Lowe. Selenium in aquatic organisms from subsurface agricultural drainage water, San Joaquin Valley, CA. *Arch. Environ. Contam. Toxicol.* 19, 496-499, 1987.

Saito, K., M. Kurosawa, K. Tatsuguchi, Y. Takagi, and I. Murakoshi. Modulation of cysteine biosynthesis in chloroplasts of transgenic tobacco over-expressing cysteine synthase (*O*-acetylserine(thiol)lyase). *Plant Physiol.* 106, 887-895, 1994.

Salt, D., M. Blaylock, N. Kumar, V. Dushenkov, B. Ensley, I. Chet, and I. Raskin. Phytoremediation: a novel strategy for the removal of toxic metals from the environment using plants. *Biotechnology* 13, 468-474, 1995.

Singh, M., N. Singh, and D.K. Bhandari. Interaction of selenium and sulfur on the growth and chemical composition of raya. *Soil Sci.* 129, 238-244, 1980.

Smith, F.W., P.M. Ealing, M.J. Hawkesford, and D.T. Clarkson. Plant members of a family of sulfate transporters reveal functional subtypes. *Proc. Nat. Acad. Sci. USA* 92, 9373-9377, 1995.

Terry, N. and A.M. Zayed. Selenium volatilization by plants. In: *Selenium in the Environment*, Frankenberger, Jr., W.T. and Benson, S., Eds., Marcel Dekker, New York, 1994.

Terry, N., C. Carlson, T.K. Raab, and A.M. Zayed. Rates of selenium volatilization among crop species. *J. Environ. Qual.* 21, 341-344, 1992.

Terry, N., A. Zayed, E. Pilon-Smits, and D. Hansen. Can plants solve the selenium problem? *Proc. 14th Annual Symposium, Current Topics in Plant Biochemistry, Physiology and Molecular Biology: Will Plants Have a Role in Bioremediation?* University of Missouri, Columbia. April 19-22, 1995.

Thompson-Eagle, E.T. and W.T. Frankenberger, Jr. Protein-mediated selenium biomethylation in evaporation pond water. *Environ. Toxicol. Chem.* 9, 1453-1462, 1990.

Thompson-Eagle, E.T. and W.T. Frankenberger, Jr. Selenium biomethylation in a saline environment. *Wat. Res.* 25, 231-240, 1991.

Thompson-Eagle, E.T., W.T. Frankenberger, Jr. and U. Karlson. Volatilization of selenium by *Alternaria alternata*. *Appl. Environ. Microbiol.* 55, 1406-1413, 1989.

Velinsky, D. and G.A. Cutter. Geochemistry of selenium in a coastal salt march. *Geochim. Cosmochim. Acta* 55, 179-191, 1991.

Weres, O., A. Jaouni, and L. Tsao. The distribution, speciation and geochemical cycling of selenium in a sedimentary environment, Kesterson Reservoir, CA, USA. *Appl. Geochem.* 4, 543-563, 1989.

Wilber, C.G. Toxicology of selenium: a review. *Clin. Toxicol.* 17, 171-230, 1980.

Wilber, C.G. *Selenium: A Potential Environmental Poison and a Necessary Food Constituent.* Charles C. Thomas Publisher, Springfield, IL, 126, 1983.

Williams, M.C. and H.F. Mayland. Selenium absorption by two-grooved milkvetch and western wheatgrass from selenomethionine, selenocystine, and selenite. *J. Range Manage.* 45, 374-378, 1992.

Wu, L., P.J. Van Mantgem, and X. Guo. Effects of forage plant and field legume species on soil selenium redistribution, leaching, and bioextraction in soils contaminated by agricultural drain water sediment. *Arch. Environ. Contam. Toxicol.* 31, 329-338, 1996.

Zayed, A. and N. Terry. Selenium volatilization in broccoli as influenced by sulfate supply. *J. Plant Physiol.* 140, 646-652, 1992.

Zayed, A. and N. Terry. Selenium volatilization in roots and shoots: effects of shoot removal and sulfate level. *J. Plant Physiol.* 143, 8-14, 1994.

Zayed, A., M. Lytle, and N. Terry. Accumulation and volatilization of different chemical species of selenium by plants. *Planta.* 206, 284-292, 1998.

Zhang, Y. and J.N. Moore. Environmental conditions controlling selenium volatilization from a wetland system. *Environ. Sci. Technol.* 31, 511-517, 1997.

Zieve, R. and P.J. Peterson. Factors influencing the volatilization of selenium from soil. *Sci Total Environ.* 19, 277-284, 1981.

Zieve, R. and P.J. Peterson. The accumulation and assimilation of dimethyl selenide by four plant species. *Planta* 160, 180-184, 1984a.

Zieve, R. and P.J. Peterson. Volatilization of selenium from plants and soils. *Sci. Total Environ.* 32, 197-202, 1984b.

Zijian, W., Z. Lihua, Z. Li, S. Jingfang, and P. An. Effect of the chemical forms of selenium on its volatilization from soils in Chinese low-selenium-belt. *J. Environ. Sci. (China)* 3(2), 113-119, 1991.

5 Metal Hyperaccumulator Plants: A Review of the Ecology and Physiology of a Biological Resource for Phytoremediation of Metal-Polluted Soils

Alan J. M. Baker, S. P. McGrath, Roger D. Reeves, and J. A. C. Smith

CONTENTS

1-56670-450-2/00/$0.00+$.50
© 2000 by CRC Press LLC

PLANT STRATEGIES TOWARD METALS

All plants take up metals to varying degrees from the substrates in which they are rooted. The concentrations in plant parts depend both on intrinsic (genetic) and extrinsic (environmental) factors and vary greatly for different species and for different metals. Baker (1981) proposed two basic strategies by which higher plants can tolerate the presence of large amounts of metals in their environment: (1) exclusion, whereby transport of metals is restricted, and low, relatively constant metal concentrations are maintained in the shoot over a wide range of soil concentrations; and (2) accumulation, whereby metals are accumulated in nontoxic form in the upper plant parts at both high and low soil concentrations. He suggested that accumulators can be characterized by a leaf:root metal concentration ratio of >1 because of the tendency to translocate metals from root to shoot, whereas in excluders the ratio is <1. An intermediate response is that of indicator plants, in which shoot metal concentrations reflect those in the soil. Whilst exclusion is more characteristic of species with both metal-tolerant and nontolerant genotypes, accumulators are commonly species restricted to metalliferous soils.

WHAT IS METAL HYPERACCUMULATION?

Peterson (1971) defined metal accumulation in two ways: (1) accumulation of an element within an organism to concentrations greater than those found in the growth medium and (2) possession of greater quantities of an element than is usual for that organism. These definitions create some problems for the interpretation of metal-accumulation patterns in plants. Only in laboratory studies using nutrient solutions is it possible to know the metal concentration of a precisely defined growth medium, while in soil it is necessary to base the discussion on total metal concentration, or one of a variety of measures of "available" metal concentration. For some metal-accumulating plant species, extreme accumulation of a metal is a normal, not an abnormal, feature of those species growing in their natural habitats.

Responses by plant species to exposure to metalliferous soils can range from phytotoxicity to survival by exclusion, with only small elevations of metal concentration (relative to the same species on nonmetalliferous soil) to survival with accumulated metal constituting a significant percentage of the plant dry matter. In the past 2 decades a number of plant species endemic to metalliferous soils have been reported to be capable of accumulating exceptional concentrations of metals such as nickel, zinc, copper, cobalt, and lead in their above-ground parts (Baker and Brooks, 1989). Such concentrations are far in excess of those normally considered to be phytotoxic. Brooks et al. (1977a) used the term *hyperaccumulators* to describe plants with Ni concentrations >1000 µg/g (0.1%) in their dried leaves, which is at least an order of magnitude greater than Ni concentrations present in nonaccumulator plants found on nickeliferous soils. Reeves (1992) further elaborated on this definition, including only those species which accumulated such concentrations when growing in their natural habitats. This criterion was also considered appropriate to specify hyperaccumulation of copper (Brooks et al., 1980) and lead (Reeves and

Brooks, 1983a), while for zinc a threshold of 10,000 μg/g (1.0 %) in dried plant material was suggested (Reeves and Brooks, 1983a) because of greater background concentrations of this metal (Table 5.1). Metal hyperaccumulation by terrestrial vascular plants was reviewed by Baker and Brooks (1989), and nickel hyperaccumulation has been discussed in detail by Reeves (1992).

TABLE 5.1

Numbers of Metal Hyperaccumulator Plants

Metal	Criterion (% in Leaf Dry Matter)	No. of Taxa	No. of Families
Cadmium	> 0.01	1	1
Cobalt	> 0.1	28	11
Copper	> 0.1	37	15
Lead	> 0.1	14	6
Manganese	> 1.0	9	5
Nickel	> 0.1	317	37
Zinc	> 1.0	11	5

Recent studies have further extended the range of species known to be capable of metal hyperaccumulation. This is an extreme evolutionary response to the presence of high metal concentrations in the soil and is not a common characteristic of terrestrial higher plants. Taxonomically, the metal hyperaccumulators identified to date account for less than 0.2% of all angiosperms. Hyperaccumulator species typically maintain high tissue metal concentrations across a wide range of soil metal concentrations, in agreement with the accumulator strategy proposed by Baker (1981). This raises important questions as to the nature of the (presumably) highly efficient metal-uptake mechanism at low soil metal concentrations, and of the processes responsible for transport and sequestration of metal in nontoxic form at higher soil concentrations. Metal concentrations in leaves, and often stems, of hyperaccumulators greatly exceed those of roots and other storage organs (Rascio, 1977; Hajar, 1987; Homer et al., 1991). Thus the roots, which are the perennial organs of many of these plants, are protected from very high metal levels, while there is the possibility of eliminating a considerable amount of accumulated metals through leaf fall (Vergnano Gambi et al., 1982; Schlegel et al., 1991). Many of the known hyperaccumulators are biennial or short-lived perennial herbs, or are shrubs or small trees.

Little is known of the mechanisms of transport of metals in hyperaccumulator plants. In view of the limited analyses of woody tissues available, it is unclear what roles xylem and phloem transport play in hyperaccumulation in leaves. The importance of nonvascular transport through laticifers, for example, is also obscure. The extent to which secondary xylem tissues become a long-term repository for accumulated metals is not known. Indeed, detailed metal budgets for individual plants are rare in the literature (see Baker et al., 1992). A research priority is therefore to gain basic information on the dynamics of metal movements, and on the sinks and their capacities in woody species capable of metal hyperaccumulation.

GEOGRAPHICAL DISTRIBUTION OF METAL HYPERACCUMULATORS

The high degree of endemism to metal-rich soils shown by hyperaccumulating taxa has been related to their survival. Such species do not compete well with nontolerant species under normal soil conditions, perhaps reflecting the metabolic "cost" of hyperaccumulation, or conceivably an elevated requirement for a particular metal compared with nonaccumulator species (Macnair and Baker, 1994; Macnair et al., Chapter 13). However, on metalliferous outcrops or on anthropogenically metal-contaminated substrata, hyperaccumulators are able to survive the strong selective pressures exerted upon other species and can become dominants, sometimes existing as nearly pure populations. Brooks et al. (1979) have drawn attention to a definite relationship in *Alyssum* species between endemism, species diversity, and proliferation on the one hand, and nickel hyperaccumulation on the other. Such a relationship is also apparent in the serpentine floras of both New Caledonia and Cuba, where an unusually large number of serpentine-endemic hyperaccumulators from several genera can be concentrated in one locality (Brooks et al., 1979; Reeves et al., 1996).

Serpentine (ultramafic) soils are generally infertile, with low NPK levels and a characteristically low Ca:Mg ratio, as well as having potentially phytotoxic concentrations of nickel (and often of chromium and cobalt), all of which have been shown to be important edaphic controlling factors (Yang et al., 1985; Proctor and Nagy, 1992; Robinson et al., 1997). The nickel-accumulating *Alyssum* species seem to be particularly well adapted to the low soil Ca:Mg ratio, still being able to acquire remarkably high Ca concentrations in their leaf tissues (Reeves et al., 1997). Studies with solution cultures have shown that the levels of Ca that are beneficial to the growth of a serpentine race of a nonaccumulator (*Silene italica*) actually depress both growth and nickel uptake by *A. bertolonii* (Gabbrielli et al., 1990).

The success of many metallophytes in overcoming the adverse edaphic conditions is such that they can exist relatively free from competition. Furthermore, the high metal concentrations of the native soils of hyperaccumulators can repress fungal growth and thus protect these plants from fungal diseases, to which they are susceptible when grown on nonmetalliferous soils (Morrison et al., 1979). Defense against fungal pathogens has been suggested as one of the most plausible reasons for the evolution of metal hyperaccumulation (Reeves et al., 1981; Boyd and Martens, 1992), at least for plants of families (such as the Brassicaceae) which are known to be very sensitive to fungal diseases. Experimental data for *Thlaspi caerulescens* and *Alyssum murale* (Hussain, 1994) have confirmed a protective role of zinc and nickel accumulated in the root cortex of seedlings in resistance against infection by the damping-off fungus, *Pythium ultimum*. Boyd and Martens (1992) and Pollard and Baker (1997) also present data to suggest that hyperaccumulated metals in leaves may play a role in defense against insect herbivory.

Two phytogeographical aspects of hyperaccumulator distribution have been noted. Serpentine endemism generally, and the occurrence of nickel hyperaccumulators in particular, are both much more marked in tropical to warm temperate parts of the world, i.e., those regions that have not suffered extensively from the effects of the Pleistocene glaciations (Reeves et al., 1983) (Figure 5.1). Within the tropics there are also marked differences in the incidence of serpentine endemism and of hyperaccumulation, related to the time of continuous availability of the metalliferous soil (Reeves et al., 1996).

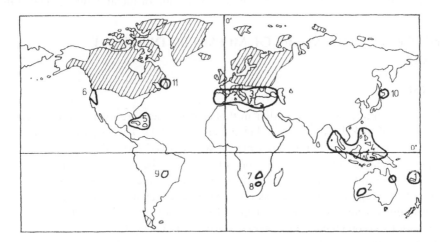

FIGURE 5.1 Global distribution of nickel hyperaccumulation. (Adapted from Brooks, R.R. *Biological Methods for Prospecting for Minerals.* John Wiley & Sons, New York, 1983.) Hatched areas indicate the limits of the ice sheets during the last glaciation. Areas and number of hyperaccumulators: 1. New Caledonia (50), 2. Australia (5), 3. S. Europe/Asia Minor (90), 4. S.E. Asia (11), 5. Cuba (128) and Dominican Republic (1), 6. U.S. (Pacific N.W. and California) (5), 7. Zimbabwe (5), 8. S. Africa (Transvaal) (4), 9. Brazil (Goiás) (11), 10. Japan (Hokkaido) (1), 11. Canada (Newfoundland) (4).

It is debated whether the endemism represented by the restriction of metal-tolerant species to such soils should be regarded as paleo-endemism or neo-endemism (Antonovics et al., 1971; Baker, 1987). In the former, originally widespread populations have become confined to severe environments through competition from other species. Neo-endemics are regarded as having developed from nontolerant precursors which have colonized metalliferous areas by the process of natural selection; in this case, more widely occurring, closely related species are often found in the surrounding area. For taxa now found exclusively on mine spoil, the neo-endemic hypothesis is favored, e.g., the metallophyte *Alyssum wulfenianum* may have the neighboring, related *A. ovirense* as precursor (Reeves and Brooks 1983a), and there are widespread species of *Alyssum, Thlaspi, Dichapetalum,* and *Brackenridgea,* for example, for which metal-accumulating taxa are recognized at subspecific rank. In genera such as *Alyssum, Thlaspi, Phyllanthus, Buxus,* and *Xylosma,* containing many metal-accumulating and nonaccumulating taxa, there is scope for the study of the geographic distribution and morphology to shed light on evolutionary relationships

(Morrison, 1980; Reeves et al., 1999). Phylogenetic questions may also be resolved using the techniques of molecular systematics, which have revealed, for example, that metal hyperaccumulation within *Thlaspi sensu lato* is restricted to a lineage consisting of *Noccaea* and *Raparia*, but does not occur in *Thlaspi sensu stricto* (Mummenhoff and Koch, 1994; Mummenhoff et al., 1997).

Whichever factors have promoted the evolution and survival of hyperaccumulating taxa, it is clear that their tolerance of stressed metalliferous environments, coupled with their exceptional ability to accumulate large quantities of metals, could be exploited in a practical way for the detoxification of metal-contaminated soils.

HYPERACCUMULATION OF METALS BY DIFFERENT TAXA

NICKEL

With the definitions set out earlier in this chapter, there are many more hyperaccumulators of nickel than of any other metal (Table 5.1). This is partly owing to the relative importance, on a global scale, of soils derived from ultramafic rocks, typically containing 0.1 to 1.0% Ni, but may also be because most effort has been expended on this group. Surface exposures of ultramafic rocks occur as continuous or narrowly separated edaphic "islands" of significant area for evolutionary purposes, especially in countries such as Cuba, Brazil (Goiás), U.S. (Pacific Northwest), Canada (Newfoundland), Italy, Greece and the Balkans, Turkey, Japan, Indonesia, the Philippines, New Caledonia, Australia (Western Australia, Queensland), South Africa (eastern Transvaal) and Zimbabwe.

At least 317 nickel-hyperaccumulating taxa are now known. They fall largely into two groups: (1) within the tropical genera and species of families such as the Violaceae and Flacourtiaceae (Order Violales), and the Buxaceae and Euphorbiaceae (Order Euphorbiales) and (2) among genera of the Brassicaceae (Order Capparales) of the northern temperate zone. In the Brassicaceae, nickel hyperaccumulators show a remarkable concentration in the genus *Alyssum* L., which contains some 48 taxa with 0.1 to 3.0% Ni, confined to Section Odontarrhena (Minguzzi and Vergnano, 1948; Brooks and Radford, 1978; Brooks et al., 1979; Vergnano Gambi et al., 1979). The geographical distribution of *Alyssum* hyperaccumulators is strongly correlated with the occurrence of nickel-rich serpentines and other ultramafic rocks of southern Europe, the eastern Mediterranean, and Asia Minor (Brooks et al., 1979); the genus shows a high degree of serpentine endemism. The genus *Thlaspi* L. (Brassicaceae) also contains a significant number of species with the ability to accumulate nickel to levels of 0.1 to 3.0% on a dry matter basis (Reeves and Brooks, 1983b); some of these are also known to be hyperaccumulators of zinc. These species have a similar distribution to *Alyssum* in central and southern Europe (although *T. caerulescens* extends more widely into northern Europe, especially on zinc-rich soils), reaching maximum diversity in Greece.

New Caledonia possesses some of the world's largest areas of ultramafic soils and is noted for its nickel flora (Figure 5.1). About 50 hyperaccumulators of nickel from 14 genera and 8 families have been identified as endemic to this island (Brooks et al., 1974, 1977a; Jaffré et al., 1976, 1979). Notable genera include *Homalium* Jacq. (Flacourtiaceae) and *Phyllanthus* L. (Euphorbiaceae).

Metal hyperaccumulators have been recognized on the basis of concentrations of metals in leaf dry matter. Little information is available on the accumulation of metals in other organs and tissues, although such analyses are needed to provide a more complete picture of metal-allocation patterns within the plant (Ernst, 1995). The majority of nickel hyperaccumulators from New Caledonia (Baker and Brooks, 1989), Indonesia (Wither and Brooks, 1977), the Philippines (Baker et al., 1992), and Cuba (Reeves et al., 1999) are trees or shrubs. The most remarkable is perhaps the New Caledonian tree *Sebertia acuminata* (Sapotaceae), which can grow to a height of >10 m and which produces a blue-green latex containing nearly 26% nickel on a dry-weight basis (or 11% on a wet-weight basis) (Jaffré et al., 1976). The woody tissues contain about 0.17% Ni compared with about 1.2% in the leaf dry matter. Similar concentrations (about 0.2%) have been reported in the wood of *Psychotria douarrei* (Rubiaceae) and *Hybanthus floribundus* (Violaceae) from New Caledonia and Western Australia, respectively. Baker et al. (1992) have described nickel hyperaccumulators from Palawan (Philippines) including *Phyllanthus "palawanensis"* (Euphorbiaceae), *Dichapetalum gelonioides* ssp. *tuberculatum* (Dichapetalaceae), *Walsura monophylla* (Meliaceae), and *Brackenridgea palustris* ssp. *foxworthyi* (Ochnaceae). The first-mentioned has woody stems that produce a jade-colored sap on cutting, and phloem tissues with >9% Ni (dry-weight basis). Analysis of herbarium specimens of taxa closely related to the Palawan plants has revealed more nickel hyperaccumulators as well as three zinc hyperaccumulators (subspecies of *D. gelonioides*). Leaves of one specimen of *D. gelonioides* ssp. *tuberculatum* from West Sumatra were found to contain 3% Zn even though the soils of the area are not thought to be metalliferous according to site notes on the herbarium sheet (Baker et al., 1992). Analysis of woody tissues has not been possible for these herbarium samples.

Morrey et al. (1992) and Anderson et al. (1997) have reported very high (up to 3.7%) Ni and elevated Cr concentrations in the tall, productive herb *Berkheya coddii* (Asteraceae) from the ultramafics of northeastern Transvaal, South Africa. This (and related) species may have a high phytoremediation potential, in view of its strong hyperaccumulation, relatively high biomass production, and its ability to grow in dense stands on nickeliferous soils. Similar potential is seen in the many hyperaccumulators in *Buxus, Leucocroton, Phyllanthus,* and *Euphorbia* from Cuba (Reeves et al., 1999).

ZINC

The range of hyperaccumulators of zinc (>1.0% dry weight) is less extensive than for nickel, possibly because of the less frequent occurrence of large exposures of zinc-rich soil and the higher threshold set in defining hyperaccumulation. It appears thus far to be limited to a few genera such as *Thlaspi* and *Cardaminopsis* in the Brassicaceae and several species in other families. Early detailed investigations of the "Galmei" (calamine) flora of western Germany and eastern Belgium by Baumann (1885) and others showed that Zn concentrations in the leaves of *Thlaspi calaminare* and *Viola calaminaria* can reach 3.5 and 1.0%, respectively. The former species, together with others regarded for many years as part of the *Thlaspi alpestre* complex, is now properly referred to as *T. caerulescens* (Ingrouille and Smirnoff, 1986). Several other *Thlaspi* species from lead/zinc-mineralized soils have been reported with up to 2.0% Zn (Reeves and Brooks, 1983a). The distribution of *T. caerulescens* in Britain and Belgium is strongly linked to lead/zinc mines (Shimwell and Laurie, 1972; Ingrouille and Smirnoff, 1986; Baker and Proctor, 1990; Baker et al., 1994) and other industrially polluted areas such as the vicinity of metal smelters (Denaeyer-De Smet and Duvigneaud, 1974). *Cardaminopsis halleri* (Brassicaceae) has also been reported as a hyperaccumulator of zinc (Ernst, 1968; Macnair et al., Chapter 13).

CADMIUM

Most plants growing over lead/zinc-mineralized soils show elevated cadmium concentrations in the range 10 to 100 μg/g. A concentration of >100 μg/g (0.01%) Cd in plant dry matter is exceptional and could be a suitable criterion for the recognition of hyperaccumulation. *Thlaspi caerulescens* has been demonstrated to accumulate Cd (occasionally) in leaf dry matter to 1000 μg/g; values above 100 μg/g generally accompany extreme zinc accumulation from mine or smelter waste.

LEAD

Hyperaccumulation of lead is particularly rare (Table 5.1). The low solubility of most lead compounds in *circum*-neutral media, and the ready precipitation of lead by sulfate and phosphate at the root systems may partly explain this. Despite this, *Thlaspi rotundifolium* ssp. *cepaeifolium* from a lead/zinc mining area of Cave del Predil (northern Italy) has been found with lead up to 8200 μg/g of dry weight (Reeves and Brooks, 1983a). *Alyssum wulfenianum* Schlecht. from the same location also contained remarkably high Pb concentrations, reaching 860 μg/g in leaf dry matter. A concentration of up to 2740 μg/g Pb was also reported in *T. caerulescens* colonizing a lead mine district in the Pennines, England (Shimwell and Laurie, 1972). A recent report of lead hyperaccumulation in the grass *Arrhenatherum elatius* (Deram and Petit, 1997) growing on calamine soils in France is of great interest as there have been no convincing reports of the phenomenon in the Poaceae. Barry and Clark (1978) recorded shoot Pb values of 130 to 11,750 μg/g in *Festuca ovina,* and Williams et al. (1977) found a mean shoot Pb value of 13,488 μg/g in pasture species growing on mining waste in the U.K. The rather erratic nature of Pb values found in this kind of environment raises the question of the relative importance of entry via the root system and direct deposition on or into the leaves (e.g., from smelter emissions and wind-blown soil particles). Careful study of this point is still required. Nevertheless, high average Pb concentrations can be achieved by uptake through the root sytem: Baker et al.(1994) found that seedlings of *T. caerulescens* exposed to a nutrient solution containing 20 μg/ml Pb contained 4500 to 7000 μg/g Pb after 21 days (roots 29,000 μg/g; shoots 280 μg/g). Even higher whole-plant concentrations can also be achieved by exposing nontolerant species to high levels of soluble lead for short periods, even though the plants may not survive the treatment. For example, addition of synthetic chelates such as EDTA (ethylenediaminetetraacetic acid) to the root medium can dramatically increase lead uptake and accumulation in *Zea mays* and *Pisum sativum*, giving shoot concentrations in excess of 1% of dry weight (Huang and Cunningham, 1996; Huang et al., 1997). Similarly, shoot concentrations of 1.5% lead by dry weight have been observed in *Brassica juncea* growing in contaminated soil amended with EDTA (Blaylock et al., 1997). These results are especially notable because they indicate that, in the presence of appropriate ligands, it is possible for lead to remain in soluble form for transport through the root system and translocation to the above-ground parts of the plant.

COPPER

Most of the hyperaccumulators of copper discovered to date are confined to Shaba Province in Zaïre (now the Democratic Republic of Congo) and the Copper Belt of northwest Zambia (Brooks et al., 1980). The copper mineralization is dispersed over some 22,000 km^2 and has provided a large area for the evolution of a multiplicity of species with the ability to accumulate and/or tolerate high Cu levels in the soil in virtual isolation, as the metal acts as a barrier to competing species. Elsewhere, copper-mineralized areas that have been subjected to detailed vegetation study cover only a few hectares, which probably explains why the known hyperaccumulators of copper are concentrated here. Some 24 hyperaccumulators of copper, several of which also hyperaccumulate cobalt, have been reported from Africa, occurring principally in advanced families such as Lamiaceae and Scrophulariaceae (Brooks and Malaisse, 1985). Particularly elevated Cu concentrations (up to 13,700 µg/g) have been found in *Aeollanthus biformifolius* De Wild. (Lamiaceae), a dwarf perennial herb which also hyperaccumulates cobalt, endemic to the southern part of the Shaban Copper Arc (Malaisse et al., 1978). Beyond Africa, there have been a few reports of high Cu concentrations in species such as *Minuartia verna* (Caryophyllaceae; 1070 µg/g) in Germany (Ernst, 1974) and *Millotia myosotidifolia* (Asteraceae; 2400 µg/g) in South Australia (Blissett, 1966). Confirmation of these reports would be helpful to ascertain whether any of these plant analyses may have been affected by dust containing nearly pure secondary copper minerals, a problem that must be kept in mind when biogeochemical work is done over copper-rich rocks and soils.

COBALT

Baker and Brooks (1989) list 26 hyperaccumulators (>1000 µg/g) of cobalt, 9 of which also hyperaccumulate copper. All are species of the vegetation of the Shaban Copper Arc, Zaïre. The majority are low-growing herbs in a range of families including Lamiaceae, Scrophulariaceae, Asteraceae, and Fabaceae. The highest concentration reported to date is 10,200 µg/g Co in *Haumaniastrum robertii* (Robyns) Duvign. et Plancke (Lamiaceae). Abnormal uptake of cobalt, occasionally exceeding 100 µg/g, has been recorded for the North American tree *Nyssa sylvatica* (Nyssaceae) from soils not known to be metalliferous (Brooks et al., 1977b).

MANGANESE

Eight hyperaccumulators of manganese were listed by Baker and Brooks in 1989. The short list remains the basis of the summary figures presented in Table 5.1 and the criterion for recognition of hyperaccumulator status is still realistically set at 10,000 µg/g (1.0%) Mn in aerial plant dry matter. To date there has been little interest in pursuing the possible exploitation of these plants from five unrelated families in phytoremediation, as manganese has not been recognized as a metal of major environmental or economic concern.

EVIDENCE FOR MULTIMETAL ACCUMULATION

Just as cotolerances have been shown to exist, it is evident from the above that the ability to accumulate unusually high levels of more than one heavy metal (coaccumulation) is present in several species, e.g., *Aeollanthus biformifolius* (Cu, Co), *Thlaspi rotundifolium* ssp. *cepaeifolium* (Pb, Zn), *T. caerulescens* (Zn, Cd, Pb). Homer et al. (1991) studied a number of *Alyssum* species in pot trials and demonstrated the ability of nickel hyperaccumulators to behave in the same way toward cobalt when this was made available, and suggested a similar mechanism of uptake for the two metals. Similarly, co-accumulation of Ni, Zn, Co, and Mn by the hyperaccumulator *Thlaspi goesingense* was attributed by Reeves and Baker (1984) to the existence of a nonspecific system of metal detoxification. Baker et al. (1994) reported that *T. caerulescens* seedlings, exposed for 21 days to nutrient solutions containing any of a variety of heavy metals at low concentration (0.5 to 25 µg/ml), were capable of taking up most of them to high total plant concentrations (>1000 µg/g). They distinguished between metals that were readily translocated to the shoot (e.g., Zn, Cd, Ni, Co, Mn) and those that became predominantly fixed at high concentrations in the root system (e.g., Pb, Fe, Al, Cu). The latter group could be found with concentrations of >10,000 µg/g in the root, but generally <600 µg/g in the shoot.

In contrast, McGrath et al. (Chapter 6) showed that differences in zinc and cadmium uptake occurred between two populations of *T. caerulescens* grown in the same soil. Both hyperaccumulated zinc but only one accumulated cadmium, suggesting strong specificity in the uptake mechanisms for the two metals.

METAL TOLERANCE AND METAL "REQUIREMENT" OF HYPERACCUMULATOR PLANTS

RELATIONSHIP BETWEEN METAL ACCUMULATION AND TOLERANCE

The precise relationship between metal accumulation and tolerance has not yet been resolved. Because hyperaccumulator plants are usually restricted in their natural distribution to metalliferous soils, it is possible to regard the hyperaccumulation trait as one form of tolerance mechanism. Some workers have proposed that there is no correlation between metal hyperaccumulation and tolerance, i.e., that they may be independent characteristics (Ingrouille and Smirnoff, 1986; Baker and Walker, 1990;

Baker et al., 1994; Lloyd-Thomas, 1995), while others state that accumulators possess a high degree of tolerance (Mathys, 1977; Reeves and Brooks, 1983b) or "hypertolerance" (Chaney et al., 1997). Screening experiments for zinc, lead, cadmium, and nickel accumulation by *T. caerulescens* populations using solution culture (Lloyd-Thomas, 1995; Chaney et al., Chapter 7) have revealed significant differences between calamine, serpentine, and alpine (nonmetallophyte) populations but no direct relationship to the metal status of the parent soils. The relationship between tolerance and accumulation may be difficult to test directly because (1) it is difficult to find populations which lack either tolerance or hyperaccumulation and (2) "tolerance" can be defined in a number of different ways. Mutagenesis of hyperaccumulator species could be one way to test for a causal relationship between hyperaccumulation and tolerance, but a mechanistic understanding of the origins of metal tolerance at the physiological and biochemical levels is also required.

The response of different plant species to particular metals varies markedly. Species from the same geographical location have evolved contrasting tolerance mechanisms toward the same metals, e.g., *Silene italica* L. and *Alyssum bertolonii* Desv. are both found on serpentine soils and are, respectively, an excluder and accumulator of nickel (Gabbrielli et al., 1990). Indeed, in most habitats with metal-rich soils it is found that hyperaccumulator species are in a minority, suggesting that metal exclusion is the more widespread strategy taxonomically.

Given that true hyperaccumulator plants are normally restricted to metalliferous soils, it is axiomatic that they must be metal tolerant, at least in the general sense of being able to grow and complete their life cycle on metal-rich soils. But it is difficult from these ecological associations alone to make inferences about tolerance mechanisms. Formally, it will not be possible to claim a causal relationship between hyperaccumulation and tolerance until the genetic basis of metal hyperaccumulation is understood. For example, the close association between nickel hyperaccumulators and ultramafic soils does not prove that this results from the nickel tolerance mechanism per se. These plants might simply be particularly well adapted to some other peculiarity of the soils, such as their nutrient imbalance (high Mg:Ca ratio, etc.). By the same token, species may be excluded from ultramafic soils because they cannot tolerate such nutrient imbalances, rather than because of their sensitivity to nickel. Nevertheless, in the largest single genus of hyperaccumulator plants, *Alyssum*, laboratory experiments have shown that the nickel-hyperaccumulator species are much more nickel tolerant than the nonaccumulator species (Homer et al., 1991; Krämer et al., 1996). Further experiments in this direction are clearly needed with other genera, but the hypothesis of an underlying, mechanistic association between metal hyperaccumulation and tolerance still seems perfectly plausible.

METAL DETOXIFICATION AND SEQUESTRATION

In hyperaccumulator plants, metal concentrations can exceed 1% of plant dry weight without any adverse effect on growth. This implies the existence of mechanisms for metal detoxification within the plant, most likely involving chelation of the metal cation by ligands and/or sequestration of metals away from sites of metabolism in the cytoplasm, notably into the vacuole or cell wall. Other possible adaptive responses include activation of alternative metabolic pathways less sensitive to metal ions, modification of enzyme structure, or alteration of membrane permeability by structural reorganization or compositional changes (Ernst et al., 1992). As yet, there is no evidence that these latter mechanisms of cellular metal tolerance are important in hyperaccumulator plants, but tolerance might be the result of a number of these processes acting collectively (Baker, 1987).

Recent analytical studies have begun to provide information on the localization of metals within hyperaccumulator plants. In the roots of plants from a population of *T. caerulescens* able to accumulate zinc and cadmium, x-ray microanalysis demonstrated that zinc accumulated principally in the vacuoles of epidermal and subepidermal cells, with smaller amounts stored in cell walls, while the apoplast was the main storage site of cadmium, with some cadmium being stored in vacuoles (Vázquez et al., 1992). Perhaps of even greater interest in hyperaccumulator plants are the sites of metal localization in the shoot, since bulk metal concentrations typically exceed those in the root, and the metal has to be translocated through the entire plant in soluble form to arrive at its final destination. In shoots of *T. caerulescens*, the highest zinc concentrations again appeared to be in the vacuoles of epidermal and subepidermal cells (Vázquez et al., 1994). This type of tissue distribution of metals may prove to be characteristic of hyperaccumulator plants, as nickel appears to be preferentially accumulated in epidermal cells of the nickel hyperaccumulators *Hybanthus floribundus* (Severne, 1974), *Senecio coronatus* (Mesjasz-Przybylowicz et al., 1994), and *Berkheya zeyheri* ssp. *rehmanii* var. *rogersiana* (Mesjasz-Przybylowicz et al., 1995).

Because it is clear that most of the metal taken up by hyperaccumulator plants must be chelated if acute metal toxicity is to be avoided, much interest has focused on identifying the intracellular ligands involved. We shall consider these in terms of the characteristic electron donor centers in the different classes of ligand.

Oxygen Donor Ligands

The feasibility of the involvement of organic acids in metal detoxification has long been recognized. Carboxylic acid anions are abundant in the cells of terrestrial plants and form complexes with divalent and trivalent metal ions of reasonably high stability. In particular, carboxylates such as malate, aconitate, malonate, oxalate, tartrate, citrate, and isocitrate are commonly the major charge-balancing anions present in the cell vacuoles of photosynthetic tissues, and they have been shown to chelate high concentrations of calcium and magnesium (Smith et al., 1996).

Analysis of metal-rich extracts from several nickel hyperaccumulators from New Caledonia has shown that nickel is predominantly bound to citrate (Lee et al., 1977, 1978) and the amount of citrate produced is strongly correlated with the accumulated nickel. Citrate also forms a 1:1 complex with aluminium in leaves of the aluminium hyperaccumulator *Hydrangea macrophylla* (Ma et al., 1997). In some other nickel-hyperaccumulating species from New Caledonia (Kersten et al., 1980), the Philippines (Homer et al., 1991), South Africa (Anderson et al., 1997), and particularly in the Mediterranean *Alyssum* species such as *A. bertolonii* (Pelosi et al., 1976), *A. serpyllifolium* subspecies (Brooks et al., 1981), and *A. troodii* (Homer et al., 1995), malate has consistently appeared as the major ligand associated with the nickel in aqueous extracts of leaf material. Malate also extracts with nickel from the root cortex, xylem fluid, and leaf sap of *A. bertolonii* (Gabbrielli et al., 1997).

Mathys (1977) found that malic acid levels were correlated with the degree of resistance to zinc, with far greater concentrations present in zinc-tolerant ecotypes. Similarly, the synthesis of mustard oils by *T. caerulescens* and of oxalate by *Silene vulgaris* was significantly greater in the Zn-resistant populations. On the basis of these findings, Mathys proposed a mechanism for zinc tolerance, whereby Zn^{2+} ions are bound by malate upon uptake into the cytoplasm, and the malate then serves as a carrier to transport the Zn^{2+} ions to the vacuole. The Zn^{2+} ions are then complexed by a terminal acceptor, possibly a sulfur-containing mustard oil in *T. caerulescens* and oxalate in *S. cucubalus*, and the released malate is able to return to the cytoplasm where it is ready to transport more Zn^{2+} ions. However, the molar ratio of S:Zn in *T. caerulescens* has been shown to be only 0.4 (Shen et al., 1997), and most of the total S would in fact be associated with protein in the plant. This makes it unlikely that mustard oils are responsible for increased tolerance and hyperaccumulation of zinc in this species.

Although the carboxylates are undoubtedly quantitatively important ligands for metal chelation in the vacuole, they tend to be present constitutively in the shoots of terrestrial plants and do not seem to account for either the metal specificity or species specificity of hyperaccumulation (Woolhouse, 1983; Ernst et al., 1992; Harmens et al., 1994). Even though the concentrations of ligands such as malate and citrate can be higher in metal-treated plants (e.g., Mathys, 1977; Thurman and Rankin, 1982; Godbold et al., 1984), this may be a general metabolic response that serves to maintain charge balance by organic acid synthesis (Osmond, 1976), rather than a specific one that accounts for tolerance toward a particular metal. This was the conclusion reached by Thurman and Rankin (1982) based on their experiments with *Deschampsia cespitosa*, which failed to show reduced metal tolerance when grown under conditions that decreased the amounts of citrate produced. Also, Qureshi et al. (1986) did not observe any correlation between malate accumulation and zinc tolerance in cell cultures of tolerant and nontolerant ecotypes of *Anthoxanthum odoratum*. The quantitative importance of the vacuole as a final repository for metals and the effectiveness of the vacuolar carboxylates in metal chelation are not in doubt, as the detailed studies of Krotz et al. (1989) and Wang et al. (1991, 1992) emphasize. Current evidence, however, suggests that metal-induced organic acid synthesis is unlikely to account for the specificity of the hyperaccumulation phenomenon.

Sulfur Donor Ligands

The sulfur donor atom in organic ligands is a considerably better electron donor than oxygen and leads to complexes of very high stability with first-row transition metals (Fraústo da Silva and Williams, 1991). Two classes of sulfur-containing ligand have been identified in plants that may play an important role in metal tolerance — metallothioneins and phytochelatins. The metallothioneins are the small cysteine-rich proteins known as metallothioneins, which are subdivided into three classes

(Robinson et al., 1993). In fungi and mammals, metallothioneins are known to be able to function in metal detoxification (Hamer, 1986; Tohayama et al., 1995), but their role in plants has been controversial (Robinson et al., 1993; Zenk, 1996). It is only recently that the protein products have been identified and purified (Murphy et al., 1997), but there is convincing evidence that expression of the *MT2a* gene correlates with copper tolerance in different ecotypes of *Arabidopsis thaliana* (Murphy and Taiz, 1995). Since the metallothioneins are evidently encoded in plants by multi-gene families (Zhou and Goldsbrough, 1995; Murphy et al., 1997), further studies of their expression patterns should provide valuable information on the involvement of specific genes in metal homeostasis.

Phytochelatins are low-molecular-weight, cysteine-rich peptides (Rauser, 1995; Zenk, 1996) now designated class III metallothioneins, as they can be regarded generically as nontranslationally synthesized metal-thiolate polypeptides (Robinson et al., 1993). They are synthesized by representatives of the whole plant kingdom upon exposure to heavy metals (Grill et al., 1987), and they are especially produced by plants growing in metal-enriched ecosystems (Grill et al., 1988). Phytochelatins are believed to be functionally analogous to the metallothioneins produced by animals and fungi (Tomsett and Thurman, 1988; Robinson et al., 1993) and consequently to be involved in cellular homeostasis of metal ions. They have the ability to bind a wide range of metals and it has been suggested by some researchers (Jackson et al., 1987; Salt et al., 1989) that phytochelatins are directly involved in heavy metal tolerance. However, there is considerable evidence to contradict this view. Metal induction of phytochelatins has been observed in both metal-resistant and metal-sensitive plants (Schultz and Hutchinson, 1988; Verkleij et al., 1991; Harmens et al., 1993). Furthermore, buthionine sulfoximine (BSO), an inhibitor of phytochelatin synthesis, has been shown not to decrease zinc tolerance (Reese and Wagner, 1987; Davies et al., 1991), while sulfur deficiency was seen to have no effect on copper tolerance of *Deschampsia cespitosa* (Schultz and Hutchinson, 1988). It is therefore questionable what exact role phytochelatins play in cellular metal-tolerance mechanisms.

There is some evidence for the involvement of phytochelatins in cadmium tolerance, but it has been argued that they are principally involved in detoxification rather than being the basis of genetically determined cadmium tolerance (Verkleij et al., 1991; de Knecht et al., 1994). This interpretation is supported by the phenotype of the *cad1* mutants of *Arabidopsis*, which have reduced phytochelatin levels and are cadmium-sensitive, but which are only slightly affected in their tolerance toward copper and zinc (Howden et al., 1995). Even if phytochelatin deficiency leads to hypersensitivity toward a particular metal, it does not follow that unusual accumulation of this metal will be related to phytochelatin production. More work is needed to establish whether this class of ligand plays any role in metal homeostasis or tolerance in the true hyperaccumulator plants.

Nitrogen Donor Ligands

Organic ligands containing nitrogen donor centers also form complexes of high stability with first-row transition metals, but with thermodynamic stability constants intermediate between those of the oxygen and sulfur donor ligands (Fraústo da Silva and Williams, 1991). In a prescient publication, Still and Williams (1980) postulated that the selectivity of the metal uptake and translocation process in nickel hyperaccumulators could be explained by chelation involving a ligand containing two nitrogen and one oxygen donor centers. This class of ligand has received relatively little attention, though Homer et al. (1995) observed in a study of bulk leaf extracts of four nickel hyperaccumulators that total amino acids changed only slightly over a wide range of nickel contents.

Earlier studies with crop plants indicated that amino acids, together with carboxylic acids, could play a significant role in metal chelation in the xylem (White et al., 1981a,b; Cataldo et al., 1988). Because root-to-shoot transfer of metals must be extremely effective in hyperaccumulator plants, Krämer et al. (1996) investigated the relationship between metal transport and xylem sap composition in species of *Alyssum* exposed to different nickel concentrations. This revealed a striking linear correlation between the concentration of nickel and histidine in the xylem of three hyperaccumulator species (*A. lesbiacum*, *A. murale*, and *A. bertolonii*). Over a range of nontoxic nickel concentrations, histidine was the only amino acid or carboxylic acid to show a significant response to metal treatment. Chemically, this is a striking vindication of the proposal of Still and Williams (1980), and in fact, at the prevailing xylem pH values, the effective stability constant for the nickel-histidine complex is higher than for any other amino acid or organic acid. Furthermore, it was shown that supplying exogenous histidine to the nonaccumulator *A. montanum*, either as a foliar spray or in the root medium, considerably increased the nickel tolerance of this species and greatly increased nickel flux through the xylem (Krämer et al., 1996). These results with *Alyssum* suggest that histidine is involved both in the mechanism of nickel tolerance and in the effective translocation of nickel to the shoot that characterizes these hyperaccumulator plants. Because nickel is complexed mainly with carboxylic acids in the shoot, the primary role of histidine may be to chelate the nickel taken up by the root cells and then to facilitate export of nickel to the shoot in the xylem. In this way, there may be a direct mechanistic link between the nickel hyperaccumulation and tolerance mechanisms, at least in the genus *Alyssum*. Further work will be required to ascertain whether metal-induced production of histidine (or perhaps other nitrogen-containing ligands) is found in other groups of nickel hyperaccumulators, or indeed in plants that hyperaccumulate other metals.

METAL "REQUIREMENT" OF HYPERACCUMULATOR PLANTS

When transplanted to soil containing only traces of the metals that they usually accumulate, hyperaccumulators display normal growth (Morrison et al., 1979; Reeves and Baker, 1984). This could indicate that there is no direct physiological requirement for elevated tissue metal concentrations and thus for their geographical distribution on mineralized or metal-enriched soils. However, when grown for long periods on some potting composts, *T. caerulescens* can become zinc deficient (McGrath, 1998). In solution culture they do show a greater requirement for some metals than nonaccumulators (Li et al., 1995; Chaney et al., 1997; Shen et al., 1997). For example, the critical solution concentrations which resulted in zinc deficiency in *T. caerulescens* grown in buffered media (<10 µm) were at least five orders of magnitude higher than those for most plant species (McGrath, 1998). Even very zinc-deficient plants grown with an external concentration of 0.1 nM (10^{-10} m) Zn, contained 300 to 500 µg/g Zn in the shoots, which is approximately 20 times the tissue concentration for the threshold of deficiency found in other plants. The reason for this high requirement may be related to the operation of strong constitutive mechanisms for sequestration and tolerance of zinc (Lloyd-Thomas, 1995). Similar stimulatory responses have been shown in the nickel hyperaccumulator *A. lesbiacum*, which when grown in solution culture (modified 0.1-strength Hoagland solution) had a growth optimum of 30 µm Ni (Krämer et al., 1996).

USE OF HYPERACCUMULATOR PLANTS IN PHYTOREMEDIATION

It has been recognized for more than 16 years that plant uptake by hyperaccumulator plants could be exploited as a biological cleanup technique for various polluted rooting media including soils, composted materials, effluents, and drainage waters. The possibility of their use in phytomining has also been demonstrated (Nicks and Chambers, 1995). However, before phytoextraction of soils is possible on a large scale, a number of important issues must be addressed. First, metal hyperaccumulator plants are relatively rare, often occurring in remote areas geographically and being of very restricted distribution in areas often threatened by devastation from mining activities. Population sizes can be extremely small. There is thus an urgent need to collect these materials, bring them into cultivation and establish a germplasm facility for large-scale production for future research and development and trials work. Secondly, the potential exploitation of metal uptake into plant biomass as a means of soil decontamination is clearly limited by plant productivity. Many of the European hyperaccumulator plants are of small biomass, although considerable natural variation exists within populations (Lloyd-Thomas, 1995; Chaney et al., 1997). In view of their infertile native habitats (and likely nutritional adaptations to these edaphic conditions), it is surprising that plants such as *Thlaspi* and *Alyssum* spp. are responsive to nutrient additions to the soil, and thus their growth potential can be enhanced by soil fertilization. Selection trials are needed to identify the fastest growing (largest potential biomass and greatest nutrient responses) and most strongly metal-accumulating genotypes. However, such a combination may not be possible and a tradeoff between extreme hyperaccumulation and lower biomass (or vice versa) may be acceptable. Selection could also identify the individuals with the deepest and most extensive and efficient root systems, and those of greatest resistance to disease. Breeding experiments are required to incorporate all these desirable properties into one plant. Future work could involve genetic engineering to further improve metal uptake characteristics, if the genes for metal accumulation can be identified and manipulated. The possibility would then exist to transfer genes for metal hyperaccumulation into a very productive (but inedible), sterile host plant. Excellent opportunities also exist through protoplast fusion techniques.

Further geobotanical exploration of metalliferous environments is needed, both in the tropics and temperate zones, to identify additional plants with potential for phytoremediation of contaminated soils. Our knowledge of hyperaccumulator plants is still fragmentary, and it is clear from the recent major discoveries in Cuba, Australia, and Brazil within the past decade that further hyperaccumulating taxa await discovery and/or recognition. The list presented in Table 5.1 will certainly prove to be incomplete, as new hyperaccumulator plants are regularly being reported. Increasing systematic effort in screening plant materials for these characteristics will most certainly reveal new hyperaccumulator plants and new potentials both for phytoremediation and biorecovery of metals.

ACKNOWLEDGMENTS

The authors wish to record their particular gratitude to the following organizations for their past and ongoing support for research into hyperaccumulator plants: U.K. Natural Environment Research Council (NERC), Royal Society of London, British Ecological Society, U.S. National Geographic Society, U.S. Army Corps of Engineers, and E.I. Du Pont de Nemours & Co, Inc.

REFERENCES

Anderson, T.R., A.W. Howes, K. Slatter, and M.F. Dutton. Studies on the nickel hyperaccumulator, *Berkheya coddii*, in Écologie des Milieux sur Roches Ultramafiques et sur Sols Metallifères. Jaffré, T., Reeves, R.D., and Becquer, T., Eds., Documents Scientifiques et Techniques, ORSTOM, Nouméa, New Caledonia, 261-266, 1997.

Antonovics, J., A.D. Bradshaw, and R.G. Turner. Heavy metal tolerance in plants. *Adv. Ecol. Res.* 7, 1-85, 1971.

Baker, A.J.M. Accumulators and excluders — strategies in the response of plants to heavy metals. *J. Plant Nutr.* 3, 643-654, 1981.

Baker, A.J.M. Metal tolerance. *New Phytol.* 106, 93-111, 1987.

Baker A.J.M. and R.R. Brooks. Terrestrial higher plants which hyperaccumulate metallic elements — a review of their distribution, ecology and phytochemistry. *Biorecovery* 1, 81-126, 1989.

Baker, A.J.M. and J. Proctor. The influence of cadmium, copper, lead and zinc on the distribution and evolution of metallophytes in the British Isles. *Plant Sys. Evol.* 173, 91-108, 1990.

Baker, A.J.M. and P.L. Walker. Ecophysiology of metal uptake by tolerant plants, in *Heavy Metal Tolerance in Plants: Evolutionary Aspects,* Shaw, A.J., Ed., CRC Press, Boca Raton, FL, 155-177, 1990.

Baker, A.J.M., J. Proctor, M.M.J. van Balgooy, and R.D. Reeves. Hyperaccumulation of nickel by the flora of the ultramafics of Palawan, Republic of the Philippines, in *The Vegetation of Ultramafic (Serpentine) Soils,* Baker, A.J.M., Proctor, J., and Reeves, R.D., Eds., Intercept Ltd, Andover, Hants., U.K., 291-304, 1992.

Baker, A.J.M., R.D. Reeves, and A.S.M. Hajar. Heavy metal accumulation and tolerance in British populations of the metallophyte *Thlaspi caerulescens* J. & C. Presl (Brassicaceae). *New Phytol.* 127, 61-68, 1994.

Barry, S.A.S. and S.C. Clark. Problems of interpreting the relationship between the amounts of lead and zinc in plants and soil on metalliferous wastes. *New Phytol.* 81, 773-783. 1978.

Baumann, A. Das Verhalten von Zinksalzen gegen Pflanzen und im Boden. *Landwirtsch. Vers. Stn.* 31, 1-53, 1885.

Blaylock, M.J., D.E. Salt, S. Dushenkov, O. Zakharova, C. Gussman, Y. Kapulnik, B.D. Ensley, and I. Raskin. Enhanced accumulation of Pb in Indian mustard by soil-applied chelating agents. *Environ. Sci. Technol.* 31, 860-865, 1997.

Blissett, A.H. Copper-tolerant plants from the Upakaringa Copper Mine, Williamstown. *Quarterly Geological Notes, Geological Survey of South Australia,* 18, 1-4, 1966.

Boyd, R.S. and S.N. Martens. The raison d'être for metal hyperaccumulation in plants, in *The Vegetation of Ultramafic (Serpentine) Soils,* Baker, A.J.M., Proctor, J., and Reeves, R.D., Eds., Intercept Ltd, Andover, Hants., U.K., 279-289, 1992.

Brooks, R.R. *Biological Methods for Prospecting for Minerals.* John Wiley & Sons, New York, 1983.

Brooks, R.R., J. Lee, and T. Jaffré. Some New Zealand and New Caledonian plant accumulators of nickel. *J. Ecol.* 62, 493-499, 1974.

Brooks, R.R., J. Lee, R.D. Reeves, and T. Jaffré. Detection of nickeliferous rocks by analysis of herbarium specimens of indicator plants. *J. Geochem. Explor.* 7, 49-77, 1977a.

Brooks, R.R., J.A. McCleave, and E.K. Schofield. Cobalt and nickel uptake by the Nyssaceae. *Taxon* 26, 197-201, 1977b.

Brooks, R.R. and F. Malaisse. *The Heavy Metal-Tolerant Flora of Southcentral Africa.* Balkema, Rotterdam, 1985.

Brooks, R.R., R.S. Morrison, R.D. Reeves, T.R. Dudley, and Y. Akman. Hyperaccumulation of nickel by *Alyssum* Linnaeus (Cruciferae). *Proc. R. Soc. London, Ser. B.* 203, 387-403, 1979.

Brooks, R.R. and C.C. Radford. Nickel accumulation by European species of the genus *Alyssum. Proc. R. Soc. London, Ser. B.* 200, 217-224, 1978.

Brooks, R.R., R.D. Reeves, R.S. Morrison, and F. Malaisse. Hyperaccumulation of copper and cobalt — a review. *Bull. Soc. Bot. Belg.* 113, 166-172, 1980.

Brooks, R.R., S. Shaw, and A. Asensi Marfil. The chemical form and physiological function of nickel in some Iberian *Alyssum* species. *Physiol. Plant.* 51, 167-170, 1981.

Cataldo, D.A., K.M. McFadden, T.R. Garland, and R.E. Wildung. Organic constituents and complexation of nickel (II), iron (III), cadmium (II), and plutonium (IV) in soybean xylem exudates. *Plant Physiol.* 86, 734-739, 1988.

Chaney, R.L., M. Malik, Y.M. Li, S.L. Brown, J.S. Angle, and A.J.M. Baker. Phytoremediation of soil metals. *Curr. Opin. Biotechnol.* 8, 279-284, 1997.

Davies, K.L., M.S. Davies, and D. Francis. The influence of an inhibitor of phytochelatin synthesis on root growth and root meristematic activity in *Festuca rubra* L. in response to zinc. *New Phytol.* 118, 565-570, 1991.

de Knecht, J.A., M. van Dillen, P.L.M. Koevoets, H. Schat, J.A.C.Verkleij, and W.H.O. Ernst. Phytochelatins in cadmium-sensitive and cadmium-tolerant *Silene vulgaris*. Chain length distribution and sulfide incorporation. *Plant Physiol.* 104, 255-261, 1994.

Denaeyer-De Smet, S. and P. Duvigneaud. Accumulation de métaux lourds dans divers écosystèmes terrestres pollués par les retombées d'origine industrielle. *Bull. Soc. Bot. Belg.* 107, 147-156, 1974.

Deram, A. and D. Petit. Ecology of bioaccumulation in *Arrhenatherum elatius* L. (Poaceae) populations — applications of phytoremediation of zinc, lead and cadmium contaminated soils. *J. Exp. Bot.* 48, Suppl., 98, 1997.

Ernst, W.H.O. Das Violetum calaminariae westfalicum, eine Schwermetallpflanzengesellschaft bei Blankenrode in Westfalen. *Mitt. Florist. Soz. Arbeitsgem.* 13, 263-268, 1968.

Ernst, W.H.O. *Schwermetallvegetation der Erde.* Fischer, Stuttgart, 1974.

Ernst, W.H.O. Sampling of plant material for chemical analysis. *Sci. Total Environ.* 176, 15-24, 1995.

Ernst, W.H.O., J.A.C. Verkleij, and H. Schat. Metal tolerance in plants. *Acta Bot. Neerl.* 41, 229-248, 1992.

Fraústo da Silva, J.J.R. and R.J.P. Williams. *The Biological Chemistry of the Elements: The Inorganic Chemistry of Life.* Clarendon Press, Oxford, 1991.

Gabbrielli, R., T. Pandolfini, O. Vergnano, and M.R. Palandri. Comparison of two serpentine species with different nickel tolerance strategies. *Plant Soil* 122, 271-277, 1990.

Gabbrielli, R., P. Gremigni, L. Bonzi Morassi, T. Pandolfini, and P. Medeghini. Some aspects of Ni tolerance in *Alyssum bertolonii* Desv.: strategies of metal distribution and accumulation, in *Écologie des Milieux sur Roches Ultramafiques et sur Sols Metallifères.* Jaffré, T., Reeves, R.D., and Becquer, T., Eds., Documents Scientifiques et Techniques, ORSTOM, Nouméa, New Caledonia, 225-227, 1997.

Godbold, D.L., W.J. Horst, J.C. Collins, D.A. Thurman, and H. Marschner. Accumulation of zinc and organic acids in roots of zinc tolerant and non-tolerant ecotypes of *Deschampsia caespitosa. J. Plant Physiol.* 116, 59-69, 1984.

Grill, E., E.-L. Winnacker, and M.H. Zenk. Phytochelatins, a class of heavy metal binding peptides from plants, are functionally analogous to metallothioneins. *Proc. Nat. Acad. Sci. USA.* 84, 439-443, 1987.

Grill, E., E.-L. Winnacker, and M.H. Zenk. Occurrence of heavy metal binding phytochelatins in plants growing in a mining refuse area. *Experientia* 44, 539-540, 1988.

Hajar, A.S.M. Comparative ecology of *Minuartia verna* (L.) Hiern and *Thlaspi alpestre* L. in the Southern Pennines, with special reference to heavy metal tolerance. Ph.D. thesis, University of Sheffield, U.K., 1987.

Hamer, D.A. Metallothionein. *Annu. Rev. Biochem.* 55, 913-951, 1986.

Harmens, H., P.R. Hartog, W.M. ten Bookum, and J.A.C. Verkleij. Increased zinc tolerance in *Silene vulgaris* (Moench) Garcke is not due to increased production of phytochelatins. *Plant Physiol.* 103, 1305-1309, 1993.

Harmens, H., P.L.M. Koevoets, J.A.C. Verkleij, and W.H.O. Ernst. The role of low molecular weight organic acids in the mechanism of increased zinc tolerance in *Silene vulgaris* (Moench) Garcke. *New Phytol.* 126, 615-621, 1994.

Homer, F.A., R.S. Morrison, R.R. Brooks, J. Clemens, and R.D. Reeves. Comparative studies of nickel, cobalt and copper uptake by some nickel hyperaccumulators of the genus *Alyssum. Plant Soil.* 138, 195-205, 1991.

Homer, F.A., R.D. Reeves, and R.R. Brooks. The possible involvement of aminoacids in nickel chelation in some nickel-accumulating plants. *Curr. Top. Phytochem.* 14, 31-37, 1995.

Howden, R., P.B. Goldsbrough, C.R. Andersen, and C.S. Cobbett. Cadmium-sensitive, cad1 mutants of *Arabidopsis thaliana* are phytochelatin deficient. *Plant Physiol.* 107, 1059-1066, 1995.

Huang, J.W. and S.D. Cunningham. Lead phytoextraction — species variation in lead uptake and translocation. *New Phytol.* 134, 75-84, 1996.

Huang, J.W.W., J.J. Chen, W.R. Berti, and S.D. Cunningham. Phytoremediation of lead-contaminated soils: role of synthetic chelates in lead phytoextraction. *Environ. Sci. Technol.* 31, 800-805, 1997.

Hussain, A. Susceptibility of metal-accumulating plants to damping off by *Pythium ultimum*. Master's thesis, University of Sheffield, U.K., 1994.

Ingrouille, M.J. and N. Smirnoff. *Thlaspi caerulescens* J. & C. Presl (*T. alpestre* L.) in Britain. *New Phytol.* 102, 219-233, 1986.

Jackson, P.J., C.J. Unkefer, J.A. Doolen, K. Watt, and N.J. Robinson. Poly(γ-glutamylcysteinyl)glycine: Its role in cadmium resistance in plant cells. *Proc. Nat. Acad. Sci. USA.* 84, 6619-6623, 1987.

Jaffré, T., R.R. Brooks, J. Lee, and R.D. Reeves. *Sebertia acuminata*: a hyperaccumulator of nickel from New Caledonia. *Science* 193, 579-580, 1976.

Jaffré, T., W.J. Kersten, R.R. Brooks, and R.D. Reeves. Nickel uptake by the Flacourtiaceae of New Caledonia. *Proc. R. Soc. London, Ser. B.* 205, 385-394, 1979.

Kersten, W.J., R.R. Brooks, R.D. Reeves, and T. Jaffré. Nature of nickel complexes in *Psychotria douarrei* and other nickel-accumulating plants. *Phytochemistry.* 19, 1963-1965, 1980.

Krämer, U., J.D. Cotter-Howells, J.M. Charnock, A.J.M. Baker, and J.A.C. Smith. Free histidine as a metal chelator in plants that accumulate nickel. *Nature* 379, 635-639, 1996.

Krotz, R.M., B.P. Evangelou, and G.J. Wagner. Relationships between Cd, Zn, Cd-peptide and organic acid in tobacco suspension cells. *Plant Physiol.* 91, 780-787, 1989.

Lee, J., R.D. Reeves, R.R. Brooks, and T. Jaffré. Isolation and identification of a citrato-complex of nickel from nickel-accumulating plants. *Phytochemistry.* 16, 1503-1505, 1977.

Lee, J., R.D. Reeves, R.R. Brooks, and T. Jaffré. The relation between nickel and citric acid in some nickel-accumulating plants. *Phytochemistry* 17, 1033-1035, 1978.

Li, Y.M., R.L. Chaney, F.A. Homer, J.S. Angle, and A.J.M. Baker. *Thlaspi caerulescens* requires over 10^3 higher Zn^{2+} activity than other plant species. *Agron. Abstr.* 1995, 261, 1995.

Lloyd-Thomas, D.H. Heavy metal hyperaccumulation by *Thlaspi caerulescens* J. & C. Presl. Ph.D. thesis, University of Sheffield, U.K., 1995.

Ma, J.F., S. Hiradate, K. Nomoto, T. Iwashita, and H. Matsumoto. Internal detoxification of Al in *Hydrangea*. Identification of Al form in the leaves. *Plant Physiol.* 113, 1033-1039, 1997.

Macnair, M.R. and A.J.M. Baker. Metal-tolerant plants: an evolutionary perspective, in *Plants and the Chemical Elements,* Farago, M.E., Ed., VCH, Weinheim, 67-85, 1994.

Malaisse, F., J. Grègoire, R.R. Brooks, R.S. Morrison, and R.D. Reeves. *Aeolanthus biformifolius* De Wild.: a hyperaccumulator of copper from Zaire. *Science.* 199, 887-888, 1978.

Mathys, W. The role of malate, oxalate and mustard oil glucosides in the evolution of zinc-resistance in herbage plants. *Physiol. Plant.* 40, 130-136, 1977.

McGrath, S.P. Phytoextraction for soil remediation, in *Plants that Hyperaccumulate Heavy Metals,* Brooks, R.R., Ed., CAB International, Wallingford, 1998, 261.

Mesjasz-Przybylowicz, J., K. Balkwill, W.J. Przybylowicz, and H.J. Annegarn. Proton microprobe and x-ray fluorescence investigations in serpentine flora from South Africa. *Nucl. Instrum. Methods Physics Res.* B89, 208-212, 1994.

Mesjasz-Przybylowicz, J., K. Balkwill, W.J. Przybylowicz, H.J. Annegarn, and D.B.K. Rama. Similarity in nickel distribution in leaf tissue of two distantly related hyperaccumulating species, in *The Biodiversity of African Plants,* van der Maesen, L.J.G., van der Burgt, X.M., and van Medenbach de Roy, J.M., Eds., Kluwer Academic Publishers, Dordrecht, 331-335, 1995.

Minguzzi, C. and O. Vergnano. Il contenuto di nichel nelle ceneri di *Alyssum bertolonii* Desv. *Atti Soc. Toscana Sci. Nat., Ser. A.* 55, 49-77, 1948.

Morrey, D.R., K. Balkwill, M.-J. Balkwill, and S. Williamson. A review of some studies of the serpentine flora of Southern Africa, in *The Vegetation of Ultramafic (Serpentine) Soils*, Baker, A.J.M., Proctor, J., and Reeves, R.D., Eds., Intercept Ltd, Andover, Hants., U.K., 147-157, 1992.

Morrison, R.S. Aspects of the accumulation of cobalt, copper and nickel by plants. Ph.D. thesis, Massey University, NZ, 1980.

Morrison, R.S., R.R. Brooks, R.D. Reeves, and F. Malaisse. Copper and cobalt uptake by metallophytes from Zaire. *Plant Soil* 53, 535-539, 1979.

Mullins, M., K. Hardwick, and D.A. Thurman. Heavy metal location by analytical electron microscopy in conventionally fixed and freeze-substituted roots of metal tolerant and non-tolerant ecotypes, *Proc. Int. Conf. Heavy Metals Environ.* Athens, CEP Consultants, Edinburgh, 43-45, 1985.

Mummenhoff, K. and M. Koch. Chloroplast DNA restriction site variation and phylogenetic relationships in the genus *Thlaspi sensu* lato (Brassicaceae). *System. Bot.* 19, 73-88, 1994.

Mummenhoff, K., A. Franzke, and M. Koch. Molecular phylogenetics of *Thlaspi* s.l. (Brassicaceae) based on chloroplast DNA restriction site variation and sequences of the internal transcribed spacers of nuclear ribosomal DNA. *Can. J. Bot.* 75, 469-482, 1997.

Murphy, A.S. and L. Taiz. Comparison of metallothionein gene expression and nonprotein thiols in ten *Arabidopsis* ecotypes. Correlation with copper tolerance. *Plant Physiol.* 109, 1-10, 1995.

Murphy, A.S., J. Zhou, P.B. Goldsbrough, and L. Taiz. Purification and immunological identification of metallothioneins 1 and 2 from *Arabidopsis thaliana*. *Plant Physiol.* 113, 1293-1301, 1997.

Nicks, L.J. and M.F. Chambers. Farming for metals? *Min. Environ. Manage.* September 1995, 15-18, 1995.

Osmond, C.B. Ion absorption and carbon metabolism in cells of higher plants, in *Encyclopedia of Plant Physiology (New Series)*, vol. 2A, Lüttge, U. and Pitman, M.G., Eds., Springer-Verlag, Berlin, 347-372, 1976.

Pelosi, P., R. Fiorentini, and C. Galoppini. On the nature of nickel compounds in *Alyssum bertolonii* Desv.-II. *Agric. Biol. Chem.* 40, 1641-1642, 1976.

Peterson, P.J. Unusual accumulation of elements by plants and animals. *Sci. Progr., Oxford* 59, 505-526, 1971.

Pollard, A.J. and A.J.M. Baker. Deterrence of herbivory by zinc hyperaccumulation in *Thlaspi caerulescens* (Brassicaceae). *New Phytol.* 135, 655-658, 1997.

Proctor, J. and L. Nagy. Ultramafic rocks and their vegetation: an overview, in *The Vegetation of Ultramafic (Serpentine) Soils*, Baker, A.J.M., Proctor, J., and Reeves, R.D., Eds., Intercept Ltd., Andover, Hants., U.K., 469-494, 1992.

Qureshi, J.A., K. Hardwick, and H.A. Collin. Malic acid production in callus cultures of zinc and lead tolerant and non-tolerant *Anthoxanthum odoratum*. *J. Plant Physiol.* 122, 477-479, 1986.

Rascio, N. Metal accumulation by some plants growing on zinc mine deposits. *Oikos.* 29, 250-253, 1977.

Rauser, W.E. Phytochelatins and related peptides. *Plant Physiol.* 109, 1141-1149, 1995.

Reese, R.N. and G.J. Wagner. Effects of buthionine sulfoximine on Cd-binding peptide levels is suspension-cultured tobacco cells treated with Cd, Zn, or Cu. *Plant Physiol.* 84, 574-577, 1987.

Reeves, R.D. Hyperaccumulation of nickel by serpentine plants, in *The Vegetation of Ultramafic (Serpentine) Soils*, Baker, A.J.M., Proctor, J., and Reeves, R.D., Eds., Intercept Ltd., Andover, Hants., U.K., 253-277, 1992.

Reeves, R.D., R.R. Brooks, and R.M. Macfarlane. Nickel uptake by Californian *Streptanthus* and *Caulanthus* with particular reference to the hyperaccumulator *S. polygaloides* Gray (Brassicaceae). *Am. J. Bot.* 68, 708-712, 1981.

Reeves, R.D., R.R. Brooks, and T.R. Dudley. Uptake of nickel by species of *Alyssum*, *Bornmuellera* and other genera of Old World Tribus Alysseae. *Taxon*. 32, 184-192, 1983.

Reeves, R.D. and R.R. Brooks. Hyperaccumulation of lead and zinc by two metallophytes from mining areas of Central Europe. *Environ. Pollut. Ser. A.* 31, 277-285, 1983a.

Reeves, R.D. and R.R. Brooks. European species of *Thlaspi* L. (Cruciferae) as indicators of nickel and zinc. *J. Geochem. Explor.* 18, 275-283, 1983b.

Reeves, R.D. and A.J.M. Baker. Studies on metal uptake by plants from serpentine and non-serpentine populations of *Thlaspi goesingense* Halácsy (Cruciferae). *New Phytol.* 98, 191-204, 1984.

Reeves, R.D., A.J.M. Baker, A. Borhidi, and R. Berazaín. Nickel-accumulating plants from the ancient serpentine soils of Cuba. *New Phytol.* 133, 217-224, 1996.

Reeves, R.D., A.J.M. Baker, and A. Kelepertsis. The distribution and biogeochemistry of some serpentine plants of Greece, in *Écologie des Milieux sur Roches Ultramafiques et sur Sols Metallifères*. Jaffré, T., Reeves, R.D., and Becquer, T., Eds., Documents Scientifiques et Techniques, ORSTOM, Nouméa, New Caledonia, 205-207, 1997.

Reeves, R.D., A.J.M. Baker, A. Borhidi, and R. Berazaín. Nickel hyperaccumulation in the serpentine flora of Cuba. *Ann. Bot.* 83, 29-38, 1999.

Robinson, B.H., R.R. Brooks, J.H. Kirkman, P.E.H. Gregg, and H. Varela Alvarez. Edaphic influences on a New Zealand ultramafic ("serpentine") flora: a statistical approach. *Plant Soil.* 188, 11-20, 1997.

Robinson, N.J., A.M. Tommey, C. Kuske, and P.J. Jackson. Plant metallothioneins. *Biochem. J.* 295, 1-10, 1993.

Salt, D.E., D.A. Thurman, A.B. Tomsett, and A.K. Sewell. Copper phytochelatins of *Mimulus guttatus*. *Proc. R. Soc. London, Ser. B* 236, 79-89, 1989.

Schlegel, H.G., J.-P. Cosson, and A.J.M. Baker. Nickel-hyperaccumulating plants provide a niche for nickel-resistant bacteria. *Bot. Acta* 104, 18-25, 1991.

Schultz, C.L. and T.C. Hutchinson. Evidence against a key role for methallothionein-like protein in the copper tolerance mechanism of *Deschampsia caespitosa* (L.) Beauv. *New Phytol.* 110, 163-171, 1988.

Severne, B.C. Nickel accumulation by *Hybanthus floribundus*. *Nature*. 248, 807-808, 1974.

Shen, Z.G., F.J. Zhao, and S.P McGrath. Uptake and transport of zinc in the hyperaccumulator *Thlaspi caerulescens* and the non-hyperaccumulator *Thlaspi ochroleucum*. *Plant Cell Environ.* 20, 898-906, 1997.

Shimwell, D.W. and A.E. Laurie. Lead and zinc contamination of vegetation in the Southern Pennines. *Environ. Pollut.* 3, 291-301, 1972.

Smith, J.A.C., J. Ingram, M.S. Tsiantis, B.J. Barkla, D.M. Bartholomew, M. Bettey, O. Pantoja, and A.J. Pennington. Transport across the vacuolar membrane in CAM plants, in *Crassulacean Acid Metabolism: Biochemistry, Ecophysiology and Evolution*, Winter, K. and J.A.C. Smith, Eds., Springer-Verlag, Berlin, 53-71, 1996.

Still, E.R. and R.J.P. Williams. Potential methods for selective accumulation of nickel (II) ions by plants. *J. Inorg. Biochem.* 13, 35-40, 1980.

Thurman, D.A. and J.L. Rankin. The role of organic acids in zinc tolerance in *Deschampsia caespitosa*. *New Phytol.* 91, 629-635, 1982.

Tohayama, H., M. Inouhe, M. Joho, and T. Murayama. Production of metallothionein in copper and cadmium resistant strains of *Saccharomyces cerevisiae. J. Ind. Microbiol.* 14, 126-131, 1995.

Tomsett, A.B. and D.A. Thurman. Molecular biology of metal tolerances of plants. *Plant Cell Environ.* 11, 383-394, 1988.

Vázquez, M.D., J. Barceló, Ch. Poschenrieder, J. Mádico, P. Hatton, A.J.M. Baker, and G.H. Cope. Localization of zinc and cadmium in *Thlaspi caerulescens* (Brassicaceae), a metallophyte that can hyperaccumulate both metals. *J. Plant Physiol.* 140, 350-355, 1992.

Vázquez, M.D., Ch. Poschenrieder, J. Barceló, A.J.M. Baker, P. Hatton, and G.H. Cope. Compartmentation of zinc in roots and leaves of the zinc hyperaccumulator *Thlaspi caerulescens* J & C Presl. *Bot. Acta.* 107, 243-250, 1994.

Vergnano Gambi, O., R.R. Brooks, and C.C. Radford. L'accumulo di nichel nelle specie italiane del genere *Alyssum. Webbia.* 33, 269-277, 1979.

Vergnano Gambi, O., R. Gabbrielli, and L. Pancaro. Nickel, chromium and cobalt in plants from Italian serpentine areas. *Acta Oecol. Plant.* 3, 291-306, 1982.

Verkleij, J.A.C., P.C. Lolkema, A.L. De Neeling, and H. Harmens. Heavy metal resistance in higher plants: biochemical and genetic aspects, in *Ecological Responses to Environmental Stress,* Rozema, J., and Verkleij, J.A.C., Eds., Kluwer, Amsterdam, 8-19, 1991.

Wang, J., B.P. Evangelou, M.T. Nielsen, and G.J. Wagner. Computer-simulated evaluation of possible mechanisms for quenching heavy metal ion activity in plant vacuoles. I. Cadmium. *Plant Physiol.* 97, 1154-1160, 1991.

Wang, J., B.P. Evangelou, M.T. Nielsen, and G.J. Wagner. Computer-simulated evaluation of possible mechanisms for sequestering metal ion activity in plant vacuoles. II. Zinc. *Plant Physiol.* 99, 621-626, 1992.

White, M.C., A.M. Decker, and R.L. Chaney. Metal complexation in xylem fluid. I. Chemical composition of tomato and soybean stem exudate. *Plant Physiol.* 67, 292-300, 1981a.

White, M.C., F.D. Baker, A.M. Decker, and R.L. Chaney. Metal complexation in xylem fluid. II. Theoretical equilibrium model and computational computer program. *Plant Physiol.* 67, 301-310, 1981b.

Williams, S.T., T. McNeilly, and E.M.H. Wellington. The decomposition of vegetation growing on metal mine waste. *Soil Biol. Biochem.* 9, 271-275, 1977.

Wither, E.D. and R.R. Brooks. Hyperaccumulation of nickel by some plants of Southeast Asia. *J. Geochem. Explor.* 8, 579-583, 1977.

Woolhouse, H.W. Toxicity and tolerance in the responses of plants to metals, in *Encyclopedia of Plant Physiology (New Series),* vol. 12C, Lange, O.L., Osmond, C. B., Nobel, P.S., and Ziegler, H., Eds., Springer-Verlag, Berlin, 245-300, 1983.

Yang, X.H., R.R. Brooks, T. Jaffré, and J. Lee. Elemental levels and relationships in the Flacourtiaceae of New Caledonia and their significance for the evaluation of the "serpentine problem." *Plant Soil.* 87, 281-291, 1985.

Zenk, M.H. Heavy-metal detoxification in higher plants. A review. *Gene.* 1179, 21-30, 1996.

Zhou, J. and P.B. Goldsbrough. Structure, organization and expression of the metallothionein gene family in *Arabidopsis. Mol. Gen. Genet.* 248, 318-328, 1995.

6 Potential for Phytoextraction of Zinc and Cadmium from Soils Using Hyperaccumulator Plants

S. P. McGrath, S. J. Dunham, and R. L. Correll

CONTENTS

1-56670-450-2/00/$0.00+$.50
© 2000 by CRC Press LLC

INTRODUCTION

Plants capable of accumulating exceptionally large concentrations of metals such as Zn, Cd, Ni, and Pb have been termed *hyperaccumulators* (Brooks et al., 1977). Unlike most plants, these species typically maintain high tissue concentrations of metals across a wide range of metal concentrations in the soil. Metal hyperaccumulation is an extreme evolutionary response to the presence of high metal concentrations in soils and is not a common characteristic among terrestrial higher plants. A number of hyperaccumulating species endemic to metalliferous soils have been reported (Baker et al., Chapter 5). Notable metal-accumulators include several members of the Cruciferae. For example, *Alyssum* and *Thlaspi* species native to serpentine soils can accumulate concentrations of Ni in excess of 2% on a dry matter basis (Brooks and Radford, 1978; Brooks et al., 1979; Reeves and Brooks, 1983a), while species of *Thlaspi* from Zn/Pb mineralized soils can accumulate Zn to more than 2%, Cd up to 0.1%, and Pb up to 0.8% (Shimwell and Laurie, 1972; Reeves and Brooks, 1983a; Reeves and Brooks, 1983b).

Plant populations colonizing metalliferous soils have developed a series of physiological and biochemical adaptations in order to overcome metal toxicities and the physical and chemical stresses frequently associated with such soils. Baker (1981) proposed two basic strategies by which plants respond to large concentrations of heavy metals in their environment: metal exclusion, whereby uptake and transport of the metal is restricted and metal accumulation, when metals are accumulated to a high degree in the upper plant parts. There is an intermediate type, that of indicator plants, in which the concentration of heavy metals mirrors that in the soils in a linear relationship. The theoretical responses of these species is shown in Figure 6.1, which differs from the conceptual diagrams of Baker (1981) in that hyperaccumulators with very high concentrations replace the "accumulator" model, which inferred that all species reach the same concentrations. This is now known to be incorrect for hyperaccumulators, as they show both massive accumulation of metals and hypertolerance (Chaney et al., 1997). Both "indicator" and possibly "excluder" species die due to phytotoxicity at lower concentrations than hyperaccumulators. Also, originally the Baker model showed the hyperaccumulator growing at low soil metal concentrations. However, we now know that hyperaccumulators have a high physiological requirement for Zn and will not grow at low soil zinc levels (McGrath, 1997; Shen et al., 1997; Küpper et al., 1999). For this reason a high cut-off point has been added at the beginning of the response curve of the "hyperaccumulator strategy" in Figure 6.1.

The potential of metal-accumulating plants to detoxify soils polluted with heavy metals was demonstrated in glasshouse studies with hyperaccumulators from western Europe and the Aegean (Baker et al., 1991). The selection of species for the trials was based on a knowledge of their performance at metal-rich sites. Little was known at the time about how hyperaccumulators would grow and accumulate metals in contaminated agricultural soils. These small-scale pot trials highlighted that *T. caerulescens* accumulated a large amount of Zn. The results of these preliminary trials were sufficiently promising to demand the appraisal of the performance of these, and other accumulator plant species, to phytoextract metals when grown under crop

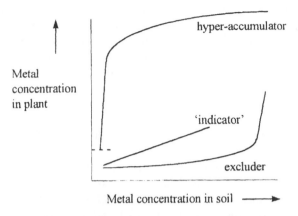

FIGURE 6.1 Conceptual response strategies of metal concentrations in plant tops in relation to increasing total metal concentrations in the soil.

conditions. The idea of using hyperaccumulator plants to phytoextract metals from industrially polluted soil was first tested on land managed by Rothamsted Experimental Station (McGrath et al., 1993). This demonstration project took place in an area that had received metal-contaminated sludges from west London 30 years ago, and is reported in detail below. The data obtained were assessed here in terms of the models of plant response to soil metal concentrations given in Figure 6.1.

Cropping with hyperaccumulator plants may prove an effective and practical means of metal extraction from superficially contaminated soils, such as those which result from the disposal to land of industrially polluted sewage sludges, Cd-containing phosphate sources, or waste materials. Our aim was to assess the performance of these hyperaccumulators to extract toxic metals under crop conditions in situations that are moderately polluted. Concentrations of all heavy metals in these soils were below 400 mg kg^{-1}, whereas the native sites where these species are endemic often have concentrations much greater than 1000 mg kg^{-1}. The objectives were (1) to measure the amounts of metals extracted by a range of potential accumulator plants and (2) to measure how the rate of removal of metals might vary with time.

MATERIALS AND METHODS

A metal-polluted field managed by Rothamsted Experimental Station (The Woburn Market Garden Experiment, situated 75 km north of London) was used. Applications of different rates of industrially contaminated sewage sludge or a sludge/straw compost were made to selected plots on this experiment over a period of 20 years, i.e., 1942 to 1961, and this resulted in increased concentrations of Zn, Cd, Cu, Ni, Cr, and Pb in the soils. The soil is a sandy loam with 9% clay (Typic Udipsamment) and a pH of 6.5 (measured in 1:2.5 soil:water). The soil pH on the plots has been maintained close to 6.5 (in water) throughout the experiment by small additions of lime (McGrath, 1984). A range of crop species, grass, and clover had been grown previously in rotation on the plots.

Nine plots were selected from the above experiment in order to provide a range of soil metal concentrations (they were plots 11, 16, 22, 31, 33, 36, 38, 39, and 40). Each main plot measured 2 x 2 m, was subdivided into 8 subplots (1 x 0.5 m), and prepared for planting by extensive roto-tilling within an area larger than the plot ($2.5 \ m^2$).

SOIL SAMPLING AND ANALYSIS

Soil samples were taken from each main plot at the beginning of the experiment. Fifteen cores were removed to plough-layer depth (23 cm), using a 3-cm diameter stainless steel semicylinder auger and bulked. The samples were air-dried and passed through a 2 mm roller mill. Total metal concentrations were determined after extraction of the soil in *aqua regia*, 4:1 v/v HCl/HNO_3 (McGrath and Cunliffe, 1985). The resulting digests were analyzed by ARL 34000 inductively coupled plasma emission spectrometer (ICP), except Cd, which was determined by graphite furnace atomic absorption.

SOWING

The species, which are all members of the Cruciferae family, and origin of seed were as follows:

1. *Thlaspi caerulescens* from Whitesike Mine, Cumbria, U.K.
2. *T. caerulescens* from Prayon, Belgium
3. *T. ochroleucum* from Thasos, Greece
4. *Cardaminopsis halleri* from a Pb/Zn smelter site in Blankenrode, Germany
5. *Cochlearia pyrenaica*, a frequent colonist of Pb/Zn mines from Derbyshire, U.K.
6. *Alyssum tenium* from Tinos, Greece
7. *A. murale* from near Thessaloniki, Greece
8. *A. lesbiacum* from Lesvos, Greece.

The last three species are known Ni hyperaccumulators. The experiment was first planted in 1990. Unfortunately, cutworm, larvae of the Turnip Moth (*Agrotis segetum*), killed most of the *T. caerulescens* and *T. ochroleucum* on all plots. A second sowing was carried out in 1991.

Seed was sown on compost then transplanted into the field at spacings appropriate for intensive cropping. Then 5 rows of 9 plants of each species were planted on each subplot, resulting in a total of over 3000 plants (9 x 8 x 45). Results for only the *Thlaspi* species and *C. halleri* are given here.

Throughout the duration of the project, the growth of the plants was monitored frequently and the plots were watered and weeded when necessary. The following basal fertilizers were applied annually: 222 g Nitrochalk 27N, supplying 150 kg N ha^{-1} and 125 g N:P:K 0:24:24, supplying 75 kg P ha^{-1} and 75 kg K ha^{-1}.

Harvesting

In 1991, alternate plants on each of the five rows were harvested after 4 months, leaving half the plants to over-winter. The plants were cut near the base and the aerial parts were removed for analysis. The stumps were left in the ground to see whether they decomposed or gave rise to new growth, as the production of further aerial growth would permit a number of successive harvests to be carried out. In early 1992, *T. ochroleucum*, *T. caerulescens*, and *C. halleri* produced new plants and the *Thlaspi* was harvested in June 1992. After harvesting in July 1992, both *C. halleri* and a small number of plants of the species *Thlaspi* produced new aerial growth. Enough plants of *C. halleri* regenerated on seven of the plots to allow a third harvest of this species only in August 1993. However, the rows of *C. halleri* could no longer be distinguished, so these were not harvested separately in 1993.

Analysis of Plant Material

All plant samples were carefully washed with water to remove any traces of soil and were then oven-dried at 80°C for 16 h and dry weights measured. The samples were finely ground in a Christy and Norris grinding mill. Plant material was digested in concentrated nitric and perchloric acids, 87% HNO_3/13% of 70% $HClO_4$ (Zhao et al., 1994). The metal concentrations in the digests were determined with an ARL 34000 ICP.

Statistical Analyses

To examine the relationship between soil-Zn or -Cd concentrations and the concentrations of these metals in the plants, the yields and metal concentrations were meaned from separate measurements made on five individual rows in 1991 and 1992, and used to calculate the extraction of each metal by the whole tops. Because of the method of harvesting in 1993, this method of averaging was not possible and data for whole subplots were used. The Genstat package was then used to perform analyses of variance and to examine the plant–soil relationships for the metals. Curve fitting progressed in the following order: (1) if curvature in the relationship between soil metal and metal concentration in plant tops was detected, a Mitscherlich curve was fitted; (2) if no curvature existed, but a strong positive linear relationship fitted, the case was described as linear; or (3) if the relationship was not linear as described in (2), no curve was fitted.

Our interpretation of this was that the Mitscherlich curve indicates hyperaccumulation, and a linear relationship could be (1) a hyperaccumulator whose concentrations were still increasing in response to increasing metals in soil or (2) an indicator or an excluder also responding to the soil gradient. In the case of no relationship, this could be either a hyperaccumulator that had reached a plateau or an excluder in the bottom part of the curve shown in Figure 6.1, or perhaps a relationship masked by field variation. The concentrations achieved can be used to distinguish between these possibilities, as hyperaccumulators should contain relatively large concentrations. Excluders should possess the lowest concentrations and these should not increase along the soil metal gradient.

RESULTS

METAL CONCENTRATIONS IN SOIL

The field plots were mostly polluted with Zn (Table 6.1), and the concentrations of this metal were the only ones that were unusually high in the accumulator plants. Total soil metal concentrations ranged from near background to in excess of European limits for Zn and Cd. Copper concentrations were close to the maximum limit, and Ni and Pb concentrations were slightly less than half the maximum limit (CEC, 1986). Chromium concentrations in the most polluted plots were under half the provisional maximum limit set by the U.K. DoE (1989).

TABLE 6.1

Metal Concentrations in the Field Plots in 1991 (mg kg^{-1})

Plot	Zn	Cd	Cu	Ni	Cr	Pb
11	124	2.82	26	20	41	40
31	157	3 36	32	20	46	47
36	206	5 22	56	24	68	63
38	269	7.77	80	29	92	81
22	288	8 43	84	29	101	157
33	327	8 36	84	30	91	213
40	340	10.23	95	30	109	109
16	406	10 52	110	35	122	122
39	444	13.61	138	35	155	136
EU maxima[a]	300	3	140	75	—	300

[a] Maximum concentrations allowed in agricultural soils receiving sewage sludge. (From CEC, *Off. J. Eur. Commun.* No L181, 6-12, 1986.)

PLANT HEALTH

In 1991, a small number of plants of *C. halleri* were infected by two closely related fungi, *Albugo candida* and *Peronospora parasitica*, the causal parasites of white rust and downy mildew, respectively. Both diseases produce marked distortion and hypertrophy (Butler and Jones, 1949; Brooks, 1953); the deformation of the stems caused by the *A. candida* prevented inflorescence, while the *P. parasitica* was responsible for a premature senescence of leaves and increased leaf turnover. The disease quickly spread to the *Thlaspi* species on adjacent plots with disease development being favored by prevailing humid and warm weather conditions. The plots were irrigated with a contact fungicide effective against both white rust and downy mildew — "Folio 575 FW," containing 75 g l^{-1} metalaxyl and 500 g l^{-1} chlorothalonil.

Two separate applications of the fungicide, at a rate of 2 l ha⁻¹, proved extremely effective and the plants recovered completely in the 2 weeks following the applications. However, the combined attack from the two fungal parasites reduced the potential biomass yield in the affected plants in 1991. Plants of *C. halleri* were again infected by white rust in July of the following year, but further applications of the fungicide contained the disease.

Cutworm (larvae of the Turnip Moth, *Agrotis segetum*) caused substantial damage to plants of the *Thlaspi* species in August 1991 by feeding on the roots. A number of plants were lost. The plots were drenched with "Spannit," active ingredient chlorpyrifos at 480 g l⁻¹, but the cutworm proved more difficult to eliminate and were found to be still present in the soil after the application of the pesticide. It is believed that the cutworm over-winters in the soil without further feeding, only emerging to feed in the spring. In order to guard against further attack, the plots were irrigated with several applications of the pesticide in May 1992 and no further losses due to cutworm damage were observed.

Biomass Production

In any season, the dry weights of *T. caerulescens*, *C. halleri*, and *T. ochroleucum* were not affected by the soil metal concentrations. *T. caerulescens* Prayon produced the greatest overall yield within this group, when they were averaged across 1991 and 1992 (Table 6.2). The dry weights of the largest plants from the Prayon and Whitesike populations of *T. caerulescens* were 28 and 31 g, respectively, indicating the maximum likely yield of individual plants under these conditions.

TABLE 6.2

Mean Yields in the Field Experiment, Averaged Across All Plots (t ha⁻¹)

Taxon	1991	1992	1993
C. halleri	2.13	8.38	1.30
T. ochroleucum	3.15	4.43	—
T. caerulescens Prayon	4.47	7.80	—
T. caerulescens Whitesike	3.62	7.45	—

Note: Least significant difference (LSD) for all species 1991 and 1992 = 1.94, LSD for *C. halleri* only for all years = 1.89

Plant–Soil Relationships

In order to emphasize similarities and differences in metal accumulation, graphs of the results for the various species have the same scale on the Y-axes (Figures 6.2 to 6.5). Each figure shows the results for both the 1991 and 1992 harvests (and 1993 for *C. halleri*).

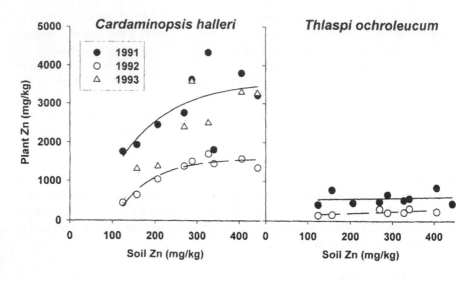

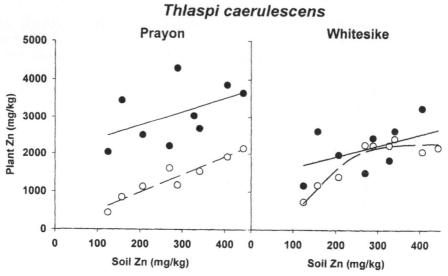

FIGURE 6.2 Zinc concentrations (mg kg⁻¹) in the shoot dry matter of field-grown plants.

Metal Concentrations in Plants

Concentrations of Zn differed between taxa (Figure 6.2). Plant response to soil Cd varied both in nature and in the amount and even between years in the same taxon (Figure 6.3). The two populations of *T. caerulescens* and *C. halleri* maintained large concentrations of Zn in their tissues over a wide range of soil metal concentrations, indicating hyperaccumulation (Figure 6.2). The effect of years was much greater on Zn concentrations in *C. halleri* and *T. caerulescens* Prayon than Whitesike (Table 6.3). The Zn concentrations in the Prayon population were greater in 1991 than those of the Whitesike population, but the latter had one particularly large value (10,625

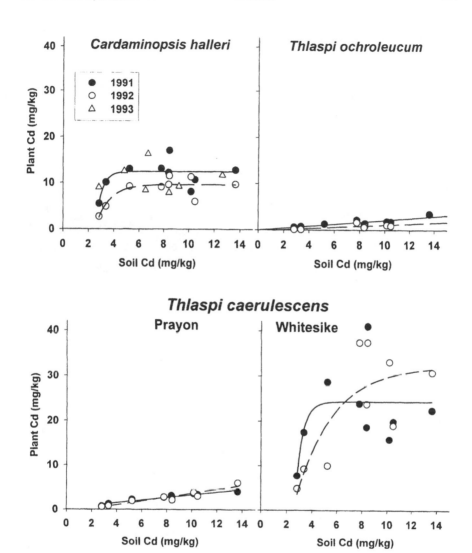

FIGURE 6.3 Cadmium concentrations (mg kg^{-1}) in the shoot dry matter of field-grown plants.

mg kg^{-1}) for one row harvested from the most polluted plot. This outlier has been omitted from the plot mean shown in Figure 6.2.

C. *halleri* (Figure 6.2) displayed similar high tissue Zn concentrations to *T. caerulescens* Prayon. Concentrations of Zn in *C. halleri* were at their lowest in 1992, but increased again in 1993 (Figure 6.2). Zinc concentrations in *T. ochroleucum* (Figure 6.2) were smaller than all other species reported here.

The largest concentrations of Cd were found in *T. caerulescens* Whitesike (Figure 6.3). The relationship between the concentrations of Cd in the plant tissues and soil was curvilinear, showing that this population is capable of accumulating large concentrations of Cd across a wide range of soil Cd concentrations, decreasing only on the least contaminated plots. Far smaller Cd concentrations were accumulated by

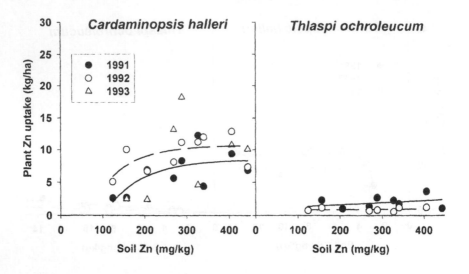

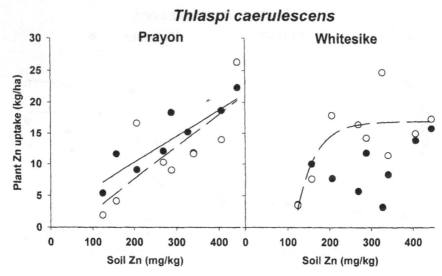

FIGURE 6.4 Zinc extractions (kg ha⁻¹) in the shoot dry matter of field-grown plants.

the other population of *T. caerulescens* from Prayon (Figure 6.3). *C. halleri* accumulated intermediate concentrations of Cd (Figure 6.3), and *T. ochroleucum* showed only small concentrations like *T. caerulescens* Prayon (Figure 6.3). *T. caerulescens* and *T. ochroleucum* maintained similar tissue concentrations of Cd in 1991 and 1992, but concentrations in *C. halleri* showed a slight decrease in 1992 (Figure 6.3). It was surprising that the Cd concentrations accumulated by *T. caerulescens* Prayon were so small, as the population originated from a Zn/Cd smelter site with a large amount of Cd in the soil, whereas Whitesike was from a Pb/Zn mine spoil area with less Cd. The Prayon population was also shown to be low in Cd by Brown et al. (1995).

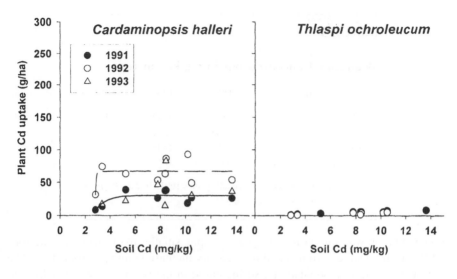

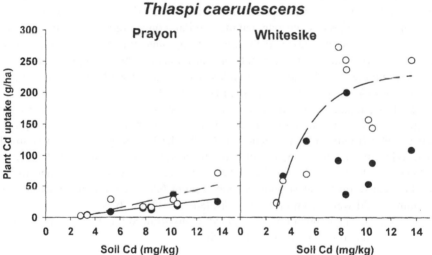

FIGURE 6.5 Cadmium extractions (g ha⁻¹) in the shoot dry matter of field-grown plants.

The largest concentrations of Ni were found in *T. caerulescens* Whitesike (10-40 mg kg⁻¹). Plant Ni concentrations were not significantly associated with soil Ni concentrations, except for *T. ochroleucum* in 1991 (data not shown). None of the species accumulated significant amounts of Ni, Cr, Cu, or Pb.

Metal Extraction

Mean Zn uptake by *T. caerulescens* Whitesike was generally similar to that of the Prayon population, but the shape of the response to soil Zn was different. Zinc extraction by the Whitesike population appeared to plateau, but Prayon appeared to be steadily increasing with increased soil Zn (Figure 6.4). Zinc uptake of *T. caer-*

TABLE 6.3

Mean Zinc Concentrations in mg kg⁻¹ in Tops

Taxon	1991	1992	1993
C. halleri	2856	1232	2533
T. ochroleucum	568	238	—
T. caerulescens Prayon	3080	1385	—
T. caerulescens Whitesike	2365	1852	—

Note: Standard error of difference, all species 1991,1992 = 239, 1993 for *C. halleri* only for all years = 371.

ulescens Prayon was similar in both years, despite smaller tissue Zn concentrations in 1992, as these were accompanied by a greatly increased biomass production. The yield of the Whitesike population also increased in the second season, resulting in greater uptake in 1992. Zinc uptake of *C. halleri* in 1991 (Figure 6.4) was less than that of *T. caerulescens*, despite having similar concentrations of Zn because it produced a smaller biomass and showed a strong plateau in relation to soil concentrations. Despite smaller tissue metal concentrations in 1992, the uptake was similar to 1991 (Figure 6.4) due to higher yields. Extraction of Zn by the nonhyperaccumulator *T. ochroleucum* was very small (Figure 6.4).

In 1991, uptake of Cd by *T. caerulescens* Whitesike was variable across the range of soil Cd concentrations in 1991, but uptake increased with increasing Cd in soil in 1992 (Figure 6.5). Extraction of Cd by the Prayon population was smaller than Whitesike (Figure 6.5) because of the lower concentrations in the tops. *C. halleri* had about half the uptake of Whitesike on most plots, but tended to have smaller uptake on the least polluted one (Figure 6.5). Again, comparatively small amounts of Cd were taken up by *T. ochroleucum* (Figure 6.5).

The Whitesike population of *T. caerulescens* showed the greatest uptake of Ni, but this amounted to only about 50 to 200 g ha⁻¹.

Soil Metal Balances

The amounts of metals extracted by the plants can be put into perspective by calculating a theoretical decrease in soil metal concentrations from the known plant maximum uptakes of Zn and Cd in 1991 and 1992. Assuming a depth of contamination of 20 cm, a soil weight of 2000 t ha⁻¹, and the maximum annual observed plant uptakes of Zn by *T. caerulescens* and *C. halleri* (33 Zn kg ha⁻¹) and the highest annual uptake of Cd (0.26 kg ha⁻¹), the soil concentrations should have decreased by 33 Zn and 0.26 Cd mg kg⁻¹ after the two harvests. However, these decreases in total concentrations are too small, in the light of sampling and analytical confidence, to be accurately quantified after 2 years.

DISCUSSION

T. caerulescens Prayon acted like an excluder of Cd but a hyperaccumulator of Zn (Figure 6.1). The population of *T. ochroleucum* used was tolerant to, but nonhyperaccumulating of, both Cd and Zn (McGrath et al., 1997). On the other hand, *T. caerulescens* Whitesike acted as an accumulator of both Cd and Zn, as did *C. halleri*.

COMPARISON WITH OTHER SPECIES

Some difficulties were encountered in growing these wild species, and substantial variation in individual plant yields and metal concentrations existed. To estimate the maximum potential removal of Zn by an optimized "crop" of hyperaccumulators, McGrath et al. (1993) made model calculations based on the uptake by the largest two rows of these plants. These estimates are similar to the maximum removals shown in Table 6.4, and a factor of 2 to 3 greater than those based on mean yields (Table 6.4). Using data from this experiment, it was calculated that 13 croppings with the Prayon population would be required to remove the excess Zn loading and bring the soil Zn concentration to within CEC (1996) limits. This compares favorable with over 800 croppings with *Brassica napus* and more than 2000 croppings with *Raphanus sativus* (McGrath et al., 1993).

TABLE 6.4

Concentrations of Zn (g t⁻¹ Dry Matter), Yields (t ha⁻¹), and Removals (kg ha⁻¹) in Selected Test Species

Year/Species	Max. Conc.	Average Conc.	Max. Yield	Average Yield	Max. Removal	Average Removal
1991						
T. caerulescens W[a]	10625	2365	5 4	4.4	57.4	10.4
T caerulescens P[a]	6328	3080	8 7	5 7	55 1	17 6
C halleri	5208	2856	2 7	2 0	14 1	5.7
1992						
T caerulescens W	2696	1852	8.6	5.0	23.2	9 3
T caerulescens P	2534	1350	13.1	5 3	33 2	7 2
C halleri	1547	1232	8.8	3 7	13 6	4.6
1993						
T caerulescens[b] W	3780	3472	9 9	4.6	37 4	16.0
C halleri[c]	3584	2533	5 5	3 2	18.1	8.1

[a] W = Whitesike; P = Prayon

[b] One plot only

[c] Seven plots

The species with the greatest capacity to extract Cd was *T. caerulescens* Whitesike, which could remove over 13 times the amount extracted by a nonaccumulating crop plant (McGrath et al., 1993).

Although the *Alyssum* species from serpentine areas showed the largest concentrations of Ni, they did not hyperaccumulate Ni because of the low soil concentrations present. However, under these conditions they did not accumulate Zn or any other divalent metal. Though serpentine species of Ni hyperaccumulators may accumulate other metals such as Co when not presented with Ni (Bernal and McGrath, 1994), they do not appear to hyperaccumulate them, perhaps indicating specificity of the uptake and/or sequestration mechanisms (Krämer et al., 1996).

EFFICIENCY OF METAL PHYTOEXTRACTION

The yields of the accumulator species and the nonaccumulating crop plants were not generally affected by the metal levels in the soil. The accumulator species investigated in the field experiment exhibited enhanced tissue concentrations of Zn and Cd, with some remarkably large values. The amounts of Ni, Cr, Cu, and Pb accumulated by these species, however, were not significant in terms of soil decontamination in the time available. *T. ochroleucum* had lower concentrations than the other species for Cd, Ni, and Zn. We attribute this to a lower transport of metals from root to shoot than *T. caerulescens*. Like many metal-tolerant but nonhyperaccumulating species, this population of *T. ochroleucum* retains much of the metal in the roots (Shen et al., 1997).

The Whitesike population of *T. caerulescens* accumulated far greater concentrations of Cd than all the other species tested. At a soil Cd concentration of 8 mg kg^{-1}, this species accumulated a maximum concentration of 40 mg kg^{-1} in the first year, with a total above-ground uptake of 100 g ha^{-1}; but this showed an overall increase in the following year because of a substantially greater yield, and Cd uptake increased to 250 g ha^{-1}.

Metal concentrations in the plants from the field experiment were generally smaller in the second year. The likely reasons were (1) the reduction in plant tissue metal concentrations was due to a growth-dilution effect, where the accumulated metals had been diluted as a result of increased plant growth; (2) that the plants take up the "most labile bioavailable" fractions of metals from the soils and that as this pool becomes smaller, uptake decreases; or (3) the root exploration may have been deeper in the well-established 2-year-old plants, thus exploiting more uncontaminated soil below 23 cm at this site (McGrath and Lane, 1989). However, because of the increased concentrations of Zn in *C. halleri* in the third year (Figure 6.2, Table 6.4), none of these can be confirmed at present.

The analysis of field material of *T. caerulescens* sampled from a site of seed collection for this study (Derbyshire, U.K.) revealed a Zn concentration of 15,100 mg kg^{-1} in the plant leaves at a soil Zn concentration of 16,200 mg kg^{-1}, giving a concentration factor (CF = concentration in plant tissue/concentration in soil) of 0.93. In contrast, the Prayon and Whitesike populations of *T. caerulescens* from this study, with maximum Zn concentrations of 5067 and 6993 mg kg^{-1}, respectively, at a soil Zn concentration of 444 mg kg^{-1}, displayed CFs of 11 and 16. *T. caerulescens*

Whitesike accumulated a maximum concentration of Cd which was 7 times that of the soil, again greater than that in the field material (CF = 3.4). It should be noted that the soils at the sites of seed collection were grossly polluted with metals, whereas the experimental field at Woburn, which received metal-contaminated sewage sludge, is moderately polluted in comparison. This may be why there is some evidence of metal concentrations not increasing to the maximum potential on these plots. Hyper-accumulators appear to have a high capacity for metal uptake (Lasat et al., 1996), but the concentration reached may depend not only on plant mechanisms, but also on the supply from the soil (McGrath et al., 1997). The soil supply may be limiting under less polluted conditions. Because concentrations even in hyperaccumulators increase with the soil concentration, e.g., Zn in *T. caerulescens* Prayon, this means that the extraction rate will not be as high as that on very polluted sites. On the other hand, the metal load to be removed is not as great in the less polluted soil.

These hyperaccumulators in their unimproved state may be of use in cleaning up marginally polluted soils. For example, if 5 t ha^{-1} is taken as the average yield of *T. caerulescens* for the three seasons, the Whitesike population of this species accumulated 150 g Cd ha^{-1} in each harvest, and *C. halleri* extracted 34 g Cd ha^{-1}. Although this is small compared to the inputs of Cd from sludge on this site, this amount is important — it is equal to decades of input to soils in many industrial countries due to phosphate fertilizer use and deposition from the atmosphere (these obviously vary according to the location and concentration of Cd in the phosphate). Small amounts of Cd from phosphates or fallout cause concern and could be extracted with a single harvest of agronomically optimized "crops" of *T. caerulescens* or *C. halleri*.

The amount of metal to be removed is an important determinant in the time needed for phytoextraction. Decontaminating soils which are very polluted with Zn would take some time, but soils just above any regulatory limits which prevent their use in agriculture could be rendered usable in just a few years (Table 6.5). So, the use of these plants for metal removal is only technically feasible in a short time period for those soils with small amounts of pollution. If more Zn and Cd can be made plant available by adding chemicals to the soil which increase metal uptake and transport within the plants, as has been shown for Pb (Huang et al., 1997), remediation could be accelerated. Thus, if Zn reaches 20,000 mg kg^{-1} and Cd 100 mg kg^{-1} in the tops and yields are increased to 10 t ha^{-1}, moderately polluted soils could be cleaned with one to two harvests (Tables 6.5 and 6.6).

COMPARISON WITH OTHER CLEANUP TECHNOLOGY

Physical and chemical methods currently available to restore metal-polluted soils can have a detrimental affect on soil physical conditions and destroy the biological activity present in the soil. Such techniques are also expensive, e.g., vitrification of the top 20 cm of contaminated soil, at an average price of $105 per tonne of soil (McNeill and Waring, 1992), would cost $273,000 per hectare, while the disposal of topsoil to 20 cm to landfill, at $45 per cubic meter, would cost $117,000 ha^{-1}. The use of hyperaccumulator plants to reclaim soils could represent both a practical and economically viable alternative strategy. Ashing harvested plant material con-

TABLE 6.5

Zinc Extraction by a Hypothetical 10 t ha⁻¹ Crop, Showing the Dependence on the Concentrations in the Tops and the Initial Soil Concentrations

Reduction Required	mg kg⁻¹ in Crop	kg ha⁻¹ Removed	% Removed in One Crop	No. Years to Target
From 400 to 300 mg kg⁻¹ ᵃ				
	1000	10	1.0	26
	2000	*20*	*1.9*	*13*
	4000	*40*	*3.8*	*7*
	10,000	100	9 6	3
	20,000	200	19 2	1
From 1000 to 300 mg kg⁻¹				
	1000	10	0 4	182
	2000	20	0.8	91
	4000	*40*	*1.5*	*46*
	10,000	*100*	*3.8*	*18*
	20,000	200	7.7	9
From 1500 to 300 mg kg⁻¹				
	1000	10	0.3	312
	2000	20	0 5	156
	4000	40	1.0	78
	10,000	*100*	*2 6*	*31*
	20,000	*200*	*5.1*	*16*
From 2000 to 300 mg kg⁻¹				
	1000	10	0.2	442
	2000	20	0.4	221
	4000	40	0.8	111
	10,000	100	1 9	44
	20,000	*200*	*3.8*	*22*

Note: Values in italics show the likely concentrations in each situation, derived from this experiment and the literature.

ᵃ Assuming 20 cm depth of contaminated soil of density 1 3.

centrates the metals, making recycling a possible option. Alternatively, the small amount of ash could be placed in a landfill at a fraction of the cost of removing or treating the polluted soil. Conventional farm practices are relatively low-cost. The average cost of growing crops (Nix, 1991) which cover different soil types, size of farm, and other such variables, and include labor, fuel, repairs, and depreciation expenses, are of the order of $1084 ha⁻¹ (Table 6.7). However, these are baseline costs which do not allow for general farm overheads or machinery (which may be preexisting). Also, the cost of the seed, any soil chemical amendments to promote metal availability, and site monitoring are not included. On the other hand, the value

TABLE 6.6

Cadmium Extraction by a Hypothetical 10 t ha⁻¹ Crop, Showing the Dependence on the Concentrations in the Tops, Soil Concentration, and Soil Target Concentration

Reduction Required	mg kg⁻¹ in Crop	kg ha⁻¹ Removed	% Removed in One Crop	No. Years to Target
From 1 to 0 2 mg kg⁻¹ ᵃ				
	10	*0 1*	*4*	*20.8*
	40	0 4	15	5
	100	1	38	2
From 10 to 3 mg kg⁻¹				
	10	*0.1*	*0 4*	*182*
	40	*0 4*	*1.5*	*46*
	100	1	3 8	18

Note: Values in italics show the likely concentrations in each situation, derived from this experiment and the literature

ᵃ Assuming 20 cm depth of contaminated soil of density 1 3

TABLE 6.7

Average Farming Costs

Operation	$ ha⁻¹
Ploughing	81
Seedbed cultivation (harrowing)	18
Fertilizer (cost) Nitrochalk	248
NPK 0:24:24	162
Fertilizer (loading and applying)	21
Drilling	23
Rolling	27
Mowing	33
Swath turning	21
Crop ashing (estimate)	450
Total	1084

From Nix, J. *Farm Management Pocketbook 1992*, 22nd ed. (pubs) Department of Agricultural Economics, Wye College, University of London, 1991, 104-1119 of any recycled metals could be a positive entry in the cost estimation (Robinson et al., 1997).

Other advantages of this "natural" technique to decontaminate soils are that the soil is not disturbed and no potentially hazardous chemicals are used. Assuming a plant yield of 20 t ha^{-1} dry matter (would be equivalent to 2 t ha^{-1} when ashed), the disposal of this volume of ash would cost approximately $70, which is a fraction of the cost of the disposal of 1 ha of the polluted soil. Alternatively, if metals in the ashed plant material were sufficiently concentrated, they could have commercial value and make recycling feasible.

Although it is recognized that many contaminated sites may also have other pollutants present, e.g., organic compounds, metal hyperaccumulator technology is a potentially valuable technique for dealing with heavy metals, which are the most difficult pollutants to remove from soils. In addition, it may be possible to combine the extraction of metals with enhanced degradation of organic pollutants in the rhizosphere.

FUTURE DEVELOPMENTS NECESSARY

The aim of this demonstration project was to ascertain the potential of cropping with hyperaccumulators as a means of soil decontamination. However, it was carried out using unselected seed from the wild. The uptakes reported are not, therefore, optimized, although some of the maximum concentrations and yields may be close to the highest potential under field conditions. Considerable variability within accumulator plants of the same species, in terms of metal accumulation and yield, was evident, e.g., the outlier for Zn in the Whitesike population. Future work is needed to (1) select optimal genotypes of these species and to initiate a program of seed multiplication, (2) determine the mechanisms of hyperaccumulation and hypertolerance, and (3) isolate the genes involved. It may then be possible to genetically engineer these traits into higher biomass crops by making them transgenic and make the processes of Zn/Cd phytoextraction more efficient.

This experiment will be continued as more information on the actual rate of decrease in soil metal concentrations with time is required. At present, we do not know whether the rate of phytoextraction will decline as the labile pool of bioavailable metals is removed from the field. If this happens, the effects of other management inputs such as acidifying agents or addition of metal-mobilizing chelates could be tested. These may increase the fraction of the total metal present that is extracted by the plants by releasing more metals from the soil matrix and promoting their transport into above-ground portions of the plants.

On the other hand, it is possible that metal mobilization may not be desirable if hyperaccumulators are shown to selectively perform "bioavailable metal stripping" (McGrath, 1998). If, for example, the small pool of bioavailable Cd is reduced, this may decrease the subsequent impact of the metal to crops and other biota to acceptable levels. McGrath et al. (1997) showed that the concentrations of "mobile" Zn (i.e., that extracted by the DIN 1995 1 M NH$_4$ NO$_3$, method) decreased after growth of *T. caerulescens*, particularly in rhizosphere soil. This decrease was up to 0.7 mg kg^{-1} in the same soil as used in this field study. Therefore, it is important to study the residual fractions of bioavailable and "fixed" forms of metals in soils following phytoextraction, and the kinetics of their reequilibration.

ACKNOWLEDGMENTS

We gratefully acknowledge financial support from DGXI of the Commission of the European Communities, the U.S. Army Corps of Engineers, and The Leverhulme Trust. IACR-Rothamsted receives grant-aided support from the Biotechnology and Biological Sciences Research Council of the U.K.

REFERENCES

Baker, A.J.M. Accumulators and excluders — strategies in the response of plants to heavy metals. *J. Plant Nutr.* 3, 643-654, 1981.

Baker, A.J.M., R.D Reeves, and S.P. McGrath. *In situ* decontamination of heavy metal polluted soils using crops of metal-accumulating plants — A feasibility study. Hinchee, R.E. and Olenbuttel, R.F., Eds., *In Situ Bioreclamation: Applications and Investigations for Hydrocarbon and Contaminated Site Remediation,* Butterworth-Heinemann, London, 600-605, 1991.

Bernal, M.P. and S.P. McGrath. Effects of pH and heavy metal concentrations in solution culture on the proton release, growth and elemental composition of *Alyssum murale* and *Raphanus sativus* L. *Plant Soil* 166, 83-92, 1994.

Brooks, F.T. *Plant Diseases,* 2nd ed., Oxford University Press, London, 109-114, 1953.

Brooks, R.R., J. Lee, R.D. Reeves, and T. Jaffré. Detection of nickeliferous rocks by analysis of herbarium species of indicator plants. *J. Geochem. Explor.* 7, 49-57, 1977.

Brooks, R.R., R.S. Morrison, R.D. Reeves, T.R. Dudley, and Y. Akman. Hyperaccumulation of nickel by *Alyssum linnaeus. Proc. R. Soc. London, Ser. B.* 203, 387-403, 1979.

Brooks, R.R. and C.C. Radford. Nickel accumulation by European species of the genus *Alyssum. Proc. R. Soc. London, Ser. B.* 200, 217-224, 1978.

Brown, S.L., R.L. Chaney, J.S. Angle, and A.J.M. Baker. Zinc and cadmium uptake by hyperaccumulator *Thlaspi caerulescens* and metal tolerant *Silene vulgaris* grown on sludge-amended soils. *Environ. Sci. Technol.* 29, 1581-1585, 1995.

Butler, E.J. and S.G. Jones. *Plant Pathology,* Macmillan and Co., London, 635-639, 1949.

Chaney, R.L., M. Malik, Y.M. Li, S.L. Brown, J.S. Angle, and A.J.M. Baker, Phytoremediation of soil metals. *Curr. Opin. Biotechnol.* 8, 279-284, 1998.

CEC: Commission of the European Communities Council Directive of 12 June 1986 on the protection of the environment, and in particular of the soil, when sewage sludge is used in agriculture. *Off. J. Eur. Commun.* No. L181, 6-12, 1986.

DoE: Department of the Environment Code of Practice for Agricultural Use of Sewage Sludge. H.M.S.O., London, 1989.

DIN (Deutsches Institut für Normung Hrsg) *Bodenbeschaffenheit, Extraktion von Spurenelementen mit Ammoniumnitratlösung.* Beuth Verlag, E DIN 19730, Berlin, 1995.

Huang, J.W., J. Chen, W.R. Berti, and S.D. Cunningham. Phytoremediation of lead-contaminated soils: role of synthetic chelates in lead phytoextraction. *Environ. Sci. Technol.* 31, 800-805, 1997.

Krämer, U., J.D. Cotter-Howells, J.M. Charnock, A.J.M. Baker, and J.A.C Smith. Free histidine as a nickel chelator in plants that hyperaccumulate nickel. *Nature.* 379, 635-638, 1996.

Küpper, H., F.J. Zhao, and S.P. McGrath. Cellular compartmentation of zinc in leaves of the hyperaccumulator *Thlaspi caerulescens*. *Plant Physiol.* 119, 305-311, 1999.

Lasat, M.M., A.J.M. Baker, and L.V. Kochian. Physiological characterization of root Zn^{2+} absorption and translocation to shoots in Zn hyperaccumulator and nonaccumulator species of *Thlaspi*. *Plant Physiol.* 112, 1715-1722, 1996.

McGrath, S.P. Metal concentrations in sludges and soil from a long-term field trial. *J. Agric. Sci. Cambridge.* 103, 25-35, 1984.

McGrath, S.P. Phytoextraction for soil remediation. In: *Plants that Hyperaccumulate Heavy Metals,* Brooks, R.R., Ed., CAB International, Wallingford, 261-287, 1998.

McGrath, S.P. and C.H. Cunliffe. A simplified method for the extraction of the metals Fe, Zn, Cu, Ni, Cd, Pb, Cr, Co and Mn from soils and sewage sludges. *J. Sci. Food Agric.* 36, 794-798, 1985.

McGrath, S.P. and P.W. Lane. An explanation for the apparent losses of metals in a long-term field experiment with sewage sludge. *Environ. Pollut.* 60, 235-256, 1989.

McGrath, S.P., C.M.D. Sidoli, A.J.M. Baker, and R.D. Reeves. The potential for the use of metal-accumulating plants for the *in situ* decontamination of metal-polluted soils. In: *Integrated Soil and Sediment Research: A Basis for Proper Protection,* Eijsackers, H.J.P., and Hamers, T., Eds., Kluwer Academic Publishers, 673-676, 1993.

McGrath, S. P., Z.G. Shen, and F.J. Zhao. Heavy metal uptake and chemical changes in the rhizosphere of *Thlaspi caerulescens* and *Thlaspi ochroleucum* grown in contaminated soils. *Plant Soil* 188, 153-159. 1997

McNeill, K.R. and S. Waring. Vitrification of contaminated soils. In: *Contaminated Land Treatment Technologies,* Rees, J.F., Ed., Elsevier Applied Science for SCI, 1992.

Nix, J. *Farm Management Pocketbook 1992,* 22nd ed. (pubs.) Department of Agricultural Economics, Wye College, University of London, 1991, 104-119.

Reeves, R.D. and R.R. Brooks. European species of *Thlaspi* L. (Cruciferae) as indicators of Ni and Zn. *J. Geochem. Explor.* 18, 275-283, 1983a.

Reeves, R.D. and R.R Brooks. Hyperaccumulation of lead and zinc by two metallophytes from mining areas of Central Europe. *Environ. Pollut.* 31, 277-285, 1983b.

Robinson, B.H., R.R. Brooks, A.W. Howes, J.H. Kirkman, and P.E.H. Gregg. The potential of the high-biomass nickel hyperaccumulator *Berkheya coddii* for phytoremediation and phytomining. *J. Geochem. Explor.* 60, 115-126, 1997.

Shen, Z.G., F.J. Zhao, and S.P McGrath. Uptake and transport of zinc in the hyperaccumulator *Thlaspi caerulescens* and the non-hyperaccumulator *Thlaspi ochroleucum*. *Plant Cell Environ.* 20, 898-906, 1997.

Shimwell, D.W. and A.E. Laurie. Lead and zinc contamination of vegetation in the Southern Pennines. *Environ. Pollut.* 3, 291-301, 1972.

Zhao, F., S.P. McGrath, and A.R. Crosland. Comparison of three wet digestion methods for the determination of plant sulphur by inductively coupled plasma atomic emission spectroscopy (ICP-AES). *Commun. Soil Sci. Plant Anal.* 25, 407-418, 1994.

7 Improving Metal Hyperaccumulator Wild Plants to Develop Commercial Phytoextraction Systems: Approaches and Progress

Rufus L. Chaney, Yin-Ming Li, Sally L. Brown, Faye A. Homer, Minnie Malik, J. Scott Angle, Alan J. M. Baker, Roger D. Reeves, and Mel Chin

CONTENTS

1-56670-450-2/00/$0 00+$ 50
© 2000 by CRC Press LLC

129

INTRODUCTION

The use of plants in environmental remediation has been called "green remediation," "phytoremediation," "botanical bioremediation," "phytoextraction," etc. This new technology is being developed for the cleanup of both soil metals and xenobiotics. Because metals cannot be biodegraded, remediation of soil metal risks has been a difficult and/or expensive goal (Chaney et al., 1995, 1997a). The general strategies for phytoremediation of soil metals is to either: (1) *phytoextract* the soil elements into the plant shoots for recycling or less expensive disposal; (2) *phytovolatilize* the soil trace elements (e.g., generation of Hg^0 or dimethylselenide which enter the vapor phase); or (3) *phytostabilize* soil metals into persistently nonbioavailable forms in the soil. The third method is usually called "*in situ* remediation" by which incorporation of soil amendments rich in Fe, phosphate, and limestone equivalent are used to transform soil Pb into forms with lower bioavailability and/or phytoavailability. Over time, soil Pb and some other elements become much less phytoavailable (or bioavailable) to organisms which consume soils; plants can contribute to this process by hastening the formation of pyromorphite, an insoluble and nonbioavailable Pb compound (e.g., Ma et al., 1993; Berti and Cunningham, 1997; Zhang et al., 1997; Brown et al., 1998).

Phytoremediation employs plants to remove contaminants from polluted soils which require decontamination under the supervision of a regulatory agency. The commercial strategy is to use phytoremediation as a lower cost alternative to current expensive engineering methods (Benemann et al., 1994; Salt et al., 1995). For every meter of soil depth removed, costs are between \$8 and \$24 million per hectare (includes disposal in a hazardous waste landfill and replacement with clean soil; Cunningham and Berti, 1993).

Soil remediation technology is needed to reverse risk to humans or the environment from metals in soil, both geochemical metal enrichment and anthropogenic soil contamination (Chaney et al., 1998b). Human disease has resulted from Cd (Nogawa et al., 1987; Kobayashi, 1978; Cai et al., 1990), Se (Yang et al., 1983), and Pb in soils. Livestock and wildlife have suffered Se poisoning at many locations with Se-rich soils (Rosenfeld and Beath, 1964; Ohlendorf et al., 1986); high soil molybdenum (Mo) first harms ruminant livestock. Soil metals have caused phytotoxicity to sensitive plants at numerous locations, especially where mine wastes and smelters caused contamination of acidic soils with Zn, Ni, or Cu (Chaney et al., 1998b). Although some of these situations can be remedied by soil amendments (e.g., Brown et al., 1998), phytoremediation offers an alternative whereby the contaminant would be removed from soils and either recycled or safely disposed. As noted below, the combination of the need to prevent adverse environmental effects of soil contaminants, and to do so at lower cost than existing technologies, has brought increased attention to phytoremediation. Improved methods for risk assessment of soil contaminants has clarified some situations where soil may be rich in an element, but no risk is observed. Among common contaminant metals, Cr^{3+} comprises a far lower risk than once assumed (Chaney et al., 1997b). Natural soils with 10,000 mg Cr kg^{-1} as Cr^{3+} do not cause risk to any organism in the environment, while contamination with Cr^{6+} can poison plants and kill soil organisms, or leach to groundwater where it could comprise risk to humans. Although soil Pb has received much attention and concern, experiments conducted to remove and replace urban soils rich in Pb did not reduce blood Pb of children very strongly, indicating that soil Pb was a smaller part of the overall Pb risk compared to interior and exterior Pb-rich paint (Weitzman et al., 1993; Chaney et al., 1998b).

As summarized in this chapter, we conclude that "hyperaccumulator" plants offer a very important opportunity to achieve economic phytoextraction to decontaminate polluted soils. The word hyperaccumulator was coined by Brooks and Reeves (Brooks et al., 1977). In order to make the definition more specific to natural systems, Reeves (1992) defined "nickel hyperaccumulator" plants as those which accumulate over 1000 mg Ni kg^{-1} dry matter in some above-ground tissue when growing in fields in which they evolved. Because plant genotypes vary somewhat in metal accumulation and in metal tolerance, using a hyperaccumulator definition of plants accumulating 10- to 100-times "normal" concentrations in plant shoots is imprecise, but this is the approximate range for hyperaccumulators of Ni, Zn, Cu, Co, etc.

As information about hyperaccumulator plants was reported in the literature, the question arose: "Can hyperaccumulator plants remove enough metal to decontaminate the soil by using simple farming technology, making hay from the biomass, and recycling the metals, if these were repeated over a number of years?" Chaney (1983) introduced this idea in a review chapter about plant uptake of metals from contaminated soils. Improved understanding of soil factors which increase or decrease metal uptake indicated that phytoextraction cropping could be managed efficiently. The unusual accumulation of specific elements could also allow the ash of the plant biomass to be recycled as an ore. In that way the value of metals in the plant biomass could offset part of the cost of soil decontamination and support "phytomining" of some elements as a commercial venture.

Phytoextraction employs plant species able to accumulate abnormally high quantities of elements from soils. Because roots use ion carriers to accumulate and translocate specific metals to their shoots, chemical and physical methods of removing metals from ores cannot be as selective as plant roots. The commercial strategy of phytomining is to concentrate metals from low-grade ores or mine and smelter wastes and then sell the ash as an alternative metal concentrate. Phytoextraction would only be applied to soils or ores that cannot be economically enriched by traditional mining and beneficial technology.

Some plant species are known to accumulate levels at least as high as 1% of the plant shoot dry matter for the elements Zn, Ni, Se, Cu, Co, or Mn, and over 0.1% Cd, but not Pb or Cr (see Table 7.1). Accumulation of other elements may also occur depending on the degree of soil contamination. Because unusual accumulation and tolerance of these elements in plant roots and shoots occurs, there is hope that the process can be applied to radionuclides (^{137}Cs, ^{60}Co, U, Am, etc.) and to other elements (Tl [Kurz et al., 1997], As) for which remediation of soils is required. The selective nature of plant accumulation of elements offers this phytoextraction technology, but we have to find plant species that can accumulate the element or radionuclide for which a soil remediation technology is required. In consideration of phytoextraction of ^{137}Cs, the presence of K required for plant growth normally

TABLE 7.1

Examples of Plant Species which Hyperaccumulate Zn, Ni, Se, Cu, Co,
or Mn to over 1% of Their Shoot Dry Matter in Field Collected Samples
(About 100-Times Higher than Levels Tolerated by Normal Crop Plants)

Element	Plant Species	Max. Metal in Leaves	Location	Ref.
Zn	*Thlaspi calaminare*[a]	39,600	Germany	Reeves and Brooks, 1983b
Cd	*T caerulescens*	1800	Pennsylvania	Li et al., 1997
Cu	*Aeollanthus biformifolius*	13,700	Zaire	Brooks et al., 1992
Ni	*Phyllanthus serpentinus*	38,100	N Caledonia	Kersten et al., 1979
Co	*Haumaniastrum robertii*	10,200	Zaire	Brooks, 1977
Se	*Astragalus racemosus*	14,900	Wyoming	Beath et al., 1937
Mn	*Alyxia rubicaulis*	11,500	N Caledonia	Brooks et al., 1981b

[a] Ingrouille and Smirnoff (1986) summarize consideration of names for *Thlaspi* species; many species and subspecies were named by collectors over many years.

inhibits uptake; a plant which could accumulate high amounts of ^{137}Cs in the presence of adequate K for maximum plant yield would be of great value in ^{137}Cs phytoextraction. Redroot pigweed (*Amaranthus retroflexus* L.) was found by Lasat et al. (1998) to accumulate higher levels of ^{137}Cs than other species evaluated; evaluation of genotypic differences in the selectivity of Cs vs. K accumulation in shoots, as well as concentration ratio (plant Cs/soil Cs), may offer substantial increase in annual Cs phytoextraction from contaminated soils. This same principle of high accumulation of the element of interest in the presence of normal soil levels of essential ions is the central theme of effective phytoextraction.

Phytovolatilization appears to be relevant to remediation of soils rich in Hg and Se, and possibly As. However, other elements do not readily form volatile chemical species in the soil environment or in plant shoots, so phytovolatilization cannot be applied to these elements. In the case of Hg, Meagher and coworkers transferred a modified gene for mercury reductase from bacteria to *Arabidopsis thaliana* (and since into other species; Rugh et al., 1996). Their research showed that transgenic plants expressing mercuric reductase can phytovolatilize Hg from test solutions and media, and their unpublished work indicates that these plants can phytovolatilize Hg from real contaminated soils. They have transferred the gene to other plant species which should be effective in field soils. Their expressed strategy is to also obtain expression in higher plants of a gene from bacteria which hydrolyzes methyl and dimethyl mercury to accompany the reductase. Organic Hg compounds are the principle source of environmental Hg poisoning because these compounds are bioaccumulated in aquatic food chains, and both predator birds and mammals are poisoned. Any potential for adverse effects of Hg^0 phytovolatilized from contaminated soils is very small compared to the reduction in risk of adverse effects by hydrolyzing any methyl mercury in soils. In the case of Se, both plants and soil microbes contribute to biosynthesis and emission of volatile Se gases, e.g., dimethyl selenide (Terry et al., 1992; Terry and Zayed, 1994; Zayed and Terry, 1994). Terry and coworkers focused their effort more on the phytovolatilization of soil Se. They found that some commercial vegetable crops are quite effective in phytovolatilization; broccoli annual Se removal was promising, although sulfate strongly inhibited Se emission (Zayed and Terry, 1994). Chapter 4 describes a wetlands system for phytovolatilization of Se in industrial wastewater. Early studies of soil phytoremediation showed that when Se reached the irrigation drainage waters, it was a co-contaminant with borate and soluble salts such that few plants can survive the combination of toxic factors (Mikkelsen et al., 1988a,b; Parker and Page, 1994; Parker et al., 1991). So it was important to remove the soil Se before it could be leached from the soil profile and require treatment in the salt rich drainage water (Bañuelos et al., 1997). Phytovolatilization offers significant opportunity to alleviate soil Se contamination while redistributing the Se to a much larger land area where the concentration would not comprise risk. Hyperaccumulator plants appear to have a significant role to play in phytoextraction of Se as well because of their ability to accumulate and phytovolatilize Se even in the presence of high levels of sulfate compared to crop plants (Bell et al., 1992).

HISTORY OF PHYTOEXTRACTION OF SOIL METALS USING METAL HYPERACCUMULATOR SPECIES

The development of phytoextraction as a soil remediation technology is being built on the earlier science of bioindicator plants and biogeochemical prospecting, and on the study of genetic or ecotypic metal tolerance by plants. Plants which occur only on mineralized or contaminated soils (endemics) have been known for centuries. Early miners searched for indicator plants before geologists knew where ores could be found (see reviews by Brooks, 1972, 1983, 1992). Bioindicator plants were important in finding uranium ores both in the U.S. (Cannon, 1955, 1960, 1971) and in the USSR (Mayluga, 1964). A review in *Science* by Cannon (1960) summarized unusual accumulation of metals by plants in this widely read forum. A book by Ernst (1974) reviewed work on metal-tolerant and hyperaccumulator plants, and it discussed their use as bioindicators of mineralization (see also Ernst, 1989). Many Se, Si, Zn, Cd, Cu, and Co accumulators were already well known before phytoremediation was conceived. Metal hyperaccumulation can be viewed as one kind of metal tolerance (Baker, 1981; Ernst, 1974, 1989). Increased interest in metal tolerance stimulated basic studies on mechanisms of tolerance, exclusion, and hyperaccumulation.

Retrospective searching for evidence of this remarkable accumulation of metals has shown that *Thlaspi calaminare* with up to 17% ZnO in the shoot ash was reported by Baumann (1885), and *Alyssum bertolonii* with up to 1% in dry matter and 10% Ni in ash was reported by Minguzzi and Vergnano (1948). These reports were based on older methods of analysis. As agronomic science showed that normal plants tolerated only about 500 mg Zn kg^{-1} dry matter or 50 to 100 mg Ni kg^{-1} dry matter, these older reports were given little credence by most researchers. However, new measurements by researchers confirmed the remarkable metal accumulation ability of a limited number of plant species, and that knowledge is the immediate predecessor of phytoextraction.

Several people played important roles in carrying forward the old information about remarkable metal accumulators, and (in the beginning) the new research using modern analytical methods which won wide appreciation of the existence of these highly unusual plants. Cannon (1960) and Mayluga (1964) did important work on biogeochemical prospecting and bioindicator plants and noted older findings in their reviews of the literature. Mayluga (1964) noted high Ni levels in *A. murale* and high levels of some other elements in specific plants. Jaffré (1992; Jaffré et al., 1997) conducted botanical research in New Caledonia, a nation which has ultramafic-derived soils as one third of its land surface, and started to find remarkable accumulators of Ni, Co, and Mn. Ernst (1968, 1974) studied the high metal accumulation in shoots of *T. alpestre* var. *calaminare*. Wild (1970) reported on Ni hyperaccumulators from Zimbabwe (since shown to be less effective than originally reported; Brooks and Yang, 1984; Baker and Brooks, 1989). Cole (1973) made an early report on an Australian Ni hyperaccumulator. Duvigneaud (1958, 1959), Duvigneaud and Denaeyer-De Smet (1963), and Denaeyer-De Smet (1970) reported studies on the Co- and Cu-accumulating plants from Africa and European Zn accumulators.

Although many researchers contributed to the recognition of metal hyperaccumulator plant species, Brooks (e.g., 1983, 1987, 1998) and Ernst (e.g., 1974), more than any other individuals, are credited with bringing this information to the attention of the wider research community. Brooks cooperated with Reeves, Baker, Jaffré, Malaise, Vergnano, and others, and validated the existence of plants that could accumulate such remarkable concentrations of generally phytotoxic elements that researchers started to examine the mechanisms of metal binding which could reduce the toxicity of these elements (e.g., Lee et al., 1977, 1978; Homer et al., 1991).

Several papers by Brooks and coworkers spread the idea of hyperaccumulators to the agricultural environmental research community. A paper on Co and Ni accumulation by *Haumaniastrum* species was published in *Plant and Soil* (Brooks, 1977). Jaffré et al. (1976) reported in *Science* their observation of a Ni-hyperaccumulating tree from New Caledonia which, when the bark was cut, expressed a latex sap that reached 25% Ni on a dry matter basis. Another especially important and innovative paper by Brooks, Lee, Reeves, and Jaffré (1977) reported the strategy of analyzing fragments of leaves from herbarium specimens to evaluate the taxonomy of metal accumulation. Because their research indicated that a number of *Alyssum* species might be able to accumulate over 1% Ni, they wanted to integrate the information collected by botanists on the occurrence of plant species into a logical database. By obtaining these samples of herbarium specimens collected at specific places on specific soils, the authors expected to find potential ore deposits that had not been found by traditional geology.

One side benefit of Ni hyperaccumulation was providing a new phenotypic characteristic which could be used by botanists to identify plant species. Normally, plant species are defined based on differences in flowering parts or leaf structures. Because of the inherent variation of plants due to many sources of environmental stress, physical differences may not always provide accurate identification of a plant species. By analyzing specimens of *Alyssum* species from many herbaria, researchers were able to test whether plant Ni concentration could be a separate indicator of plant species. During this exercise, reexamination of the botanical specimen often found misidentification of species when only a few specimens of one species had low or high Ni levels compared to the bulk of samples of that species analyzed. In this way, researchers found that one "tribe" of the *Alyssum*, the *Odontarrhena*, achieved hyperaccumulator Ni levels when they were growing on serpentine soils (Brooks et al., 1977; Brooks and Radford, 1978; Brooks et al., 1979; Reeves et al., 1983). Other *Alyssum* species (e.g., *A. montanum, A. serpyllifolium* subsp. *serpyllifolium*) were also endemic on serpentine soils, but did not accumulate high levels of Ni, nor did most other species endemic on such soils. The use of Ni hyperaccumulation as a biomarker for this genetic ability of some *Alyssum* species supported a significant step forward in understanding of the taxonomic relationships of this complex genus, including the definition of two subspecies of *A. serpyllifolium* (*A. serpyllifolium* subsp. *malacitanum* Rivas Goday; proposed *A. malacitanum*; Dudley, 1986a), and *A. serpyllifolium* subsp. *lusitanicum* (proposed *A. pintodasilvae*; Dudley, 1986b) which were very like the parent species in most taxonomic characters but hyperaccumulated Ni (Brooks et al., 1981a). Many other exciting kinds of knowledge have been developed from these initial investigations of hyperaccumulation of metals by serpentine species, species endemic on Zn/Pb rich soils in Europe, or Cu/Co rich soils in Africa.

The original focus of these researchers was to identify either a bioindicator species of new ore bodies, or a taxonomic tool for species identification to improve phylogenetic understandings. Still, as this information was reported in the literature, other applications were evident to other researchers. Chaney and coworkers suggested the possibility of using hyperaccumulator species (e.g., *Arenaria patula, Alyssum bertolonii*) for the phytoextraction of metals from contaminated soils. The model of using *A. bertolonii* as a metal extraction crop compared to corn was reported in the initial publications of Chaney and coworkers on phytoextraction (Chaney et al., 1981a,b; Chaney, 1983).

"Revival Field" — Art Helps Spread the Phytoremediation Meme

Two field experiments testing metal phytoextraction were begun in 1991, the test at Rothamsted (Baker et al., 1994), and a field test at St. Paul, MN by Chin, Chaney, Homer, and Brown. The Minnesota opportunity for us to conduct research to characterize the potential of phytoextraction arose when Mel Chin, an artist from New York, contacted Chaney. Chin had accepted a commission to create an art work for the 20th anniversary of Earth Day (in 1992), and while considering alternatives, read of metal-tolerant cell cultures in the Whole Earth Catalog. With that information, he thought that such plants might be used to decontaminate polluted soil, and wanted to use this idea in his art work. While searching for plants to use, and for a site where he could install the art work on a hazardous soil, he was referred to Chaney for information on plant metal accumulation. Although the *Datura innoxia* which he sought tolerated Cd (see Jackson et al., 1984), there was no evidence that this species had the ability to accumulate metals to achieve decontamination of polluted soils. Chaney brought to Chin's attention the information on natural hyperaccumulator plants, and Chin obtained further information on these plants from the literature and from R.D. Reeves and A.J.M. Baker (personal communications). Chin submitted a grant proposal for his art work ("Revival Field") to the National Endowment for the Arts (NEA). The proposal won approval of all the review committees but was rejected by the Chair of the NEA. This political rejection of the proposal became a "cause" to the art community and news of the rejection was reported in many of the largest U.S. newspapers, and even in the news section of *Science* magazine (Anon., 1990). This half-page note rapidly spread the concept of phytoremediation using hyperaccumulators to the research community where others began to investigate the promise of phytoextraction. The press reports encouraged the Chair of the NEA to meet with Chin and others in the art community; the grant was subsequently awarded and the field experiment begun.

In the St. Paul Revival Field, Chin cooperated with Chaney and Baker to study hyperaccumulator plant species from the germplasm collections that Baker accumulated over years of searching for these species. The art work/field experiment was designed as a circle sectioned into quadrants by walkways, representing a rifle sight focused on Earth, but also incorporated a randomized complete block field experiment which tested the effect of sulfur addition to lower soil pH, as well as the form of nitrogen (NO_3-N which can raise rhizosphere pH vs. NH_4-N which can lower rhizosphere pH) which was expected to affect Zn and Cd uptake. Five plant species were grown on the plots in a split-plot arrangement: "Prayon" *Thlaspi caerulescens*; "Palmerton" *Silene vulgaris*; "Parris Island" Romaine lettuce (*Lactuca sativa* L. var. *longifolia*); a Cd-accumulator corn (*Zea mays* L.) inbred, FR-37; and the Zn/Cd-tolerant "Merlin" red fescue (*Festuca rubra*).

Table 7.2 shows the effect of the treatments on Zn, Cd, and Pb in *T. caerulescens*, and in lettuce in the 1993 crop. The initial soil condition was highly calcareous with 25 mg Cd, 475 mg Zn, and 155 mg Pb kg^{-1} soil. The soil at the test field had become

---

TABLE 7.2

Effect of Sulfur and Nitrogen Fertilizer Treatments on Metals in Shoots of *Thlaspi caerulescens* and Lettuce Grown at St. Paul, MN in 1993

Treatment		Soil	Metals in Thlaspi			Metals in Lettuce		
S	N	pH	Cd	Zn	Pb	Cd	Zn	Pb
				(mg kg^{-1} dry weight)				
0	NH$_4$	7 4	9.6	1360	0 5	5 3	58	0.8
0	NO$_3$	7.5	9 4	1260	4.6	4.5	64	0.8
1	NH$_4$	6.7	11 7	3100	1 9	7.8	86	2.1
1	NO$_3$	6.8	8 0	2060	1 5	7 5	77	1 7

metal enriched by land application of ash from a sewage sludge incinerator during a period when the sludge from St. Paul, MN was highly contaminated with Cd from a Cd-Ni battery manufacturer. Lime was used in dewatering the sludge, which prevented the sulfur treatment from acidifying the soil as much as had been expected. This study demonstrated the ability of soil acidification to increase uptake of Zn and Cd, and the ability of *T. caerulescens* to grow well when plant competition was limited by weed control, and when appropriate fertilizers were provided. The very low concentration of Pb in the plants confirmed the experience of most researchers — that Pb hyperaccumulation was not likely to be possible when adequate phosphorus was provided to improve biomass yield.

During the same period, Brown et al. (1994; 1995a,b) characterized the ability of *T. caerulescens* to hyperaccumulate and hypertolerate Zn and Cd. The first study (Brown et al., 1995a) was a nutrient solution evaluation of metal uptake in relation to metal concentration in the nutrient solution. By using the Fe-chelate FeEDDHA, the test system avoided the confoundment which results when Zn displaces Fe from FeEDTA used in other studies of metal tolerance by this species (see Parker et al., 1995). The research compared a widely studied plant species, tomato (*Lycopersicon esculentum*), and the Palmerton Zn-tolerant strain of *S. vulgaris* with *T. caerulescens*, finding that as Zn concentration increased in the nutrient solution, the tomato and bladder campion suffered Zn phytotoxicity at lower solution Zn and much lower plant Zn than required to injure the *T. caerulescens*. Figure 7.1 shows the shoot Zn concentration vs. solution Zn concentration. These findings were interpreted as evidence that *T. caerulescens* reaches high shoot Zn by tolerating higher Zn in shoots rather than accumulating Zn more effectively at lower Zn activity in the rhizosphere (Chaney et al., 1995, 1997a). Other investigators have confirmed the importance of Zn and Cd tolerance in the hyperaccumulation of Zn by *T. caerulescens* (Tolrà et

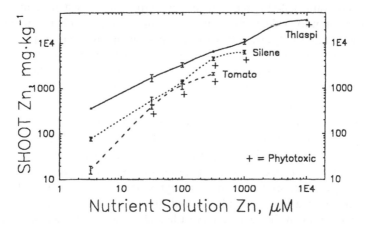

FIGURE 7.1 Shoot Zn concentration and phytotoxicity from nutrient solution with 3 to 10,000 FM Zn; "Rutgers" tomato, "Palmerton" *Silene vulgaris*, and "Prayon" *Thlaspi caerulescens* were grown in half-strength Hoagland solution with strong Fe chelate and low maintained phosphate. (From Brown et al., 1995a. *Soil Sci. Soc. Am. J.* 59: 125-133.)

al., 1996a,b; Shen et al., 1997; Mádico et al., 1992). This pattern was a sharp contrast to the higher uptake at low Ni supply found for *Alyssum* hyperaccumulator species by Morrison et al. (1980). Vázquez et al. (1992; 1994) reported that leaf Zn and Cd were primarily stored in vacuoles, confirming a prediction of Ernst (1974).

The importance of following effective agronomic practices in phytoextraction was illustrated by the contrasting observations from Revival Field and a study by Ernst (1988), who examined the harvest of Zn and Cd on a Zn smelter contaminated site in The Netherlands where both metal-tolerant grasses and some Zn hyperaccumulators grew. Because the hyperaccumulators were very low to the ground in the natural environment, he concluded that these species would not be harvestable and thus not provide useful phytoextraction — this outcome would be expected in the absence of weed control and fertilization to optimize the production of biomass of the hyperaccumulator. In the wild, *T. caerulescens* is often nearly covered by grasses. In an effort to improve growth conditions of hyperaccumulator species, Chaney and his colleagues (unpublished) controlled weeds and supplied fertilizers to increase yield of the hyperaccumulator species, thus allowing them to test the genetic potential of the species rather than the field collection of wild plants. Other comparisons of *T. caerulescens* with crop plants have been made under invalid conditions. For example, Ebbs and Kochian (1997) remarked about the high yield of *Brassica juncea* at Zn supplies which did not cause Zn phytotoxicity to this species, but as seen in Figure 7.1, a Zn supply which allows a typical plant species such as tomato or *B. juncea* to survive does not allow expression of the genetic potential of *T. caerulescens*. At Zn supplies which give zero yield of crop plants, *T. caerulescens* reaches over 2% Zn on a dry matter basis, and some genotypes contained over 2.5% Zn at harvest with no evidence of yield reduction. As illustrated by the studies of genotypic variation in nutrient efficiency of crop plants by Gabelman and Gerloff (e.g., Gerloff, 1987), the test system must allow expression of the genetic potential of a species in order to find strains with higher or lower ability to accumulate a nutrient from soils.

In an attempt to evaluate the utility of *T. caerulescens* in phytoextraction of Zn and Cd, Li et al. (1997) tested this species at a Zn smelting site in Palmerton, PA. They believed that with the ability to adjust soil pH, and the higher level of soil contamination present at this area, it would be possible to better evaluate the genetic potential of this species. At Palmerton, Zn smelting for 80 years caused extensive contamination of soils in the community adjacent to the smelter. The levels of Zn and Cd in lawns and vegetable gardens in the more highly contaminated parts of the village were as high as 10,000 mg Zn and 100 mg Cd kg^{-1} dry soil. Such soils caused severe Zn phytotoxicity to garden crops and lawn grasses unless soils were limed heavily and fertilized well. Many homeowners gave up on growing lawns due to the strong Zn toxicity to Kentucky bluegrass cultivars (Chaney, 1993). Li et al. (1997) reported that lower soil pH favored Zn and Cd accumulation in *T. caerulescens* shoots, and that the second harvest had about double the Zn and Cd concentration of the first harvest. Levels of 20 g Zn kg^{-1} shoots and 200 mg Cd kg^{-1} shoots were obtained in the field.

Additional laboratory studies conducted by Chaney and his colleagues showed a wide range in Zn tolerance and in Cd:Zn ratio in shoots of different *T. caerulescens*

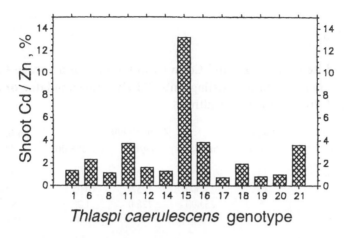

FIGURE 7.2 Cd:Zn ratio in shoots of several *Thlaspi caerulescens* genotypes harvested from Zn and Cd contaminated soils of Revival Field-2 at Palmerton, PA in 1997. Although genotypes differed in Zn accumulation, all accumulated between 10 and 20 g Zn/kg. Remarkable variation in Cd accumulation and Cd:Zn ratio was observed indicating that such high Cd-accumulating genotypes might be useful for rapid phytoremediation of Cd-contaminated soil which cause adverse health effects in subsistence consumers of rice or tobacco grown on the soil.

genotypes grown on the same treatments. Figure 7.2 shows the Cd:Zn ratio of different genotypes. Although the Prayon genotype performed well with near 20 g Zn and 200 mg Cd/kg dry shoots, one of the high Cd:Zn genotypes accumulated nearly 1800 mg Cd kg[-1] dry shoots with 18 g Zn kg[-1]. As shown in Table 7.3, we believe the identification of such genotypes offers a completely different soil remediation opportunity than was recognized earlier. Further, using chelator-buffered nutrient solutions, they showed that Zn hyperaccumulator strains of *T. caerulescens* had a remarkable Zn requirement (Chaney, 1993; Li et al., 1995).

One of the significant challenges of adapting hyperaccumulators to practical phytoextraction is the small size of many of these species. The small size or rosette growth pattern precludes mechanical harvesting, increasing the cost of annual harvest. For the Zn/Cd accumulator system, Reeves et al. (personal communication) found that the tall tree which accumulates Zn does not accumulate Cd. Brewer et al. (1997) made somatic hybrids between *T. caerulescens* and canola, using selection for Zn tolerance. Nonrosette form hybrids with high Zn tolerance were recovered, but were sterile. If somatic hybridization does not provide a method to combine higher harvestable yield with effective Zn + Cd hyperaccumulation, biotechnology may be the only likely route to such plants. Genes involved in hyperaccumulation have not yet been reported to have been cloned, but it is only a matter of time before the genes required for this phenotype are cloned and their mechanisms characterized. It is clear that very many opportunities remain for discovery of critical biochemical, plant physiological, agronomic, soil science, and engineering information that will increase the efficiency of development of reliable phytoextraction systems.

TABLE 7.3

Estimated Removal of Zn and Cd in Crop Biomass of a Forage Crop (Corn), Compared to an Existing Zn + Cd Hyperaccumulator or an Improved Phytoextraction Cultivar

Crop	Yield (t/ha)	Zn in Shoots (mg/kg)	(kg/ha)	(% of Soil)	Zn in Ash (%)
Corn, normal	20	25	0.5	0.005	0.025
Corn, toxic	10	500	5.0	0.05	0.50
Thlaspi	5	25,000	125 0	1 25	25.0
Remed crop	20	25,000	500 0	5.0	25 0

Crop	Yield (t/ha)	Cd in Shoots (mg/kg)	(kg/ha)	(% of Soil)	Cd in Ash (%)
Corn, normal	20	0.5	0.010	0.005	0 0005
Corn, toxic	10	5	0.05	0.025	0.005
Thlaspi	5	250	1.25	0.62	0 40
Super-Cd *Thlaspi*	5	2500	12.5	6 2	4.0
Remed crop	20	250	5 0	2.5	0 40
Super-Cd crop	20	2500	50.0	25.0	4 0

Note: Presume the soil contains 5000 mg Zn/kg and 50 mg Cd/kg dry weight (or 10,000 kg Zn/ha@15 cm or 100 kg Cd/ha@15 cm) Crop is presumed to have 10% ash of the dry matter.

PHILOSOPHY OF SOIL METAL PHYTOEXTRACTION

Since the introduction of the concept of phytoextraction, different models have been put forward for the development of practical phytoextraction systems. Research on the mechanism of hyperaccumulation has been encouraged by all parties because it seems evident that improved understanding of the fundamental processes which natural metal hyperaccumulator plants use to achieve this phenotype should provide ideas on how to develop successful commercial systems. We believe that a fuller understanding could be used to breed improved hyperaccumulators and develop agronomic management practices to give high yields of metal-rich shoot biomass with higher annual metal removal than the wild parents. Alternatively, the metal hyperaccumulator genes required for this phenotype could be cloned and transferred to high yielding crop plants; using a direct gene transfer technique, the combination of yield and metal concentration could be obtained. Other researchers felt that study of mutants of *Arabidopsis* might reveal the genes involved in the hyperaccumulator phenotype. It is likely that many different approaches will contribute to improved understanding of metal hyperaccumulation.

Developers of soil metal phytoextraction must also recognize that soil remediation occurs only when someone agrees to pay for the process. In a market economy system, remediation occurs when a governmental unit orders the remediation, or when the landowner or company offering phytoextraction service decides they can make sufficient profit to conduct the remediation. In the U.S., the Superfund program can require companies to remediate contaminated soils, and the method to be used is negotiated among the parties and the community involved. If no "responsible parties" are found for the contaminated site, the EPA can select the method for remediation and use tax monies for the field work. DOE and DOD have a number of contaminated properties that require remediation as a federal government obligation. In some cases, when the sale of a property for reuse in industry is limited by soil contaminants, a party voluntarily remediates the site to achieve the sale (such sites are often called brownfield sites). Phytoextraction must fit within the market economy or it will not supplant soil removal and replacement, the traditional engineering approach to soil remediation. This understanding influenced the selection of phytoextraction of metals with higher potential economic value in our research program.

Our group looked at the natural metal hyperaccumulators and sought methods to improve them for practical systems for phytoextraction. We initially examined the case for phytoextraction of Zn and Cd because Cd is known to comprise risk to humans who consume rice grown on contaminated soils for a lifetime (Chaney and Ryan, 1994). We chose to study *T. caerulescens* because seed had been collected by Baker, and earlier research indicated promise (Baker and Brooks, 1989). We have considered the recovery of metals from the ash of the plant shoots as a value to offset the cost of soil remediation. The matrix of plant ash generally has little interaction with high concentrations of shoot metals in comparison with ores or soils. When plant ash has 10 to 40% Zn, Cu, Ni, or Co, it is expected that recovery of the metals by standard metallurgical methods would be readily achieved. Extraction of some metals from their ores is very difficult because of the presence of high levels of Fe, Mn, and Si (e.g., lateritic Ni ores). Biomass energy or pyrolysis is nearly cost effective for energy production, and could be an important source of energy if petroleum were more expensive (Agblevor et al., 1995; MacDougall et al., 1997). Biomass ash and pyrolysis char offer metal recovery after energy recovery; biomass burn facilities will require effective emission controls to collect the ash for recycling. Because of the value of recoverable metals in ash of hyperaccumulator plants, we believe that use of metal hyperaccumulator plants would give much lower costs for remediation of metal-contaminated soils than possible with the engineering alternatives (MacDougall et al., 1997).

A quite different approach involves crops grown to produce a marketed crop that would provide the profit potential required in agriculture. Lee and Chen (1992) considered growing cut flowers to phytoextract Cd from contaminated soil in Taiwan; the flowers would be sold, not posing a Cd risk to the purchasers, and the remainder of the plants removed to achieve phytoextraction of Cd from contaminated rice paddies. Some of the flowers they tested reached over 70 mg Cd kg^{-1} dry matter, but the annual removal was not high enough to support economic soil remediation. The removal of Cd in crop parts, which are not marketed, was also considered by McLaughlin and by Chaney and coworkers in 1997. McLaughlin (personal communication) suggested that growing canola could remove Cd which had been added by high Cd phosphate fertilizers to acidic farmland. The market pays for the seed oil, and the Cd accumulated in the unused seed meal and/or stover could be burned and recovered for marketing or disposal. Alternatively, removal of potato shoots with high Cd concentration due to high soil chloride concentrations (McLaughlin et al., 1994) could phytoextract soil Cd at little additional cost compared to potato production alone.

In order to develop agronomic practices and improved cultivars which produced flax and nonoilseed sunflower with lower Cd concentrations, Li et al. (1997) characterized genetic differences within these crops for grain Cd. Upon reconsideration, it seemed possible that breeding higher grain Cd genotypes might give cultivars with good seed oil properties and support phytoextraction over time. For phytoremediation of surface soils, the rooting pattern of flax is shallow while that of sunflower is deep; thus, we considered the use of flax cultivars with high Cd accumulation as a crop plant which could achieve phytoextraction as a side benefit of normal crop production. If the linseed oil extracted from the flax seed gave sufficient profit to grow the crop, the seed meal could be burned to recover or give efficient disposal of the Cd. This model for phytoextraction is based on the commercial production of a crop to market some flower or oil which either has no metal risk to consumers, or is not consumed, with metal phytoextraction a byproduct of the normal economic crop production. The payment for soil remediation would not have to be large to pay for the remediation if the dominant value is obtained from the seed oil. This model also fits some concepts of phytoextraction of radionuclides. As found for heavy metals, little or no ^{137}Cs enters the extracted oil of canola seed. An experiment conducted to characterize alternative crops which could be grown on the contaminated land surrounding Chernobyl expected to find that biodiesel could be produced safely on the land; when the oil was found to be not contaminated with ^{137}Cs, it was used as food oil (Chaney et al., 1993, unpublished data). As in the flax model, if the seed meal or shoots accumulate sufficient amounts of radionuclide, or sufficiently concentrated levels of a radionuclide, the reduction in cost of disposal compared to disposal of the entire soil might make this an economic technology.

TYPICAL CROPS WILL NOT REMOVE ENOUGH METALS TO SUPPORT PHYTOREMEDIATION

Table 7.3 shows the potential crop removal of Zn and Cd by a typical crop plant, corn (*Zea mays* L.), compared to the Zn + Cd hyperaccumulator plant *T. caerulescens*. Corn silage can have a high biomass yield (e.g., 20 t ha^{-1}), but because corn has only normal Zn tolerance, yield is reduced substantially (e.g., 50%) when the whole shoots contain 500 mg Zn kg^{-1} (Chaney, 1993). Under normal conditions, corn shoots contain only about 20 to 50 mg Zn kg^{-1}, so the removal of Zn from soils by even typical forage crops is small, and grains remove even less than the forage (Chaney, 1973). Because of this understanding, it was clear in 1973 that normal crop removal of Zn and Cd from soil would not rapidly remove these metals applied in biosolids or wastewater effluents used on cropland.

We believe that the use of element hyperaccumulator species would give much lower costs for remediation of metal-contaminated soils than possible with the engineering alternatives. Scientists have long wondered how hyperaccumulator plants were selected by evolutionary processes. Research has recently provided evidence that the unusual accumulation of metals gives these plants the ability to limit predation by chewing insects and plant disease caused by microbes; high metals in the leaves defend the plant (Boyd and Martens, 1994; Boyd et al., 1994; Pollard and Baker, 1997). At each location where evolution of such plants occurred, there are many plant species that are tolerant of high levels of soil metals by exclusion of metals. However, the small subset of hyperaccumulators took advantage of their high metal tolerance to increase their ability to compete for survival.

DEVELOPMENT OF A TECHNOLOGY FOR PHYTOREMEDIATION OF TOXIC METALS IN SOIL

In our judgement, if evolution has selected plants which can both tolerate and hyperaccumulate an element, X (some element for which plant species are known that hyperaccumulate the element), it is likely that one could develop a technology for phytoremediation of contaminated or mineralized soils rich in X, and recycle X commercially along with biomass energy production. Such a research and development program needs to cover the total system:

1. Collection of plant genetic diversity so that improved phytoremediation cultivars can be selected and/or bred.
2. Valid comparison of genotypic differences in yield and X hyperaccumulation.
3. Breeding improved plant cultivars which are effective in metal-rich field.
4. Identification of soil and plant management practices needed to attain high yields and high metal concentrations in the biomass including tillage, fertilization, soil conditioners, pH adjustment, herbicides, etc.
5. Identification of methods to plant, grow, harvest, and market the biomass.
6. Selection of methods to economically recover the metals from the biomass (e.g., a method to burn the biomass which retains the metals in a form that can be sold as a high grade metal ore), and the biomass energy may be used for power generation.
7. Identification of methods to recover the metals from the ash.
8. Identification of farming systems that allow use of this technology to produce jobs and profits for growers as well as smelters.

We have used this approach to try to develop a commercially useful technology for phytoextraction of Zn and Cd, but in the absence of taller genotypes of *T. caerulescens*, our technology is only cost effective in developing countries where hand-harvest of biomass could be a normal production practice. We also used this approach in development of technology for phytoextraction of soil Ni and Co, and obtained a utility patent for the product of our research and development efforts (Chaney et al., 1998a). In situations where the available hyperaccumulator species are too small to afford economic practices, biotechnology may be needed to bring the hyperaccumulator phenotype together with high biomass characteristics. This may be especially necessary when one is trying to develop plants to selectively accumulate one and not especially accumulate a closely related ion, such as ^{137}Cs vs. K. Use of biotechnology to hasten development of phytoextraction crops is discussed in more detail below.

DEVELOPING COMMERCIAL PHYTOREMEDIATION TECHNOLOGY FOLLOWING THE PARADIGM OF DOMESTICATION AND BREEDING OF IMPROVED HYPERACCUMULATOR PLANT SPECIES

Select Plant Species

Select hyperaccumulator plant species that have a high likelihood for domestication to make a commercial phytoextraction system. Select plant species that would be expected to grow and accumulate commercially useful levels of a hyperaccumulated metal, designated X, from metal X-rich soils in the climatic zone where soils rich in X occur. In selecting species for development, all characteristics of known hyperaccumulators would be reviewed, and a few species would be selected for germplasm collection and evaluation. As noted in Chaney et al. (1997a), an effective phytoremediation crop should attain well in excess of 1% of X (on the order of 100-fold normal levels tolerated by plants), with a dry biomass yield of 20 t/ha or greater under optimum production conditions. For some contaminants, the concentration ratio (concentration in plant dry matter divided by concentration in dry soil) is a stronger consideration than simple hyperaccumulation. For example, if ^{137}Cs or ^{90}Sr are accumulated to much higher concentrations in the ash of a plant than in the soil being remediated, the amount of radiological waste which must be appropriately disposed is small enough to justify this approach. Where soil elements will be recycled for value, the economics of recovery of X from the plant ash will affect the value of the ash and the overall economics of phytoextraction. At different stages in phytoremediation, plants with higher tolerance and hyperaccumulation would be most effective, and at other stages, genotypes with high plant:soil concentration ratios would be most effective. As with crop plants, cultivars with different properties may be needed to obtain an effective overall technology. Preliminary soil and plant studies would be conducted to learn if the selected plant species actually hyperaccumulate from soils which might be available for commercial planting, and possess the ability to grow outside of their area of collection.

Collect Seeds

Collect seeds of plants from the wild, bioassay their utility in phytoextraction, and breed improved cultivars for commercial phytoextraction system. Based on confirmation that high enough X concentration was accumulated in the collected plant shoots to justify investment of funds to develop improved genotypes for commercial use, a genetic collection would be undertaken. In general, when one is domesticating a new crop, there are no germplasm collections held by public agencies from which one may request seeds. Trained staff would visit natural sites where the plants grow and obtain seeds that represent the genetic diversity available in the species. Genetic diversity is needed to allow breeding of improved cultivars in which characteristics found in different strains can be brought together into a desired phenotype. Each seed sample would represent a single plant from a specific place in the natural ecosystem with differing ecosystem properties (soil properties, plant growth characteristics, moisture, slope, aspect, etc.) The collected seed would be cleaned, dried, and managed to preserve seed quality (appropriate seed storage methods may have to be determined experimentally for previously unstudied species).

After the germplasm is collected, its potential for phytoremediation must be evaluated. It should be recognized that most hyperaccumulator plants occur naturally on soils with metal toxicity and severe infertility, and may have seasonal drought. So, the potential of collected genotypes to perform phytoextraction cannot be identified from measurements at the point of collection. Both greenhouse and field testing are useful. The field testing is needed to see how the plants perform in the environment for which the selection and commercial production will be made. Each genotype collected would be grown in replicated field plots in at least two environments (soils or locations with a range of properties). The seed would be processed to allow mechanical planting. The plant density would be somewhat lower than expected to be used in production fields so that the genetic potential of the collected germplasm could be fully evaluated. Selection criteria may focus on improved metal tolerance and hyperaccumulation, improved concentration ratio, or others relevant to commercial uses of improved cultivars.

The goals of such a germplasm evaluation testing program are to estimate the ability of the genotype to achieve high yields on fertilized soil rich in X, and to accumulate high X levels in harvestable plant shoots. The constancy of the test field/soil should be verified by use of check genotypes throughout the blocks. For the genetic evaluation, soil management practices near optimum for yield and X hyperaccumulation should be used which have been identified by preliminary studies. Such practices could involve management of soil pH, phytoavailable levels of N, P, K, Ca, Mg, etc. Because the goal is to reach high biomass with high X concentrations, soil conditions should be managed to allow high yields. Agronomic testing would determine if other than optimal biomass production conditions should be considered to maximize annual phytoextraction of X.

Depending on the outcome of preliminary studies of X uptake by these species in test fields, the germplasm evaluation testing plots would be harvested during flowering at the time when biomass metals appear to be at a maximum. Separate experiments (see below) would be used to identify optimum harvesting time. Yield would be measured, and a representative sample of each experimental unit collected for analysis. The biomass would be dried, ground, ashed, and analyzed by ICP-atomic emission spectrometry for both the metal(s) of commercial interest (X) and plant nutrients. The nutrient analysis could identify any potential plant nutritional limitations of the test system, or any unusual nutrient levels in the species and genotypes under development.

Based on the outcome of these replicated plots to assess performance of each collected genotype, specific strains would be selected as genetic parents in a breeding program. If possible, genetic crosses would be made to combine characteristics that would be useful in a commercial cultivar (yield, maximum annual phytoextraction, perennial performance, disease resistance, accumulation of other elements which might have value or reduce the value of the biomass ash).

During the field-testing program, separate research would be conducted to learn how to force flowering of the test species. A rapid breeding program to generate commercially useful cultivars (which would be protected by the Plant Variety Protection Act) would greatly benefit from being able to control the flowering date so that crosses can be made according to the program needs rather than the annual weather cycle.

Presuming that the species being studied are from a region with a significant change in day length during the year, are perennial species which flower in the spring, and have mature dry seed in the early summer (or are annual plants sown in the spring, with mature seed in the fall), day length should play a role in stimulating flowering. However, the plants may have to undergo vernalization (growth under cold soil and environment conditions) before the day length stimulus can induce flowering. Drought can be another flowering signal for species. In some cases, plant hormones can substitute for the vernalization treatment. Otherwise, forcing of flowering will require a vernalization (cold) treatment at short-day, followed by long-day, and optimum growth conditions. Such studies would identify the controls on flowering of the species selected for breeding.

Basic genetic studies may be needed to characterize specific plant characters which may be important for the commercial development of the species studied. The inheritance of metal accumulation, biomass yield, plant height, regrowth from cut stems, retention of leaves on the plant during air drying, and other characters may require more detailed evaluation, depending on the outcome of the research. At a minimum, genotypes differing widely in ability to hyperaccumulate or hypertolerate X would be crossed to determine the inheritance of metal hyperaccumulation and/or tolerance. Although the progenies will need to be tested in the field in any case, study in nutrient solutions may allow characterization of variation in X accumulation ratio, translocation to shoots, and/or X tolerance, which could be combined by breeding to build a more effective phytoextraction cultivar.

Presuming that the methods developed to assess the phytoremediation potential of the collected germplasm is a valid bioassay of genetic ability of the species, a breeding program would be conducted in which yield, harvestability, metal concentration in the biomass, metal tolerance by the plant, and other characteristics normally considered by a professional breeder would be combined. The progenies would be tested to characterize the outcome and individuals selected for subsequent improvement. If the species does become of commercial value, a plant breeding program would be a part of the commercial development program so that more of the potential genetic diversity is used to breed improved cultivars for a wider range of environments.

Identify Soil Management Practices

Identify soil management practices to improve the efficiency of phytoextraction. Research has identified a number of soil management practices which may affect the efficiency of phytoextraction of elements such as Zn and Cd. However, different elements may be affected differently by pH, fertilization, and other management practices. Because soil pH is known to affect plant uptake of most trace elements (Zn, Cd, Ni, Co, Mn) from soils, studies should be conducted to evaluate the independent effect of soil pH, and soil cations on yield and X uptake. It was noted above that soil phosphate levels needed for high yields inhibit Pb uptake from soils. Because yield is so important to the success of phytoextraction, nutrient interactions which affect either yield or X concentration in the shoots should be identified by experiments.

Other soil fertility management may affect yield, X hyperaccumulation, or hypertolerance. Some wild plants suffer yield reduction at soil levels of specific nutrients considered to be below optimal for crop plants (e.g., many Australian native species are very efficient in accumulating phosphate and can be poisoned by levels of P fertilizer commonly used to obtain maximum yield elsewhere). Thus, X-rich soils should be used in pot and/or field tests in determining the amount of N, P, K, Ca, Mg, Zn, Cu, Mn, B, and Mo that supports production of the maximum value of X in the plants grown. Changing soil fertility may affect the distribution of X within the plant parts, so plant tissues from such tests should be analyzed separately. If stems accumulated higher levels of X under one soil management regime than another, production practices could consider that information.

Changing soil fertility or pH may increase the uptake or toxicity of other element(s) in the soil such that biomass yield is reduced without increasing X concentration in the biomass. In particular, if soil acidification favors hyperaccumulation of X (as found for Zn and Cd; Brown et al., 1995a), making the soil too acidic could induce Al or Mn phytotoxicity. Thus, it may be necessary to improve the Al or Mn tolerance of a phytoextraction cultivar so that it can be grown at increasingly acidic soil pH to maintain a high annual phytoextraction level. As the plants deplete the soil reserve of X, by lowering pH the phytoavailability could be increased to maintain economic phytoextraction if the improved cultivar had sufficient tolerance of ions solubilized at lower soil pH, especially Al and Mn.

Soil management practices may also affect the depth from which roots can efficiently extract X. Some pH-modifying soil amendments and nutrients will improve plant performance and affect rooting density only to the depth in which they are incorporated. Thus the effect of the depth of incorporation of some of the amendments on rooting depth and X hyperaccumulation may need to be evaluated to develop the most effective phytoextraction technology.

Develop Crop Management Practices

Developing crop management practices to improve phytoextraction efficiency. Domestication of a wild plant species into a commercially useful crop is usually a difficult undertaking. In practice, all aspects of cropping must be considered, and those evaluated which could affect the economics of phytoextraction using the improved cultivars.

Effect of planting practices

Depending on the nature of the seeds of the species involved, commercial planting using mechanical equipment may require pelletization of the seed or other novel techniques to obtain the plant density desired in the field. The ability of commercially available planters to establish the desired stand density would be tested. The effect of seeding depth, seed bed moisture, and any requirement of light exposure for germination would be tested.

The effect of stand density

The density of plants in a field usually influences the biomass yield per plant (and per hectare) and may affect element concentration in the biomass. Little or no data have been published on this relationship for hyperaccumulator plants. Density (plants/hectare) is expected to be an important management variable, and it is possible that genetic improvement may need to take into account the relationship between stand density and costs of phytoextraction. Replicated field experiments are required with a number of promising genetic lines to assess the effect of plant density on yield and X concentration. It is also likely that the effect of stand density on yield of X will interact with the harvest date.

Annual/perennial management

The optimum fundamental crop management approach must be identified for the plant species to be domesticated. If the plant is perennial in nature, it may be most economical to establish a perennial planting, and then harvest it annually. Alternatively, it may be more profitable to reseed each year. These choices could depend on the seasonal availability of soil water or irrigation at production locations. As the results of evaluating the many genotypes collected from natural sites are obtained, experiments would be designed to evaluate whether annual, biennial, or perennial crop management is more cost effective with the plants selected for development.

Weed control practices

In modern agriculture, weed control chemicals are used to reduce weed competition with crops for soil nutrients, moisture, or light. Development of a new domesticated crop will require identification of weed control practices. Depending on the plant species selected for development, existing herbicide data can be examined to identify promising chemicals, and the actual tolerance of the new crop to promising chemicals determined by experiment. The overall combination of agronomic practices interacts with the nature of soil and herbicide to control effectiveness. Furthermore, it is possible that management of soil pH and fertility in an attempt to maximize annual X phytoextraction will provide metal toxicity stress for weeds and crop plants and limit weed competition.

Harvest schedule and methods

Methods to harvest the crop would be evaluated. Although the conceptual model has been that of "hay making" (using solar drying to reduce production costs), it is not clear what approach and/or equipment can be used to collect the biomass from the field so it can be processed to recover energy and X, or if the weather will support drying when the crop needs to be cut. Thus, it will be necessary to experimentally cut the hay, allow it to dry and metabolize in the field, "lift" the hay, and probably bale the biomass so it can be handled mechanically at the X recovery facility. It is unlikely that use of energy to dry the field moist crop could be cost effective. Thus, the retention of X-rich leaves during field drying must be evaluated for the genotypes collected so that effective methods for harvest may be identified.

Seed management

With any new crop, the methods to harvest, process, and store seeds must be established. Genetic selection to improve phytoextraction needs to consider the effect on seed quality and survival during storage. Poor seed quality may cause problems with regard to domestication of a plant species due to the rapid decline of germination percentage of the seeds. Methods for seed increase, with mechanical harvest of the seed, should be developed as part of the development of commercial practices. An example of the difficulty which short longevity of seed quality may cause during domestication of a plant species can be found in the rapid decline of germination percentage of the seed of *Berkheya coddii* reported by Anderson et al. (1997).

Process the Biomass

Processing the biomass to produce energy and valuable ash in a form which can be used as ore or disposed safely at low cost. Recovery of energy by biomass burn or pyrolysis could help make phytoextraction cost effective. Strategies for biomass processing to recover the hyperaccumulated element need to consider the normal markets for X and the smelting or other technologies used in traditional industrial practices.

Several methods are available to process biomass, which recovers the ash while recovering energy. The Department of Energy has been conducting R&D to develop biomass energy technologies using more efficient methods such as gas turbine generators. They found that the KCl in biomass tended to accumulate on turbine blades and eventually harm the equipment when the biomass included leafy materials (Agblevor et al., 1995). Thus, it is likely that external firing for a gas turbine generator would be necessary. Traditional incinerators have been demonstrated to be able to process biomass and produce an ash that can be recovered with the usual pollution control devices. The intentional burning of plant biomass to both produce energy and release metals for recovery will require demonstration that the pollution control equipment is highly effective in recovering the metals. In this regard, plant species which accumulate Si, such as rice (*Oryza sativa*), horsetail (*Equisetum arvense*), and many other species, would cause significant problems in element recovery from the plant ash. Not only may the Si dilute the metals in the ash substantially, the presence of high Si levels may cause formation of a glass or slag from which recovery of metals may be very difficult.

Alternatively, pyrolysis may be used to produce gaseous or liquid hydrocarbon fuels, and a mixture of ash and "char" which has characteristics similar to charcoal (Agblevor et al., 1995). Such a "char" may have further value in pyrometallurgical processing of the metals.

Recovery of value from X in the biomass could use different smelting methods. If the ash has appropriate properties, it may be possible to dissolve X with acids, and "electrowin" the metals from the acid solution. On the other hand, if the ash has a high enough concentration, it may have more value for use in a pyrometallurgical smelter or for alloy production. These processing methods are tolerant of many of the other constituents in ash from plant biomass. Efficient and cost-effective methods for element recovery from the biomass ash should be able to be identified using established geochemical and metallurgical testing methods to characterize the effect of possible biomass processing methods on the value of X obtained by the overall technology.

Develop Commercial Systems

Development of commercial systems for production of metal rich biomass for recovery of energy and metals. For commercial phytomining/phytoremediation to be successful, soils must be found which contain high enough levels of X to support growing the improved cultivars for profit. Additionally, growers must produce the crop using production practices identified by the R&D to reliably maximize yield and metals concentration in the biomass. Where soil decontamination is required by a regulatory agency, cost savings may control the selection of the technology to be used; phytoextraction appears to be highly cost effective if plants can reach even 1% element in shoots (Cunningham and Berti, 1993). Areas of mineralized or contaminated soils rich enough in X to justify phytoextraction must exist for the plants to be useful.

Properties of different X-rich soils may affect the response to different fertilizers, yield potential, and X phytoavailability differently. Thus, some testing protocol needs to be developed which predicts the phytoextractability of X from the specific fields available to be cropped. There is no reason to expect testing methods developed for element deficiency to be useful in predicting phytoavailability to a hyperaccumulator plant.

It seems likely that soil tests logically related to the usual soil fertility tests used in agricultural production should provide an important management tool for phytoextraction. Based on chemical analysis for X quantity and phytoavailability in the soil, and soil fertility characteristics found to optimize yield and X accumulation, it should be possible to estimate the quantities of soil amendments required to bring a field into phytoextraction crop production. Such soil tests would have to be validated by field testing.

Practical phytoextraction cropping for metal recycling and biomass energy might become a commercial system depending on the value of the metal recovered in the biomass ash (Robinson et al., 1997), density of metal-rich soils, and the profit potential compared to other crops the landowner could grow, among other complex economic criteria. Government requirements or payments to phytoextract contaminants from soils must support a market-based system or no phytoextraction will occur. Whether local farmers or phytoremediation companies do the cropping, it seems likely that the biomass processor or ore purchaser may strongly affect the location of production because of the cost of hauling biomass or ash from sites of production to sites where energy and metals are recovered. Given adequate information on how to produce the crop as an economic agricultural technology, farmers can play an important role in environmental remediation, using the low-cost agricultural technologies usually considered fundamental to the whole concept of phytoremediation. Phytoextraction is fundamentally an agricultural technology. Connecting this agricultural production system to metal recovery technologies will be a challenge for researchers and managers of phytoextraction enterprises.

POTENTIAL NEGATIVES FOR PHYTOEXTRACTION WITH HYPERACCUMULATOR SPECIES

We believe that every environmental remediation technology has some characteristic which limits adoption by some users. For high value land, the time required to achieve phytoextraction to regulatory limits may cost so much that removal and replacement of the soil is more cost effective than phytoextraction, and small areas of contamination will usually be removed more cost effectively than phytoextracted. For Pb-contaminated urban soils where children are at risk for acute Pb toxicity from soil ingestion, it is not appropriate to keep children exposed for the number of years required for effective phytoextraction. For soils contaminated with metals for which plants have evolved hyperaccumulation, and which are large arable land areas, phytoextraction will often be highly cost effective. The volatility of the metal marketplace is viewed by some as a negative of phytoextraction with recycling of metals from the biomass; perhaps storage of the biomass or the biomass ash will provide flexibility to producers. Some have been concerned about dispersal of the metal hyperaccumulators into "normal" soils; for *T. caerulescens*, the finding of a remarkably high Zn requirement for growth confirmed that the hyperaccumulator genotypes are endemic to contaminated soils (Li et al., 1995). These genotypes cannot grow on normal soils, and they cannot compete with nonhyperaccumulator species on soils with background phytoavailable Zn levels.

An important potential adverse aspect of phytoremediation is the exposure of wildlife to plant biomass which is rich in elements which could cause toxicity. On the positive side, most highly contaminated sites are barren because of metal toxicity and support no wildlife. Also, the nature of the plants domesticated for phytoextraction will affect the potential exposure of wildlife. *T. caerulescens* does not produce large seeds which could sustain a wildlife population, and the metals which this species accumulates are not biomagnified in terrestrial food chains. Large mammals are not at risk because their range is larger than a phytoextraction production field. The wildlife at risk from phytoextraction appear to be small mammal herbivores which would live on the contaminated site and have access only to a monoculture of hyperaccumulator plant managing to contain over 2% Zn. If the feeding tests of Pollard and Baker (1997) and Boyd and Martens (1994) are a model for response of primary consumers of plant biomass, the consumers will take great effort to avoid consuming these plants, and not try to live on the site. Predators of these primary consumers are not expected to be at any risk because little of the ingested Zn and Cd are retained in bodies of small mammals. In practice, the management of such contaminated soils to produce high yields of biomass will help reduce erosive redistribution of the metals, and by evapotranspiration, reduce the potential for leaching of soil metals. Because roots alter the phytoavailability of metals in their rhizosphere, hyperaccumulators achieve this outcome without appreciably changing the risk of metal leaching in the soil. After remediation of a contaminated soil has been completed, raising the pH to normal levels will allow establishment of diverse vegetation as part of the remediated safe ecosystem. This endpoint, a remediated diverse ecosystem, was the goal of Revival Field and of our research and development program for phytoextraction using hyperaccumulator plants.

ACKNOWLEDGMENT

We gratefully acknowledge our many colleagues who have discussed soil remediation and hyperaccumulator plants with us over the many years. The research community developing phytoremediation has shared ideas to help each other for many years: Jim Ryan, Scott Cunningham, Robert Brooks, Ilya Raskin, Bill Berti, Mike Blaylock, Joe Pollard, Dick Lee, Mike McLaughlin, Gary Bañuelos, Norman Terry, David Parker, Lee Tiffin, Mike White, Meredith Leech, Bob Wright, Carrie Green, R.B. Corey, K.-Y. Kuang, and others too numerous to list.

REFERENCES

Agblevor, F.A., S. Besler, and A.E. Wiselogel. 1995. Fast pyrolysis of stored biomass feedstocks. *Energy Fuel* 9:635-640.

Anderson, T.R., A.W. Howes, K. Slatter, and M.F. Dutton. 1997. Studies on the nickel hyperaccumulator, *Berkheya coddii*. 261-266. In T. Jaffré, R.D. Reeves, and T. Becquer (Eds.) *The Ecology of Ultramafic and Metalliferous Areas* (Proc. Second Intern. Conf. on Serpentine Ecology, Noumea, New Caledonia, July 31-Aug. 5, 1995.) ORSTOM, New Caledonia.

Anon. 1990. NEA dumps on science art. *Science* 250:1515.

Baker, A.J.M. 1981. Accumulators and excluders S-strategies in the response of plants to heavy metals. *J. Plant Nutr.* 3:643-654.

Baker, A.J.M. and R.R. Brooks. 1989. Terrestrial higher plants which hyperaccumulate metal elements. A review of their distribution, ecology, and phytochemistry. *Biorecovery* 1:81-126.

Baker, A.J.M., S.P. McGrath, C.M.D. Sidoli, and R.D. Reeves. 1994. The possibility of *in situ* heavy metal decontamination of polluted soils using crops of metal-accumulating crops. *Res. Conserv. Recyc.* 11:41-49.

Bañuelos, G.S., H.A. Ajwa, B. Mackey, L. Wu, C. Cook, S. Akahoue, and S. Zambruzuski. 1997. Evaluation of different plant species used for phytoremediation of high soil selenium. *J. Environ. Qual.* 26:637-646.

Baumann, A. 1885. Das Verhalten von Zinksalzen gegen Pflanzen und im Boden. *Landwirtsch. Vers.-Stan.* 31:1-53.

Beath, O.A., H.F. Eppsom, and C.S. Gilbert. 1937. Selenium distribution in and seasonal variation of type vegetation occurring on seleniferous soils. *J. Am. Pharmacol. Assoc.* 26:394-405.

Bell, P.F., D.R. Parker, and A.L. Page. 1992. Contrasting selenate-sulfate interactions in selenium-accumulating and nonaccumulating plant species. *Soil Sci. Soc. Am. J.* 56:1818-1824.

Benemann, J.R., R. Rabson, J. Tavares, and R. Levine (and workshop participants). 1994. Summary Report of a Workshop on Phytoremediation Research Needs. DOE-EM-0224, 24.

Berti, W.R. and S.D. Cunningham. 1997. In-place inactivation of Pb in Pb-contaminated soils. *Environ. Sci. Technol.* 31:1359-1364.

Boyd, R.S. and S.N. Martens. 1994. Nickel hyperaccumulated by *Thlaspi montanum* var. *montanum* is acutely toxic to an insect herbivore. *Oikos* 70:21-25.

Boyd, R.S., J.J. Shaw, and S.N. Martens. 1994. Nickel hyperaccumulation as a defense *Streptanthus polygaloides* (Brassicaceae) against pathogens. *Am. J. Bot.* 81:294-300.

Brewer, E.P., J.A. Saunders, J.S. Angle, R.L. Chaney, and M.S. McIntosh. 1997. Somatic hybridization between heavy metal hyperaccumulating *Thlaspi caerulescens* and canola. *Agron. Abstr.* 1997:154.

Brooks, R.R. 1972. *Geobotany and Biogeochemistry in Mineral Exploration.* Harper & Row, London.

Brooks, R.R. 1977. Copper and cobalt uptake by *Haumaniastrum* species. *Plant Soil* 48:541-544.

Brooks, R.R. 1983. *Biological Methods of Prospecting for Minerals.* John Wiley & Sons, New York.

Brooks, R.R. 1987. *Serpentine and Its Vegetation.* Dioscorides Press, Portland, OR.

Brooks, R.R. 1992. Geobotanical and biogeochemical methods for detecting mineralization and pollution from heavy metals in Oceania, Asia, and The Americas. 127-153. In B. Markert (Ed.) *Plants as Biomonitors and Indicators for Heavy Metals in the Terrestrial Environment.* VCH Publishers, Weinheim.

Brooks, R.R. (Ed.) 1998. *Plants that Hyperaccumulate Heavy Metals.* CAB International, Wallingford, 379.

Brooks, R.R. and C.C. Radford. 1978. Nickel accumulation by European species of the genus *Alyssum. Proc. Roy. Soc. Lond.* B200:217-224.

Brooks, R.R., A.J.M. Baker, and F. Malaisse. 1992. Copper flowers: the unique flora of the copper hills of Zaïre. *National Geographic Res. Explorer.* 8:338-351.

Brooks, R.R., J. Lee, R.D. Reeves, and T. Jaffré. 1977. Detection of nickeliferous rocks by analysis of herbarium specimens of indicator plants. *J. Geochem. Explor.* 7:49-57.

Brooks, R.R., S. Shaw, and A. Asensi Marfil. 1981a. Some observations on the ecology, metal uptake, and nickel tolerance of *Alyssum serpyllifolium* subspecies from the Iberian peninsula. *Vegetatio* 45:183-188.

Brooks, R.R., J.M. Trow, J.-M. Veillon, and T. Jaffré. 1981b. Studies on manganese-accumulating Alyxia from New Caledonia. *Taxon* 30:420-423.

Brooks, R.R. and X.-H. Yang. 1984. Elemental levels and relationships in the endemic serpentine flora of the Great Dyke, Zimbabwe and their significance as controlling factors for this flora. *Taxon* 33:392-399.

Brown, S.L., R.L. Chaney, J.S. Angle, and A.J.M. Baker. 1994. Zinc and cadmium uptake by *Thlaspi caerulescens* and *Silene vulgaris* in relation to soil metals and soil pH. *J. Environ. Qual.* 23:1151-1157.

Brown, S.L., R.L. Chaney, J.S. Angle, and A.J.M. Baker. 1995a. Zinc and cadmium uptake of *Thlaspi caerulescens* grown in nutrient solution. *Soil Sci. Soc. Am. J.* 59:125-133.

Brown, S.L., J.S. Angle, R.L. Chaney, and A.J.M. Baker. 1995b. Zinc and cadmium uptake by *Thlaspi caerulescens* and *Silene vulgaris* grown on sludge-amended soils in relation to total soil metals and soil pH. *Environ. Sci. Technol.* 29:1581-1585.

Brown, S.L., C.L. Henry, R.L. Chaney, and H. Compton. 1998. Bunker Hill Superfund Site: Ecological restoration program. *Proc. 15th Nat. Mtg. Am. Soc. Surface Mining and Reclamation* (May 17-22, 1998, St. Louis, MO).

Cai, S., Y. Lin, H. Zhineng, Z. Xianzu, Y. Zhaolu, X. Huidong, L. Yuanrong, J. Rongdi, Z. Wenhau, and Z. Fangyuan. 1990. Cadmium exposure and health effects among residents in an irrigation area with ore dressing wastewater. *Sci. Total Environ.* 90:67-73.

Cannon, H.L. 1955. Description of indicator plants and methods of botanical prospecting for uranium deposits on the Colorado Plateau. In Contributions to the geology of uranium. *Bull. U.S. Geol. Surv.* 1030:399-515.

Cannon, H.L. 1960. Botanical prospecting for ore deposits. *Science* 132:591-598.

Cannon, H.L. 1971. The use of plant indicators in ground water surveys, geologic mapping, and mineral prospecting. *Taxon* 20:227-256.

Chaney, R.L. 1973. Crop and food chain effects of toxic elements in sludges and effluents. 120-141. In *Proc. J. Conf. on Recycling Municipal Sludges and Effluents on Land.* Nat. Assoc. St. Univ. and Land Grant Coll., Washington, D.C.

Chaney, R.L. 1983. Plant uptake of inorganic waste constituents. 50-76. In J.F. Parr, P.B. Marsh, and J.M. Kla (Eds.) *Land Treatment of Hazardous Wastes.* Noyes Data Corp., Park Ridge, NJ.

Chaney, R.L. 1993. Zinc phytotoxicity. 135-150. In A.D. Robson (Ed.) *Zinc in Soils and Plants.* Kluwer Academic Publishers, Dordrecht.

Chaney, R.L. and J.A. Ryan. 1994. Risk Based Standards for Arsenic, Lead and Cadmium in Urban Soils. (ISBN 3-926959-63-0) DECHEMA, Frankfurt, 130.

Chaney, R.L., S.B. Hornick, and L.J. Sikora. 1981a. Review and preliminary studies of industrial land treatment practices. 200-212. In *Proc. Seventh Annual Research Symposium on Land Disposal of Municipal Solid and Hazardous Waste and Resource Recovery.* EPA-600/9-81-002b.

Chaney, R.L., D.D. Kaufman, S.B. Hornick, J.F. Parr, L.J. Sikora, W.D. Burge, P.B. Marsh, G.B. Willson, and R.H. Fisher. 1981b. Review of information relevant to land treatment of hazardous wastes. Report to EPA-Solid and Hazardous Waste Research Division, 476.

Chaney, R.L., S. Brown, Y.-M. Li, J.S. Angle, F. Homer, and C. Green. 1995. Potential use of metal hyperaccumulators. *Min. Environ. Manage.* 3(3):9-11.

Chaney, R.L, M. Malik, Y.M. Li, S.L. Brown, E.P. Brewer, J.S. Angle, and A.J.M. Baker. 1997a. Phytoremediation of soil metals. *Curr. Opin. Biotechnol.* 8:279-284.

Chaney, R.L., J.A. Ryan, and S.L. Brown. 1997b. Development of the U.S.-EPA limits for chromium in land-applied biosolids and applicability of these limits to tannery by-product derived fertilizers and other Cr-rich soil amendments. 229-295. In S. Canali, F. Tittarelli, and P. Sequi (Eds.) *Chromium Environmental Issues.* Franco Angeli, Milano, Italy [ISBN-88-464-0421-1].

Chaney, R.L., J.S. Angle, A.J.M. Baker, and Y.-M. Li. 1998a. Method for phytomining of nickel, cobalt and other metals from soil. U.S. Patent 5,711,784.

Chaney, R.L., S.L. Brown, and J.S. Angle. 1998b. *Soil Root Interface: Food Chain Contamination and Ecosystem Health.* In M. Huang et al. (Eds.) Soil Science Society of America, Madison, WI, 279-311.

Cole, M.M. 1973. Geobotanical and biogeochemical investigations in the sclerophyllosis woodland and shrub associations of the Eastern goldfields area of Western Australia, with particular reference to the role of *Hybanthus floribundas* (Lindl.) F. Muell. as a nickel indicator and accumulator plant. *J. Appl. Ecol.* 10:269-320.

Cunningham, S.D. and W.R. Berti. 1993. Remediation of contaminated soils with green plants: an overview. *In Vitro Cell Dev. Biol.* 29P:207-212.

Denaeyer-De Smet, S. 1970. Considerations sur l'accumulation du zinc par les plantes poussant sur sols calaminaires. *Bull. Inst. R. Sci. Nat. Belg.* 46:1-13.

Dudley, T.R. 1986a. A new nickeliferous species of *Alyssum* (Cruciferae) from Portugal: *Alyssum pintodasilvae* T.R. Dudley. *Feddes Repert.* 97:135-138.

Dudley, T.R. 1986b. A nickel hyperaccumulating species of *Alyssum* (Cruciferae) from Spain: *A. malacitanum* (Rivas Goday) T.R. Dudley comb. et stat. nov. *Feddes Repert.* 97:139-141.

Duvigneaud, P. 1958. La vegetation du Kaatanga et de ses sols metalliferes. *Bull. Soc. R. Botan. Belg.* 90:127-286.

Duvigneaud, P. 1959. Plantes cobaltophytes dans le Haut Katanga. *Bull. Soc. R. Botan. Belg.* 91:111-134.

Duvigneaud, P. and S. Denaeyer-De Smet. 1963. Cuivre et vegetation au Katanga. *Bull. Soc. R. Bot. Belg.* 96:93-231.

Ebbs, S.B. and L.V. Kochian. 1997. Toxicity of zinc and copper to Brassica species: implications for phytoremediation. *J. Environ. Qual.* 26:776-781.

Ernst, W. 1968. Zur Kenntis der Soziologie und Ökologie der Schwermetallvegetation Groflbritanniens. *Ber. Dtsch. Bot. Ges.* 81:116-124.

Ernst, W.H.O. 1974. *Schwermetallvegetation der Erde.* Fisher, Stuttgart, Germany.

Ernst, W.H.O. 1988. Decontamination of mine sites by plants: an analysis of the efficiency. 305-310. In *Proc Intern. Conf. Environmental Contamination (Venice).* CEP Consultants, Ltd., Edinburgh.

Ernst, W.H.O. 1989. Mine vegetation in Europe. 21-37. In A.J. Shaw (Ed.) *Heavy Metal Tolerance in Plants: Evolutionary Aspects.* CRC Press, Boca Raton, FL.

Gerloff, G.C. 1987. Intact-plant screening for tolerance of nutrient-deficiency stress. *Plant Soil* 99:3-16.

Homer, F.A., R.D. Reeves, R.R. Brooks, and A.J.M. Baker. 1991. Characterization of the nickel-rich extract from the nickel hyperaccumulator *Dichapetalum gelonioides. Phytochemistry* 30:2141-2145.

Ingrouille, M.J. and N. Smirnoff. 1986. *Thlaspi caerulescens* J. & C. Presl. (*T. alpestre* L.) in Britain. *New Phytol.* 102:219-233.

Jackson, P.J., E.J. Roth, P.R. McClure, and C.M. Naranjo. 1984. Selection, isolation, and characterization of cadmium-resistant Datura innoxia suspension cultures. *Plant Physiol.* 75:914-918.

Jaffré, T. 1992. Floristic and ecological diversity of the vegetation on ultramafic rocks in New Caledonia. 101-107. In A.J.M. Baker, J. Proctor, and R.D. Reeves (Eds). *The Vegetation of Ultramafic (Serpentine) Soils.* Intercept, Andover, Hampshire, U.K.

Jaffré, T., R.R. Brooks, J. Lee, and R.D. Reeves. 1976. *Sebertia acuminata*: a hyperaccumulator of nickel from New Caledonia. *Science* 193:579-580.

Jaffré, T., R.D. Reeves, and T. Becquer (Eds.) 1997. *The Ecology of Ultramafic and Metalliferous Areas* (Proc. Second Intern. Conf. on Serpentine Ecology, Noumea, New Caledonia, July 31-Aug. 5, 1995). ORSTOM, New Caledonia.

Kersten, W.J., R.R. Brooks, R.D. Reeves, and T. Jaffré. 1979. Nickel uptake by New Caledonian species of *Phyllanthus. Taxon* 28:529-534.

Kobayashi, J. 1978. Pollution by cadmium and the itai-itai disease in Japan. 199-260. In F.W. Oehme (Ed.) *Toxicity of Heavy Metals in the Environment.* Marcel Dekker, New York.

Kurz, H., R. Schulz, and V. Römheld. 1997. Phytoremediation of thallium and cadmium from contaminated soils — possibilities and limitations. 120-132. In *Proc. Intern. Seminar on Use of Plants for Environmental Remediation* (Sept. 20, 1997, Tokyo, Japan). Council for Promotion of Utilization of Organic Materials, Tokyo.

Lasat, M.M., M. Fuhrman, S.D. Ebbs, J.E. Cornish, and L.V. Kochian. 1998. Phytoextraction of radiocesium-contaminated soil: evaluation of cesium-137 bioaccumulation in the shoots of three plant species. *J. Environ. Qual.* 27:165-169.

Lee, D.-Y. and Z.-S. Chen. 1992. Plants for cadmium polluted soils in Northern Taiwan. 161-170. In D.C. Adriano, Z.-S. Chen, and S.-S. Yang (Eds.) *Biogeochemistry of Trace Elements. Sci. Technol. Lett.,* London, U.K.

Lee, J., R.D. Reeves, R.R. Brooks, and T. Jaffré. 1977. Isolation and identification of a citrato-complex of nickel from nickel-accumulating plants. *Phytochemistry* 16:1503-1505.

Lee, J., R.D. Reeves, R.R. Brooks, and T. Jaffré. 1978. The relation between nickel and citric acid in some nickel-accumulating plants. *Phytochemistry* 17:1033-1035.

Li, Y.M., R.L. Chaney, F.A. Homer, J.S. Angle, and A.J.M. Baker. 1995. *Thlaspi caerulescens* requires over 103 higher Zn^{2+} activity than other plant species. *Agron. Abstr.* 1995:261.

Li, Y.M., R.L. Chaney, K.Y. Chen, B.A. Kerschner, J.S. Angle, and A.J.M. Baker. 1997. Zinc and cadmium uptake of hyperaccumulator *Thlaspi caerulescens* and four turf grass species. *Agron. Abstr.* 1997:38.

Ma, Q.Y., S.J. Traina, T.J. Logan, and J.A. Ryan. 1993. *In situ* lead immobilization by apatite. *Environ. Sci. Technol.* 27:1803-1810.

MacDougall, R., F. Zoepfl, and R.L. Chaney. 1997. Phytoremediation and bioenergy — a new, dual-use opportunity for industrial crops. *United Bioenergy Commercialization Assoc. Bull.* 4(2):1-4.

Mádico, J., C. Poschenreider, M.D. Vázquez, and J. Barceló. 1992. Effects of high zinc and cadmium concentrations on the metallophyte *Thlaspi caerulescens* J. et C. Presl. (Brassicaceae). *Suelo Planta* 2:495-504.

Mayluga, D.P. 1964. Biogeochemical Methods of Prospecting. Acad. Sci. Press, Moscow, 1963; Translated Consultants Bureau, NY, 205.

McLaughlin, M.J., L.T. Palmer, K.G. Tiller, T.W. Beech, and M.K. Smart. 1994. Increasing soil salinity causes elevated cadmium concentrations in field-grown potato tubers. *J. Environ. Qual.* 23:1013-1018.

Mikkelsen, R.L., G.H. Haghnia, A.L. Page, and F.T. Bingham. 1988a. Influence of selenium, salinity, and boron on alfalfa tissue composition and yield. *J. Environ. Qual.* 17:85-88.

Mikkelsen, R.L., A.L. Page, and G.H. Haghnia. 1988b. Effect of salinity and its composition on the accumulation of selenium by alfalfa. *Plant Soil* 107:63-67.

Minguzzi, C. and O. Vergnano. 1948. Il contenuto di nichel nelle ceneri di *Alyssum bertolonii* (Nickel content of the ash of Alyssum bertolonii (in Italian). *Atti Soc. Toscana Sci. Nat. Pisa, Mem. Ser. A* 55:49-74 (*Chem. Abstr.* 44:9003h).

Morrison, R.S., R.R. Brooks, and R.D. Reeves. 1980. Nickel uptake by *Alyssum* species. *Plant Sci. Lett.* 17:451-457.

Nogawa, K., R. Honda, T. Kido, I. Tsuritani, and Y. Yamada. 1987. Limits to protect people eating cadmium in rice, based on epidemiological studies. *Trace Subst. Environ. Health* 21:431-439.

Ohlendorf, H.M., J.E. Oldfield, M.K. Sarka, and T.W. Aldrich. 1986. Embryonic mortality and abnormalities of aquatic birds: apparent impacts by selenium from irrigation drain water. *Sci. Total Environ.* 52:49-63.

Parker, D.R. and A.L. Page. 1994. Vegetation management strategies for remediation of selenium contaminated soils. 327-342. In W.T. Frankenberger, Jr. and S. Benson (Eds.) *Selenium in the Environment.* Marcel Dekker, New York.

Parker, D.R., A.L. Page, and D.N. Thomason. 1991. Salinity and boron tolerances of candidate plants for the removal of selenium from soils. *J. Environ. Qual.* 20:157-164.

Parker, D.R., R.L. Chaney, and W.A. Norvell. 1995. Equilibrium computer models: applications to plant nutrition research. 163-200. In R.H. Loeppert, A.P. Schwab, and S. Goldberg (Eds.). *Chemical Equilibrium and Reaction Models.* Soil Science Society of America Special Publ. No. 42. Soil Sci. Soc. Am./Am. Soc. Agron., Madison, WI.

Pollard, A.J. and A.J.M. Baker. 1997. Deterrence of herbivory by zinc hyperaccumulation in *Thlaspi caerulescens* (Brassicaceae). *New Phytol.* 135:655-658.

Reeves, R.D. 1988. Nickel and zinc accumulation by species of *Thlaspi* L., *Cochlearia* L., and other genera of the Brassicaceae. *Taxon* 37:309-318.

Reeves, R.D. 1992. The hyperaccumulation of nickel by serpentine plants. 253-277. In A.J.M. Baker, J. Proctor, and R.D. Reeves (Eds). *The Vegetation of Ultramafic (Serpentine) Soils.* Intercept Ltd., Andover, Hampshire, U.K.

Reeves, R.D. and R.R. Brooks. 1983a. Hyperaccumulation of lead and zinc by two metallophytes from mining areas of Central Europe. *Environ. Pollut.* A31:277-285.

Reeves, R.D. and R.R. Brooks. 1983b. European species of *Thlaspi* L. (Cruciferae) as indicators of nickel and zinc. *J. Geochem. Explor.* 18:275-283.

Reeves, R.D., R.R. Brooks, and T.R. Dudley. 1983. Uptake of nickel by species of *Alyssum, Bornmuellera* and other genera of Old World Tribus Alysseae. *Taxon* 32:184-192.

Robinson, B.H., A. Chiarucci, R.R. Brooks, D. Petit, J.H. Kirkman, P.E.H. Gregg, and V. De Dominicis. 1997. The nickel hyperaccumulator plant *Alyssum bertolonii* as a potential agent for phytoremediation and phytomining of nickel. *J. Geochem. Explor.* 59:75-86.

Rosenfeld, I. and O.A. Beath. 1964. *Selenium: Geobotany, Biochemistry, Toxicity and Nutrition.* Academic Press, New York.

Rugh, C.L., H.D. Wilde, N.M. Stack, D.M. Thompson, A.O. Summers, and R.B. Meagher. 1996. Mercuric ion reduction and resistance in transgenic *Arabidopsis thaliana* plants expressing a modified bacterial merA gene. *Proc. Natl. Acad. Sci. USA* 93:3182-3187.

Salt, D.E., M. Blaylock, P.B.A. Kumar, V. Dushenkov, B.D. Ensley, I. Chet, and I. Raskin. 1995. Phytoremediation: a novel strategy for the removal of toxic metals from the environment using plants. *Bio/Technology* 13:468-474.

Shen, Z.G., F.J. Zhao, and S.P. McGrath. 1997. Uptake and transport of zinc in the hyperaccumulator *Thlaspi caerulescens* and the non-hyperaccumulator *Thlaspi ochroleucum.* *Plant Cell Environ.* 20:898-906.

Terry, N. and A.M. Zayed. 1994. Selenium volatilization in plants. 343-367. In W.T. Frankenberger, Jr., and S. Benson (Eds.) *Selenium in the Environment.* Marcel Dekker, New York.

Terry, N., C. Carlson, T.K. Raab, and A.M. Zayed. 1992. Rates of selenium volatilization among crop species. *J. Environ. Qual.* 21:341-344.

Tolrá, R.P., C. Poschenreider, and J. Barceló. 1996a. Zinc hyperaccumulation in *Thlaspi caerulescens.* I. Influence on growth and mineral nutrition. *J. Plant Nutr.* 19:1531-1540.

Tolrá, R.P., C. Poschenreider, and J. Barceló. 1996b. Zinc hyperaccumulation in *Thlaspi caerulescens.* II. Influence on organic acids. *J. Plant Nutr.* 19:1541-1550.

Vázquez, M.D., J. Barceló, C. Poschenreider, J. Mádico, P. Hatton, A.J.M. Baker, and G.H. Cope. 1992. Localization of zinc and cadmium in *Thlaspi caerulescens* (Brassicaceae), a metallophyte than hyperaccumulate both metals. *J. Plant Physiol.* 140:350-355.

Vázquez, M.D., C. Poschenreider, J. Barceló, A.J.M. Baker, P. Hatton, and G.H. Cope. 1994. Compartmentation of zinc in roots and leaves of the zinc hyperaccumulator *Thlaspi caerulescens* J&C Presl. *Bot. Acta* 107:243-250.

Weitzman, M., A. Aschengrau, D. Bellinger, R. Jones, J.S. Hamlin, and A. Beiser. 1993. Lead-contaminated soil abatement and urban children's blood lead levels. *J. Am. Med. Assoc.* 269:1647-1654.

Wild, H. 1970. Geobotanical anomalies in Rhodesia. 3. The vegetation of nickel-bearing soils. *Kirkia* 7(Suppl.):1-62.

Yang, G., S. Wang, R. Zhou, and S. Sun. 1983. Endemic selenium intoxication of humans in China. *Am. J. Clin. Nutr.* 37:872-881.

Zayed, A.M. and N. Terry. 1994. Selenium volatilization in roots and shoots: effects of shoot removal and sulfate level. *J. Plant Physiol.* 143:8-14.

Zhang, P., J.A. Ryan, and L.T. Bryndzia. 1997. Pyromorphite formation from goethite adsorbed lead. *Environ. Sci. Technol.* 31:2673-2678.

8 Physiology of Zn Hyperaccumulation in *Thlaspi caerulescens*

Mitch M. Lasat and Leon V. Kochian

CONTENTS

INTRODUCTION

Thlaspi caerulescens J&C Presl has been reported to have a great potential for the extraction of Zn from metalliferous soils (Reeves and Brooks, 1983; Brown et al., 1994; Brown et al., 1995a). The identification of this and several other metal hyper-accumulator plant species (Brooks et al., 1977) demonstrates that the genetic potential exists for successful phytoremediation of contaminated soils. In support of this, Brown et al. (1995b) reported that *T. caerulescens* grown in hydroponic medium accumulated more than 25,000 μg Zn g^{-1} in shoots before any shoot biomass reduction occurred. Despite its ability to accumulate high levels of zinc and other metals in the shoot, the use of *T. caerulescens* for commercial scale soil remediation is severely limited by its small size and slow growth (Ebbs et al., 1997). The transfer of zinc-hyperaccumulating properties from *T. caerulescens* into high biomass producing plants has been previously suggested as a potential avenue for making phytoremediation a commercial technology (Brown et al., 1995a). Progress in this area, however, is hindered by a lack of understanding of the fundamental mechanisms involved in zinc transport and accumulation in roots and shoots.

The physiology of Zn accumulation in hyperaccumulator plants is not well understood. Several reports have dealt with general aspects of plant Zn uptake and translocation (for reviews, see Kochian, 1991, 1993). Studies on the concentration-dependent kinetics of root Zn absorption showed that Zn uptake follows Michaelis-Menten kinetics with apparent K_m values of 3 and 1.5 μM for barley (*Hordeum distichon* L.) and maize (*Zea mays* L.), respectively (Veltrup, 1978; Mullins and

Sommers, 1986). This suggests that Zn transport into the cytosol is via a protein-mediated transport system with a fairly high affinity for zinc. In addition to uptake into roots, another important characteristic of the hyperaccumulator *T. caerulescens* is its ability to transport zinc from the root and to accumulate it to a high level in the shoot. Currently, however, there is little basic information regarding mechanisms of Zn transport and accumulation in the shoot.

In this study, we employed radiotracer flux techniques to characterize root $^{65}Zn^{2+}$ fluxes and compartmentation, as well as Zn translocation to the shoot in hydroponically grown seedlings of *T. caerulescens* and a related species, *T. arvense*, which is not a Zn hyperaccumulator.

ZINC ACCUMULATION IN ROOTS AND SHOOTS

To investigate the physiology of Zn hyperaccumulation in *T. caerulescens*, seedlings of both *Thlaspi* species were grown hydroponically. Seeds of *T. arvense* and *T. caerulescens* were surfaced sterilized in 0.5% NaOCl and immersed in drops of 0.7% (w/v) agarose placed on a 1-mm nylon mesh. The mesh was floated on a nutrient solution containing macronutrients (mM): Ca^{2+}, 0.8; K^+, 1.2; Mg^{2+}, 0.2; NH_4^+, 0.1; NO_3^-, 2.0; PO_4^{3-}, 0.1; SO_4^{2-}, 0.2, and micronutrients (μm): BO_3^{2-}, 12.5; Cl^-, 50; Cu^{2+}, 0.5; Fe^{3+}-EDDHA, 10.0; MoO_4^{2-}, 0.1; Mn^{2+}, 1.0; Ni^{2+}, 0.1; Zn^{2+}, 1.0. The solution was buffered at pH 5.5 with 1 mM Mes-Tris. The mesh was covered with black polyethylene beads to minimize illumination of the growth solution. Seedlings were grown in a growth chamber at 24°C/15°C (light/dark: 16:8 h) under light intensity of 300 μmol photons m^{-2} s^{-1}. In order to use *T. caerulescens* and *T. arvense* seedlings at a similar developmental stage, experiments were conducted with *T. caerulescens* seedlings that were approximately 10 days older than *T. arvense*, as the growth rate of *T. caerulescens* seedlings was considerably slower than *T. arvense*.

The 22-day-old seedlings grown in normal nutrient medium (containing 1 μm Zn^{2+}) were transferred to 5-l black plastic tubs filled with nutrient solution containing 1, 25, 50, or 100 μm Zn^{2+}. After 10 days, plants were harvested and roots were desorbed (to remove Zn associated with the root apoplasm) for 15 min in a solution containing 5 mM Mes-Tris buffer (pH 6.0), 5 mM $CaCl_2$, and 100 μm $CuCl_2$, and separated from shoots. Shoots and roots were oven-dried at 65°C for 3 days. Dried shoot and root material was ground, and a 0.2-g aliquot was digested overnight in 1 ml of HNO_3 at 120°C. Subsequently, samples were dissolved in 0.75 ml of $HNO_3:HClO_4$ (1:1, v/v) and incubated at 220°C until dry. Samples were redissolved in 10 ml of 5% HNO_3 and analyzed for elemental composition with a trace analyzer emission spectrometer (model ICAP 61E, Thermo-Jarrell, Waltham, MA).

After 10 days of growth in solutions containing different Zn^{2+} concentrations (1, 25, 50, or 100 μm), *T. arvense* accumulated more Zn^{2+} in roots, whereas most of the Zn was translocated to the shoots of *T. caerulescens* (Table 8.1). From the growth solution containing 100 μm Zn^{2+}, *T. caerulescens* shoots accumulated 4.6-fold more Zn compared with *T. arvense*, whereas 1.9-fold more Zn accumulated in roots of *T. arvense* than in *T. caerulescens*. Accumulation of elevated zinc levels in shoots was not toxic to *T. caerulescens*. Conversely, *T. arvense* grown in solutions

TABLE 8.1

Zinc Accumulation in *T. arvense* and *T. caerulescens* Seedlings Exposed for 10 Days to Different Zn^{2+} Levels in the Growth Solution

	T. arvense		*T. caerulescens*	
Zn Conc.	Roots	Shoots	Roots	Shoots
(μm)	($\mu mol\ Zn^{2+}$ g dry wt^{-1})		($\mu mol\ Zn^{2+}$ g dry wt^{-1})	
1	—	3	—	10
25	54 ± 6	12 ± 2	19 ± 1	20 ± 3
50	54 ± 8	14 ± 2	25 ± 3	40 ± 5
100	90 ± 11	15 ± 3	48 ± 8	70 ± 3

Note: Tubs containing 22-day-old seedlings grown on nutrient media containing 1 μm Zn were refilled with nutrient solution containing 1, 25, 50, or 100 μm Zn After an additional 10 days of growth, zinc concentration in roots and shoots was determined by ICP-ES The results are presented as means ±SE (n = 8-27).

containing >1 μm zinc showed severe chlorosis (data not shown), despite reduced accumulation of zinc in leaves.

Significant differences between zinc accumulation in roots and shoots of the two *Thlaspi* species were also obtained in time-course studies conducted with $^{65}Zn^{2+}$. One day prior to the uptake experiment, four seedlings of each *Thlaspi* species were bundled together and immersed with roots in an aerated-pretreatment solution consisting of 2 m*M* Mes-Tris (pH 6.0) and 0.5 m*M* $CaCl_2$. Bundles of *Thlaspi* seedlings were immersed with roots in 1 l of aerated uptake solution containing 10 μm $ZnCl_2$ labeled with $^{65}Zn^{2+}$ (1.4 μCi). Following different uptake periods, one bundle of each species was harvested and roots desorbed for 15 min and separated from shoots. Then, both roots and shoots were weighed and ^{65}Zn accumulation measured using a Packard Model 5530 Gamma Counter.

After 3 hours of incubation in the radioactive uptake solution, twice as much ^{65}Zn accumulated in *T. caerulescens* roots compared with *T. arvense* (Figure 8.1 inset). However, after 24 h, comparable amounts of radiozinc accumulated in the roots of the two species, and after 96 h, 29% more zinc accumulated in roots of *T. arvense* compared to *T. caerulescens* (Figure 8.1A). In contrast, zinc translocation to the shoot was 10-fold greater in *T. caerulescens* (Figure 8.1B).

INVESTIGATION OF PLASMA MEMBRANE AND TONOPLAST TRANSPORT IN ROOTS

To investigate zinc influx across the plasma membrane of root cells, the concentration-dependent kinetics of short-term Zn uptake in roots were studied. Roots of four

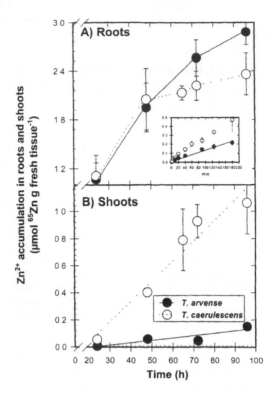

FIGURE 8.1 Time-course of $^{65}Zn^{2+}$ accumulation in (A) roots and (B) shoots of the two *Thlaspi* species. Roots of intact seedlings were immersed in an uptake solution containing 10 µm $^{65}Zn^{2+}$. After incubation periods shown, roots were desorbed, excised, blotted, and both roots and shoots weighed and gamma activity measured.

intact *T. caerulescens* and *T. arvense* seedlings were immersed in individual 80-ml Plexiglas wells filled with pretreatment solution. One minute before the addition of $^{65}ZnCl_2$ (0.08 µCi), zinc was added as $ZnCl_2$ to each uptake well to yield final concentrations between 0.5 and 100 µm. After a 20-min uptake period, the radioactive uptake solution was vacuum-withdrawn and roots desorbed for 15 min. After desorption, seedlings were harvested, and roots excised, blotted, weighed, and ^{65}Zn measured.

Kinetics for root-$^{65}Zn^{2+}$ influx were characterized by a nonsaturating curve which could be graphically resolved into a saturable and a linear component (data not shown). The linear component was shown to be zinc remaining in the root apoplasm following desorption. The saturable component which displayed Michaelis-Menten kinetics was due to zinc influx across the root cell plasma membrane (Figure 8.2). The apparent K_m values for the saturable components were 8 and 6 µm for *T. caerulescens* and *T. arvense*, respectively, while V_{max} values for Zn^{2+} influx were 4.5 times greater in *T. caerulescens* (270 vs. 60 nmol Zn^{2+} g^{-1} fresh weight h^{-1}).

To investigate subcellular localization of absorbed ^{65}Zn, we conducted an efflux (compartmentation) study with roots of the two *Thlaspi* species. Seedlings of *T. arvense* and *T. caerulescens* were immersed with roots in uptake wells filled with

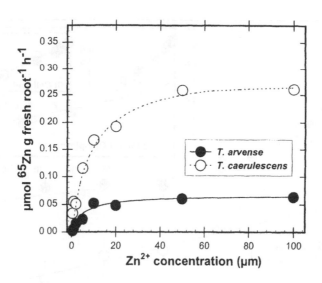

FIGURE 8.2 Concentration-dependent kinetics of $^{65}Zn^{2+}$ influx into roots of the two *Thlaspi* species. Roots were immersed in an uptake solution containing $^{65}Zn^{2+}$ at concentrations shown. After 20 min, roots were desorbed, excised, blotted, weighed, and gamma activity measured.

80 ml of aerated uptake solution containing 2 mM Mes-Tris (pH 6.0), 0.5 mM CaCl$_2$, and 20 μm ZnCl$_2$ labeled with $^{65}Zn^{2+}$ (1.4 μCi l^{-1}). After 24 h, the radioactive uptake solution was removed, roots were quickly rinsed with deionized water, and uptake wells were refilled with an efflux solution containing 2 mM Mes-Tris (pH 6.0), 0.5 mM CaCl$_2$, and 20 μm unlabeled ZnCl$_2$. At various time intervals, a 1-ml aliquot of the solution was collected and ^{65}Zn efflux from the roots was measured. Following the removal of each aliquot, wells were drained and refilled with fresh efflux solution. After 6 h, roots were excised, blotted, weighed, and ^{65}Zn measured. Efflux of $^{65}Zn^{2+}$ from roots was determined by monitoring the appearance of radiozinc in efflux solution over time.

Efflux curves could be dissected into three linear components which are traditionally interpreted to represent efflux from three cellular compartments in series. The linear efflux phase for time points between 180 and 360 min was considered to represent ^{65}Zn efflux from the vacuole (Figure 8.3A). Subtraction of this component from total efflux data yielded a curve that was interpreted to represent ^{65}Zn efflux from the cytoplasm and cell wall (Figure 8.3B). The linear phase for this efflux curve observed for Zn^{2+} efflux between 30 and 60 min was interpreted as Zn efflux from the cytoplasm. Efflux data from the cell wall (Figure 8.3C) was obtained after subtracting the cytoplasmic efflux (the straight line in Figure 8.3B) from the data points plotted in Figue 8.3B. The two major pieces of information obtained from this analysis were (1) estimates of $^{65}Zn^{2+}$ accumulation in the respective subcellular compartment at the end of the radioactive uptake period determined from the intersection of extrapolated linear components with the y-axis and (2) estimates of $^{65}Zn^{2+}$ efflux rates ($t_{1/2}$, half-time) calculated from the slopes of the linear, first-order kinetic components. Based on this analysis, similar amounts of ^{65}Zn were

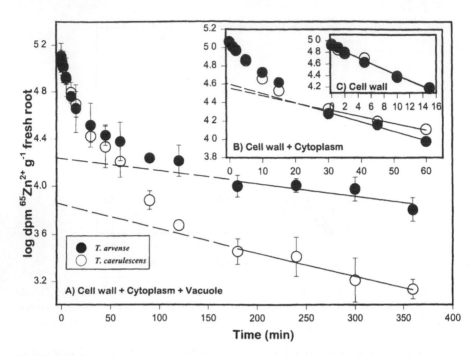

FIGURE 8.3 Efflux (compartmental) analysis of root ^{65}Zn accumulation. After a 24-h incubation period in an uptake solution containing 20 μm ^{65}Zn^{2+}, radiozinc efflux from roots of the two *Thlaspi* species was monitored over a subsequent 6-h period. Dashed lines represent regressions of linear portion of each curve extrapolated to the y-axis. The curve shown in B was derived by subtracting the linear component in A from the data points in A. The curve in C was similarly derived from the curve in B.

accumulated in the cell wall and cytoplasm of *T. arvense* and *T. caerulescens* roots (Table 8.2). However, approximately 2.4-fold more ^{65}Zn was accumulated in the root vacuole of *T. arvense* than in *T. caerulescens*. Furthermore, the rates of ^{65}Zn efflux from the cell wall and cytoplasm were similar in *T. arvense* and *T. caerulescens* roots, yet the half-time for ^{65}Zn vacuolar efflux in *T. caerulescens* was approximately 72% smaller than in *T. arvense*, indicating that Zn accumulated in root-cell vacuoles of *T. arvense* was transported back out of the vacuole much more slowly than in *T. caerulescens* (Table 8.2).

ZINC UPTAKE IN LEAVES

To investigate the ability of the *Thlaspi* species to accumulate zinc in leaf cells, we conducted uptake studies with leaf sections of the two species. Leaves of *T. arvense* and *T. caerulescens* seedlings were gently abraded with carborundum powder to remove the cuticle and to facilitate infiltration of the radioactive uptake solution into the tissue. Leaves were subsequently washed with water to remove any adhering powder. Leaf blades were cut into 10- to 20-mm^2 sections and subsequently immersed in an aerated uptake solution containing 1 mM ^{65}ZnCl$_2$ (1.4 μCi l^{-1}), 2

TABLE 8.2

Intracellular Zn Compartmentation and Zinc Efflux Rates ($t_{1/2}$) from Roots of *T. arvense* and *T. caerulescens*

	T. arvense			*T. caerulescens*		
	Root (dpm)	$^{65}Zn^{2+}$ (%)	$t_{1/2}$ (min)	Root (dpm)	$^{65}Zn^{2+}$ (%)	$t_{1/2}$ (min)
Cell wall	87,100	61	6	87,100	66	6
Cytoplasm	39,800	27	30	38,300	29	38
Vacuole	17,000	12	260	7100	5	150

Note: Estimates of ^{65}Zn accumulation in the subcellular compartments were determined from the intersection of extrapolated linear components shown in Figure 8.3 with the y-axis Estimates of ^{65}Zn efflux rates ($t_{1/2}$, half-time) were calculated from the slope of the linear first-order kinetic components shown in Figure 8 3

mM Mes-Tris (pH 6.0), and 0.5 mM CaCl$_2$. Following incubation periods of up to 48 h, $^{65}Zn^{2+}$ uptake was terminated by removing the leaf sections and immersing them in an aerated desorption solution containing 100 μm ZnCl$_2$, 5 mM CaCl$_2$, and 5 mM Mes-Tris (pH 6.0). After 15 min of desorption, leaf sections were harvested, blotted, weighed, and ^{65}Zn measured.

From an uptake solution containing 1 mM $^{65}Zn^{2+}$, approximately 2.5-fold more zinc accumulated in leaf sections of *T. caerulescens* at the end of a 48-h uptake period (Figure 8.4).

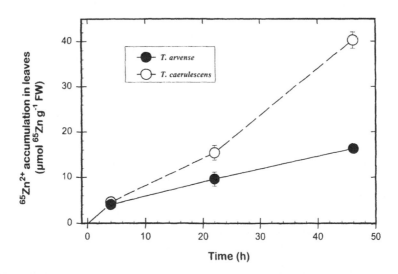

FIGURE 8.4 Time course of $^{65}Zn^{2+}$ accumulation in leaves of the two *Thlaspi* species. Leaf sections were immersed in an uptake solution containing 1 mM $^{65}Zn^{2+}$. After exposures of up to 48 h, leaf sections were desorbed, blotted, weighed, and gamma activity measured.

MODEL FOR Zn HYPERACCUMULATION IN
T. caerulescens

Thlaspi caerulescens, which can accumulate zinc in shoots to concentrations exceeding 4% dry weight (Chaney et al., 1993), represents an intriguing model system for studying physiological mechanisms of metal hyperaccumulation. In plants grown for 10 days in a nutrient solution supplemented with Zn at concentration as high as 100 μm, 5-fold more Zn accumulated in shoots of *T. caerulescens* than in *T. arvense*. Moreover, zinc accumulation in *T. caerulescens* leaves was not associated with any symptom of toxicity, whereas *T. arvense* leaves showed severe chlorosis after growth in solution containing 25, 50, or 100 μm Zn. These results indicate that *T. caerulescens* has the capacity not only to hyperaccumulate zinc in the leaves, but also to tolerate a high level of the heavy metal, which is toxic to the nonaccumulator species *T. arvense*. Interestingly, however, after 10 days of growth, more zinc accumulated in the roots of *T. arvense* compared to *T. caerulescens*. To investigate the kinetics of Zn accumulation in shoots and roots, we conducted a time-course uptake study with the two *Thlaspi* species. Following root exposure to radioactive uptake solution, translocation of ^{65}Zn to the shoot was detectable after 24 h. Accumulation of ^{65}Zn in the shoot was linear for up to 96 h. At all time points, more ^{65}Zn was accumulated in shoots of *T. caerulescens* than in shoots of *T. arvense*. In roots, however, Zn accumulation followed a different pattern. In the first 3 h, approximately twice as much radiozinc accumulated in the root cells of *T. caerulescens* compared to *T. arvense*. After 24 h, similar amounts of ^{65}Zn accumulated in the roots of the two species. Furthermore, after 72 h of accumulation, more radiozinc was measured in the root of *T. arvense*. The observation that initially more ^{65}Zn accumulated in roots of *T. caerulescens* suggests that this species maintains a greater unidirectional Zn influx than does *T. arvense*.

To characterize zinc influx across the root cell plasma membrane (the entry point for Zn accumulation in the plant), we conducted a concentration-dependent uptake study. The kinetics of unidirectional zinc influx into root cells of *T. caerulescens* and *T. arvense* were dominated by a single saturable component (Figure 8.2). The similar K_m values suggest that in both *Thlaspi* species transport across the root cell plasma membrane is mediated by proteins with similar Zn^{2+} affinities. The significantly larger V_{max} value for Zn^{2+} influx into roots of *T. caerulescens* indicates that there might be a higher expression of Zn^{2+} transporters in *T. caerulescens* roots, resulting in a deployment of more transporters in the root-cell plasma membrane. These results indicate that an important feature of Zn hyperaccumulation in *T. caerulescens* is enhanced zinc influx into the root symplasm. The observation that more zinc accumulated in roots of *T. arvense* after 72 h indicates that, over time, a greater proportion of the absorbed zinc is translocated from the root to the shoot of *T. caerulescens*. In contrast, most of the absorbed zinc remains sequestered in *T. arvense* root with little metal translocation to the shoot.

To investigate the sequestration of zinc in root cells of the two *Thlaspi* species in more detail, we conducted a compartmentation analysis (Figure 8.3). It is possible that reduced Zn uptake into the root symplasm along with lower rates of translocation to the shoot in *T. arvense* may be due to enhanced binding to the root cell walls;

storage of zinc in the root apoplasm has been previously suggested by Peterson (1969). However, in the current study, we found that similar amounts of zinc were accumulated in the root cell walls of the two *Thlaspi* species (Table 8.2). Comparable amounts of Zn were also accumulated in the root-cell cytoplasm of *T. arvense* and *T. caerulescens*, and we found similar rates of Zn^{2+} efflux across the root-cell plasma membrane. Although like amounts of zinc were accumulated in the cell walls and cytoplasm of the two *Thlaspi* species, 2.4-fold more zinc was accumulated in the vacuole of *T. arvense*. Furthermore, the rate of vacuolar efflux was significantly smaller in *T. arvense* (Table 8.2). These results suggest that zinc is sequestered in the vacuole of *T. arvense* and made unavailable for translocation to the shoot. A model for zinc transport and compartmentation in the root cells of *T. arvense* and *T. caerulescens* is proposed in Figure 8.5.

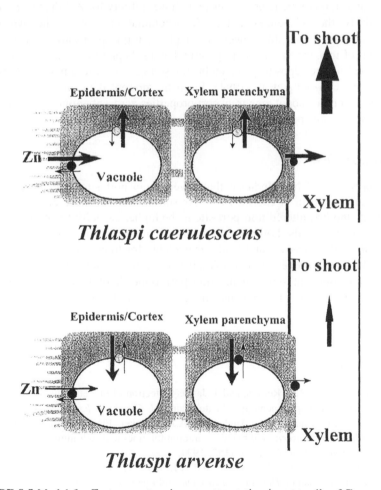

FIGURE 8.5 Model for Zn transport and compartmentation in root cells of *T. arvense* and *T. caerulescens*. The bold arrows denote a large Zn influx or efflux, while the thin arrows indicate a small Zn flux.

Thlaspi caerulescens accumulated about fivefold higher concentrations of Zn in the xylem sap compared with *T. arvense* (data not shown). These findings indicate that the leaf cells in *T. caerulescens* are bathed with xylem solution containing high levels of zinc. Thus, we hypothesized that another important transport site involved in zinc hyperaccumulation in *T. caerulescens* could be reabsorption of xylem-borne Zn into leaf cells with subsequent storage in the vacuole. To test this hypothesis, we investigated the time-dependent kinetics of ^{65}Zn accumulation in leaf sections of the two *Thlaspi* species. At the end of a 48-h uptake period, from a solution containing 1 mM ^{65}ZnCl$_2$, a dramatic (2.5-fold) increase in Zn accumulation was observed in *T. caerulescens* leaves compared to *T. arvense* (Figure 8.4). One mechanism which could account for the enhanced Zn accumulation in *T. caerulescens* leaves is a stimulated Zn^{2+} transporter operating across the leaf-cell plasma membrane with different kinetic properties (e.g., lower affinity for Zn) than in *T. arvense*. Alternatively, the differences in leaf Zn accumulation between the two *Thlaspi* species could be due to differences in tonoplast Zn transport into the leaf vacuole. In support of this, the vacuoles of epidermal and subepidermal leaf cells have been documented as the repository organelle for storage of large amounts of Zn in *T. caerulescens* (Vázquez et al., 1994). From the present data, however, we cannot make any inference about zinc transport properties across the tonoplast in leaf cells of the two *Thlaspi* species.

CONCLUSIONS

The data presented here indicate that several transport sites are involved in Zn hyperaccumulation in *T. caerulescens*. As summarized in the model presented in Figure 8.5, the first altered transport site is the higher capacity for Zn influx into *T. caerulescens* root cells. Following entry into symplasm, zinc is sequestered in the root vacuole of *T. arvense* and made unavailable for translocation to the shoot. In *T. caerulescens*, however, symplasmic zinc is readily available for loading into the xylem and subsequent long-distance transport to the shoot. Our data also indicate a greater ability for zinc transport into the leaf cells of *T. caerulescens* compared to *T. arvense*.

REFERENCES

Brooks, R.R., J. Lee, R.D. Reeves, and T. Jaffré. Detection of nickelferous rocks by analysis of herbarium specimens of indicator plants. *J. Geochem. Explor.* 7, 49-58,1977.

Brown, S.L., R.L. Chaney, J.S. Angle, and A.J.M. Baker. Phytoremediation potential of *Thlaspi caerulescens* and bladder campion for zinc- and cadmium-contaminated soil. *J. Environ. Qual.* 23, 1151-1157, 1994.

Brown, S.L., R.L. Chaney, J.S. Angle, and A.J.M. Baker. Zinc and cadmium uptake by hyperaccumulator *Thlaspi caerulescens* and metal tolerant *Silene vulgaris* grown on sludge amended soils. *Environ. Sci. Technol.* 29, 1581-1585, 1995a.

Brown, S.L., R.L. Chaney, J.S. Angle, and A.J.M. Baker. Zinc and cadmium uptake by hyperaccumulator *Thlaspi caerulescens* grown in nutrient solution. *Soil Sci. Soc. Am. J.* 59, 125-133, 1995b.

Chaney, R.L. Zinc Phytotoxicity, in *Zinc in Soil and Plants,* Robson, A.D., Ed., Kluwer Academic Publishers, Dordrecht, The Netherlands, 1993.

Ebbs, S.D., M.M. Lasat, D.J. Brady, J.E. Cornish, R. Gordon, and L.V. Kochian. Phytoextraction of cadmium and zinc from a contaminated site. *J. Environ. Qual.* 26, 1424-1430, 1997.

Kochian, L.V. Mechanisms of micronutrient uptake and translocation in plants, in *Micronutrients in Agriculture,* Mortvedt, J.J., Ed., Soil Science Society of America Book Series, Madison, WI, 1991.

Kochian, L.V. Zinc Absorption from hydroponic solutions by plant roots, in *Zinc in Soil and Plants,* Robson, A.D., Ed., Kluwer Academic Publishers, Dordrecht, The Netherlands, 1993.

Mullins, G.L. and L.E. Sommers. Cadmium and zinc influx characteristics by intact corn (*Zea mays* L.) seedlings. *Plant Soil* 96, 153-164, 1986.

Peterson, P.J. The distribution of zinc-65 in *Agrostis tenuis* Sibth. and *A. stolonifera* L. tissues. *J. Exp. Bot.* 20, 863-875, 1969.

Reeves, R.D. and R.R. Brooks. Hyperaccumulation of lead and zinc by two metallophytes from mining areas of central Europe. *Environ. Pollut.* A31, 277-285, 1983.

Vázquez, M.D., C. Poschenrieder, J. Barceló, A.J.M. Baker, P. Hatton, and G.H. Cope. Compartmentation of zinc in roots and leaves of the zinc hyperaccumulator *Thlaspi caerulescens* J&C Presl. *Bot. Acta* 107, 243-250, 1994.

Veltrup, W. Characteristics of zinc uptake by barley roots. *Physiol. Plant.* 42, 190-194, 1978.

9 Metal-Specific Patterns of Tolerance, Uptake, and Transport of Heavy Metals in Hyperaccumulating and Nonhyperaccumulating Metallophytes

Henk Schat, Mercè Llugany, and Roland Bernhard

CONTENTS

INTRODUCTION

The possibility of effective phytoremediation of heavy metal-contaminated soil depends on the availability of plant varieties with high rates of accumulation and tolerance of the metal(s) to be extracted. High levels of tolerance to specific heavy metals are known to occur in wildtype plant populations from heavy metal-enriched substrates such as ore outcrops or mine-waste deposits. A minority of the species concerned, the so-called hyperaccumulators, exhibit extremely high rates of foliar metal accumulation. It may be argued that hyperaccumulators are insufficiently productive in terms of harvestable above-ground biomass to be useful in phytoremediation, although this argument does not seem to apply to all of them. Plant species from metalliferous soil ("metallophytes") possess desirable properties

1-56670-450-2/00/$0.00+$.50
© 2000 by CRC Press LLC

which might be transferable to more productive species by means of genetic engineering or interspecific crossing. Moreover, hyperaccumulating and nonhyperaccumulating metallophytes exhibit highly variable patterns of metal specificity with regard to uptake, root-to-shoot transport, and tolerance, both inter- and intraspecifically. This variation, which could be useful in phytoremediation, has not been fully explored, and the underlying genetic and physiological mechanisms are far from understood. Further studies may be expected to provide insights with possible implications for phytoremediation technology.

This chapter summarizes the present knowledge of the mechanisms and genetics of heavy metal tolerance in hyperaccumulating and nonhyperaccumulating metallophytes. Special attention will be paid to the relationships between tolerance and the rates of metal uptake and metal root-to-shoot transport, as well as to the patterns of metal specificity of these properties. Possible implications for phytoremediation will be discussed.

PLANT ADAPTATION TO METALLIFEROUS SOILS

Strongly heavy metal-enriched substrates are hostile to plant growth. Nevertheless, even extremely metal-toxic soils eventually become colonized. Vegetations on metalliferous soil often remain free of trees and maintain a low species diversity for very long periods. Younger metallophyte vegetations often show a remarkably constant floristic composition within wide geographic ranges, more or less irrespective of the metal composition of the soil. Many pioneer metallophytes are common and widely distributed, both on metalliferous and nonmetalliferous soil ("facultative metallophytes," such as *Agrostis capillaris*, *A. canina*, *A. stolonifera*, *Festuca rubra*, *F. ovina*, *Deschamsia cespitosa*, *D. flexuosa*, *Holcus lanatus*, *Anthoxanthum odoratum*, and *Silene vulgaris* in western and central Europe; Ernst, 1974). Very old, undisturbed heavy metal vegetations in parts of the world that remain free from pleistocene glaciations usually exhibit a much higher floristic diversity and may accommodate considerable numbers of "strict metallophytes," often with very limited geographic distributions ("edaphic endemics;" Brooks, 1987; Brooks et al., 1985).

The colonization of metalliferous substrates involves evolutionary adaptation. Plants of facultative metallophyte populations from metalliferous soil almost invariably exhibit higher levels of metal tolerance than those of normal soil populations. This apparently adaptive variation has been demonstrated to be genetically based (see below). The question of why only a limited number of species become adapted to metalliferous soil has not been definitively answered. When sown on metalliferous substrate, seed collections of nonmetallophytes usually do not produce survivors, whereas those of nonadapted populations of facultative metallophytes often produce survivors with high levels of metal tolerance in considerable frequencies (0.1 to 0.5%; Gartside and McNeilly, 1974). This observation suggests that nonmetallophyte populations are usually devoid of sufficiently tolerant mutants. The reasons for this are not clear, the more so because it appears to be possible to select stable, highly tolerant cell lines from suspension or callus cultures of many nonmetallophytes, suggesting that the appropriate mutability should be present (e.g., Gilissen and Van Staveren, 1986; Jackson et al., 1984). However, the basic "constitutive" level of tolerance seems to be higher in metallophytes than in nonmetallophytes. This could reduce the amount of genetic change necessary to produce the tolerance levels required to survive and reproduce on metalliferous soil. Moreover, it might allow survival and growth on less toxic transitional soil, which would increase the number of individuals that are effectively screened for higher tolerance by natural selection and, consequently, enhance the chances for evolutionary change (Baker and Proctor, 1990).

In general, the transmission genetics of heavy metal tolerance in facultative metallophytes seem to be relatively simple. All the larger data sets available thus far demonstrate that metal tolerance, in so far as it segregates in crosses between plants from metalliferous and normal soil, depends basically on a small number (usually one or two) of major genes: e.g., Cu tolerance in *Mimulus guttatus* (Macnair, 1983); Cu, Zn, and Cd tolerance in *S. vulgaris* (Schat and Ten Bookum, 1992a; Schat et al., 1993; 1996); and arsenate tolerance in *Holcus lanatus* (Macnair et al., 1992). However, particularly in the case of Cu tolerance and arsenate tolerance, these major genes merely control the occurrence of any tolerance, relative to the nontolerant parent plant, rather than its degree. The latter seems to be strongly affected by "modifiers," which could be described as additive hypostatic factors affecting the penetrance in the phenotype of the genes that produce the tolerance themselves (Macnair 1983; Schat and Ten Bookum, 1992a; Macnair et al., 1992; 1993). The degree of intergenic variation among tolerant plants of different geographic origin seems to be low. In *S. vulgaris*, for example, Schat et al. (1996) found only two major gene loci for Cu tolerance, two for Zn tolerance, and one or two for Cd tolerance among a total of four mine populations from Germany and one from Ireland. In all cases, the plants from the Irish population and those from all of the German populations, even though they represented different subspecific taxa (ssp. *maritima* and ssp. *vulgaris*, respectively), appeared to have the tolerant genotype in common at one of the loci concerned. This clearly suggests independent "parallel evolution" in different local ancestral populations. It also suggests that evolution of tolerance might critically depend on the availability of appropriate genetic variation at no more than one or two specific gene loci.

MULTIPLE TOLERANCE AND COTOLERANCE

Metalliferous soils are often enriched in combinations of different heavy metals and, therefore, local metallophyte populations often exhibit combined tolerance to different metals. Such combined tolerance could either rely on less specific mechanisms that pleiotropically confer tolerance to different metals ("cotolerance") or on combinations of independent metal-specific mechanisms ("multiple tolerance"). Many authors have reported tolerance to metals that do not seem to be present at toxic levels in the soil at the population site (e.g., Gregory and Bradshaw, 1965; Allen and Sheppard, 1971; Hogan and Rauser, 1979; Verkleij and Bast-Kramer, 1985; Schat and Ten Bookum, 1992b; Von Frenckell-Insam and Hutchinson, 1993). The occurrence of these apparently nonfunctional tolerances has been interpreted as circumstantial evidence of co-tolerance. However, very little direct genetic evidence is available. Schat and Vooijs (1997a) have demonstrated independent genetic control of Cu, Zn, and Cd tolerance in *S. vulgaris*. In agreement with this, Cu and Zn tolerance (Macnair, 1993) and Cu and Ni tolerance (Tilstone and Macnair, 1997) have been shown to be under independent genetic control in *M. guttatus*. Likewise, Brown and Brinkman (1992) demonstrated independent phenotypic variation of Zn and Pb tolerance in *F. ovina*. On the other hand, several nonfunctional tolerances in *S. vulgaris*, such as those to Ni and Co in zinc-mine populations (Schat and Ten Bookum, 1992b) and to Ag and Hg in copper mine populations, cosegregated consistently with a particular Zn tolerance gene and with both Cu tolerance genes, respectively (Schat and Vooijs, 1997a; H. Schat, unpublished). In so far as they can be compared with their functional analogs in other metallophyte populations, these apparent cotolerances are of a low level. For example, Ni tolerance in serpentine plants, both hyperaccumulating and nonhyperaccumulating, is often much higher (e.g., Krämer et al., 1996) and not associated with any appreciable Zn tolerance (Ernst, 1972; Schat and Vooijs, 1997a). Thus, each heavy metal, when present at a toxic concentration, seems to provoke unique adaptations which are strongly, though not necessarily completely, metal-specific.

MECHANISMS OF TOLERANCE

The physiological and biochemical mechanisms of metal tolerance are poorly understood. In general, tolerance in unadapted and metal-adapted plants might depend on qualitatively different mechanisms. Copper tolerance in the nonmetallophyte *Arabidopsis thaliana*, for example, depends on the degree of Cu-induced expression of the genes encoding the metallothionein MT2 (Murphy and Taiz, 1995; Murphy et al., 1997) but not on phytochelatin (PC) synthesis (Howden and Cobbett, 1992), although Cu activates phytochelatin synthase and binds to PCs *in vivo* (Verkleij et al., 1989). Also, adaptive Cu tolerance in mine populations of *S. vulgaris* does not depend on PC synthesis (Schat and Kalff, 1992; De Vos et al., 1992; Schat and Vooijs, 1997b). However, high-level Cu tolerance can be demonstrated within several minutes upon exposure by measuring net potassium efflux from roots (De Vos et al., 1991), which precludes a decisive role for any Cu-induced MT gene transcription as well.

Normal constitutive Cd tolerance in *Arabidopsis* has been shown to depend on PC synthesis (Howden and Cobbett, 1992; Howden et al., 1995a,b) but, except in PC-deficient mutants, not on MT expression (P. B. Goldsbrough, personal communication). However, studies of Cd-hypersensitive fission yeast mutants have shown that the capacity for PC synthesis as such is not sufficient to produce a normal level of Cd tolerance. The transport of PC-Cd complexes into the vacuoles, which is mediated by an ATP-binding cassette-type transporter (Ortiz et al., 1992), and the stabilization of Cd-PC complexes by incorporation of acid-labile sulfide, which presumably takes place in the vacuole (Ortiz et al., 1992; Speiser et al., 1992b), are also required for this. There are reasons to believe that these processes may be equally important in higher plants (Vögeli-Lange and Wagner, 1989; Speiser et al.,

1992a). Thus, adaptive high-level Cd tolerance might be conceived to depend on (combined) enhancements of PC synthesis itself, vacuolar compartmentation of PC-Cd complexes, or increased incorporation of acid-labile sulfide into Cd-PC complexes.

Normal Cd tolerance in unadapted *S. vulgaris* appeared to depend on PC synthesis. This is because treatment with buthionine sulfoximine (BSO), which inhibits the synthesis of the PC precursor γ-glutamylcysteine, caused hypersensitivity to Cd (De Knecht et al., 1992). However, Cd-adapted mine plants of this species exhibited a much lower PC synthesis under Cd exposure, and their Cd tolerance was not reduced by BSO treatment (De Knecht et al., 1992). The former must be due to a lower *in vivo* PC-synthase activity, because the PC breakdown rates were the same in nontolerant and tolerant plants (De Knecht et al., 1995). The maximum Cd-inducible activities, the K_m values for glutathione, and the activation constants for Cd (the Cd concentrations that induced 50% activation) of PC synthase in crude root protein fractions were also the same in both plant types. There were neither differences in the rates of Cd uptake nor the *in vivo* availability of reduced glutathion under Cd exposure (De Knecht et al., 1992; 1995). The *in vivo* acid-labile sulfide contents and the chain length distributions of the PC-Cd complexes were identical as well (De Knecht et al., 1994). Thus, adaptive Cd tolerance in *S. vulgaris* must be due to an alternative cellular sequestration system, which apparently reduces the availability of cellular Cd for PC-synthase activation *in vivo*.

The mechanisms of normal constitutive Zn tolerance have not been elucidated thus far. PC synthesis appeared to be unimportant in *Arabidopsis* (Howden and Cobbett, 1992) and unadapted *S. vulgaris* (H. Harmens, unpublished), although Zn is certainly capable of activating PC synthase *in vivo* (Grill et al., 1987; Harmens et al., 1993a). Binding of Zn to PCs *in vivo* has never been demonstrated, however. There are some indications that MTs might be involved (Robinson et al., 1996). Adaptive high-level Zn tolerance in *S. vulgaris* is associated with decreased PC synthesis (Harmens et al., 1993a) but constitutively increased levels of malate and citrate in the leaves, such as in Zn-tolerant populations of many other species (Ernst, 1975; Mathys, 1977). Increased organic acid concentrations as such cannot explain increased tolerance, however (Harmens et al., 1994). Recent results of *in vitro* experiments showed a large tolerance-correlated difference in Mg-ATP-dependent Zn uptake by tonoplast vesicles isolated from nontolerant and Zn-tolerant *S. vulgaris* (Verkleij et al., 1998), suggesting that the capacity or affinity of Zn-transporting systems at the tonoplast may be decisive.

Except for the metalloid arsenic (see below), the mechanisms of constitutive and adaptive tolerances to other heavy metals are largely unknown.

HYPERACCUMULATION

The present knowledge of adaptive metal tolerance in plants, such as summarized above, is almost completely based on research on nonhyperaccumulating metallophytes. There is much less information on the patterns of metal specificity, the genetics, and the mechanisms of metal tolerance and metal accumulation in hyperaccumulating species. This might relate to the fact that hyperaccumulators are more often strictly bound to metalliferous soil types, which limits the possibilities for genetic analyses and relevant intraspecific physiological and biochemical comparisons. Normal soil populations of the Zn hyperaccumulating facultative metallophyte *Thlaspi caerulescens* seem to have much lower degrees of Zn tolerance than those of "calamine" soils (Ingrouille and Smirnoff, 1986), suggesting that hyperaccumulators are not inherently tolerant to high soil metal concentrations, not even with regard to the (preferentially) hyperaccumulated metal(s). As yet unpublished studies in the authors' laboratory confirmed this point of view (Table 9.1). Of the three *T. caerulescens* populations compared, the one from normal soil (Le) was no more tolerant to Zn and Cd than the *Silene* population from normal soil (Am). The calamine population (La), on the other hand, was even more tolerant to these metals than the calamine *Silene* population (Bl). Populations of Ni-hyperaccumulating species are invariably highly tolerant to Ni (e.g., Gabbrielli et al., 1990; Homer et al., 1991; Bernal and McGrath, 1994; Krämer et al., 1996; Table 9.1), but virtually all the species tested (mainly *Alyssum* sp.) were endemic to serpentine soil, which precludes comparisons with normal soil populations.

TABLE 9.1

Tolerance, Expressed as the External EC_{50} for Root Growth (the Concentration in the Test Solution which Inhibits Root Elongation by 50%) in a 4-Day Test

Species (Population)	EC_{50} (µm Me^{2+})				
	Cu	Zn	Cd	Ni	Co
Silene vulgaris (Am)	3 (2)	140 (14)	20 (3)	15 (4)	110 (10)
S. vulgaris (Im)	170 (8)	1800 (35)	170 (10)	50 (6)	240 (15)
S. vulgaris (Bl)	3 (2)	1800 (30)	110 (8)	50 (7)	220 (14)
Thlaspi caerulescens (Le)	0.5 (0.4)	160 (36)	8 (7)	150 (25)	250 (20)
T. caerulescens (La)	1 (1)	2300 (200)	230 (20)	600 (40)	200 (20)
T. caerulescens (Mo)	1 (1)	1200 (150)	60 (9)	800 (55)	300 (25)
T. arvense (Am)	0.5 (1)	20 (7)	1 (2)	40 (7)	15 (15)
Alyssum bertolonii (Pi)	0.5 (1)	440 (42)	15 (5)	570 (100)	500 (30)
A. argenteum (Bo)	n.t.	360 (n.t.)	n.t	440 (n t)	n.t
A. saxatilis (Fi)	n.t	230 (n.t)	n.t.	25 (n t)	n.t.

Note: Populations originated from nonmetalliferous soil at Amsterdam (Am), The Netherlands; a copper mine near Imsbach (Im); a zinc mine near Blankenrode (Bl), Germany; nonmetalliferous soil near Lellingen (Le), Luxemburg; calamine ore waste near La Calamine (La), Belgium; serpentine sites at Pieve S. Stefano (Pi), Bobbio (Bo), and Monte Prinzera (Mo), Italy; and nonmetalliferous soil near Firenze, (Fi), Italy. The plant-internal EC_{50}s (the total amount of plant metal per unit of total plant dry weight, expressed as µmol g^{-1}, at 50% root growth reduction) is given in brackets (root systems were desorbed with ice-cold $Pb(NO_3)_2$ prior to plant harvest)

n.t = not tested

Hyperaccumulators, just like nonhyperaccumulating metallophytes, may show relatively high levels of tolerance to metals that are not present at toxic concentrations in soil at the sites of population origin. Examples of this phenomenon are Zn tolerance in the serpentine *Thlaspi* population (Mo) and, though to a much lower degree, in the serpentine endemics *Alyssum bertolonii* and *A. argenteum*, as well as Ni tolerance in the calamine *Thlaspi* population (La), and Co tolerance in the serpentine *Thlaspi* population (Mo) and *A. bertolonii* (Table 9.1). These seemingly nonfunctional tolerances might represent the pleiotropic byproducts of functional tolerances to Zn or Ni. A common property of the metals Zn, Ni, and Co is a relatively high preference for oxygen- or nitrogen-based coordination over sulfur-based coordination. This clearly distinguishes them, as a group, from highly sulfhydryl-reactive metals such as Cd and Cu (Smith and Martell, 1989), and might provide a basis for (mutual) cotolerances, such as demonstrated in *S. vulgaris* (see above). Future analysis of intraspecific crosses in *T. caerulescens* should clarify this. However, with the possible exception of Cd tolerance in the serpentine *Thlaspi* population (Table 9.1), nonfunctional tolerances to Cu and Cd do not seem to occur. Therefore, it is not possible to maintain that hyperaccumulators would exhibit a degree of inherent nonspecific tolerance to all heavy metals (compare Baker et al., 1994).

The mechanisms of metal tolerance in hyperaccumulators have not been widely studied. Krämer et al. (1996) suggested a role for Ni-inducible accumulation of free histidine in the tolerance and hyperaccumulation of Ni in *Alyssum*. Although their results are strongly suggestive of a critical role for xylem histidine in plant-internal Ni transport, they are less conclusive with regard to the role for free histidine in the cellular Ni tolerance mechanism. Moreover, Ni-exposed *A. bertolonii* exclusively showed increased xylem-histidine concentrations. The root and leaf tissue concentrations were often even lower than in nonmetallophytes (H. Schat and M. Llugany, unpublished). The precise relationships between tolerance and hyperaccumulation mechanisms are unclear. Hyperaccumulation has been considered to represent a tolerance mechanism as such (e.g., Brooks, 1987), but this seems to be problematic from a logical point of view. Moreover, tolerance and hyperaccumulation apparently exhibit a high degree of independent variation in *Thlaspi* (see above) and seem to have clearly different patterns of metal specificity, both in *Thlaspi* and *Alyssum* (Gabbrielli et al., 1991; see below).

METAL UPTAKE, TRANSPORT, AND TOLERANCE: CORRELATED PATTERNS OF METAL SPECIFICITY?

Evidently, a reduced rate of uptake into the plant body would provide a simple explanation for increased tolerance to the metal(s) concerned ("avoidance" *sensu* Levitt, 1980). For example, arsenate tolerance in mine populations of a number of grasses appeared to be produced by reduced arsenate uptake through suppression of the high-affinity phosphate uptake system (Meharg and Macnair, 1991; 1992a,b). Apart from arsenate tolerance, there are no clear-cut examples of avoidance among heavy metal-adapted higher plant populations. Even when the tolerant plants take up less metal than the nontolerant ones, such as in the case of Cu- and nontolerant

S. vulgaris, the difference is far from sufficient to explain the difference in tolerance (De Vos et al., 1991; Schat and Kalff, 1992). It seems that such a reduced Cu uptake in Cu-tolerant plants represents a consequence, rather than a primary cause of tolerance (Strange and Macnair, 1991).

On the other hand, variation in constitutive tolerance among nonmetal-adapted nonhyperaccumulating plants might be more often determined by differential rates of uptake (e.g., Leita et al., 1993; Yang et al., 1995, 1996). For example, the external EC_{50} values for Zn, Cd, and Co were much higher (7- to 20-fold) in nonadapted *S. vulgaris* (population Am) than in *Thlaspi arvense*. The plant-internal EC_{50}s, however, were similar, or much less different (Table 9.1), suggesting that the differences in tolerance are strongly related to different uptake rates. In contrast with this, nonadapted *T. caerulescens* (population Le) tolerated much higher rates of accumulation, particularly of Ni and Zn, than nonadapted *S. vulgaris* (population Am) did, although the external EC_{50}s for Zn and Cd were similar or even lower in the former species (Table 9.1). This could be taken to suggest that hyperaccumulators, as compared to nonhyperaccumulators, might inherently tolerate higher rates of accumulation rather than higher external concentrations of particular metals.

It has been suggested that differential heavy metal tolerance in nonmetallophytes would be based on differential root-to-shoot transport rather than differential uptake. Metals are often more strongly retained in the root system in relatively tolerant species (e.g., Leita et al., 1993; Yang et al., 1996), although there are marked exceptions to this "rule" (Yang et al., 1995). Also, nonhyperaccumulating metallophytes might show tolerance-correlated differences in the degree of metal retention in the root system. For example, tolerance to Cd and Zn in mine populations of *S. vulgaris* is associated with a strongly reduced root-to-shoot translocation of these metals (Baker, 1978; Verkleij and Prast, 1989; De Knecht et al., 1992; Harmens et al., 1993b). The same has been found for Cu tolerance in *Minuartia hirsuta* (Ouzounidu et al., 1994). On the other hand, high-level Cu tolerance in *S. vulgaris*, *Lotus purshianus*, and *M. guttatus* is not consistently associated with reduced root-to-shoot transport of this metal (Lolkema et al., 1984; Lin and Wu, 1994; Harper et al., 1997).

Tolerance-correlated decreases of metal root-to-shoot transport in metal-adapted nonhyperaccumulating metallophyte populations are certainly not the primary cause of increased tolerance. Harmens et al. (1993b) reported split-root experiments with Zn- and nontolerant *S. vulgaris* in which individual root systems were divided over two compartments with either different or equal Zn concentrations. In all cases, the responses of individually exposed parts of the root system were exclusively dependent on the Zn concentration in their own compartment, irrespective of the Zn burden of the shoot and the other part of the root system, both in tolerant and nontolerant plants. This clearly proves that Zn-imposed root growth inhibition is due to a direct effect of Zn on the roots themselves and that high-level tolerance, which is particularly manifest from the root growth response, cannot be explained by increased retention of Zn in the root system. Additional split-root experiments with *S. vulgaris* populations have unambiguously shown that this holds also true for Cd tolerance (De Knecht, 1994). Thus, increased retention in the roots in metal-adapted metallophytes seems to be a mere consequence of the operation of a high-level tolerance mechanism in the roots, rather than a primary mechanism of tolerance as such. Tolerance-correlated shifts in metal uptake or transport in nonhyperaccumulating metallophytes, whenever they occur, seem to be as metal specific as tolerance itself (Schat and Ten Bookum, 1992b; H. Schat and R. Vooijs, unpublished).

Hyperaccumulators seem to exhibit a high degree of metal specificity with regard to foliar accumulation. Most of the hyperaccumulators described thus far are more or less restricted to serpentine soil and hyperaccumulate nickel (e.g., *Alyssum* sp.). A lower, but still considerable number of species preferentially occur on calamine soil (e.g., *Thlaspi* sp., *Cardaminopsis halleri*) and primarily hyperaccumulate Zn and occasionally Cd, Pb, and Ni. Some hyperaccumulators of Ni and Zn are also found on nonmetalliferous soil. Also under these conditions, they may show extraordinarily high foliar metal contents. Hyperaccumulation of Cu and Co is confined to a low number of species from the Shaban copper belt in southeastern Congo (Baker and Brooks, 1989). However, much of the evidence concerning the metal-specific patterns of foliar accumulation in hyperaccumulators is based on comparisons of metal contents in field-collected plant materials which are likely to be biased by strongly differential soil metal compositions at the sites of origin. Hyperaccumulation of Ni and Zn in serpentine and calamine hyperaccumulators, even in slightly enriched substrates and nutrient solutions, has been demonstrated amply in short-term laboratory experiments (e.g., Baker et al., 1994; Krämer et al., 1996; Lasat et al., 1996). Hyperaccumulation of Cu in presumed Cu hyperaccumulators from Shaba, on the other hand, was not reproducible under controlled conditions (Köhl et al., 1997). At any rate, extensive laboratory studies, which would allow accurate comparisons of the patterns of metal specificity of uptake and transport in different hyperaccumulators, are hardly available to date. As yet unpublished studies in the authors' laboratory have revealed very distinctive metal preference patterns in calamine and serpentine *Thlaspi* and serpentine *Alyssum*. With regard to metal uptake, calamine *Thlaspi* (population La) exclusively showed a strongly increased uptake of Zn, particularly at lower external concentrations (Figure 9.1A). The uptake of Ni (Figure 9.2A), Cd (Figure 9.3A), Co (Figure 9.4A), and Cu (data not shown) was not higher than in nonhyperaccumulating metallophytes and nonmetallophytes. Zn uptake, even at low external Zn concentrations, was not suppressed in the presence of high concentrations of Cd, Ni, or Co and, conversely, high external Zn concentrations did not suppress the uptake of Cd, Ni, and Co (data not shown). On the other hand, in agreement with earlier studies (Gabbrielli et al., 1991; Homer et al., 1991; Bernal and McGrath, 1994), *A. bertolonii* exhibited a strongly increased uptake of Ni, Zn, and, to a lower degree, Cd, and Co (Figures 9.1A to 9.4A), but not of Cu (data not shown), at lower external concentrations. Ni uptake was almost completely suppressed (about 90%) in the presence of equimolar external concentrations of either Zn, Cd, or Co, even in the lower concentration range, whereas high external Ni concentrations did not suppress the uptake of Ni, Cd, and Co. For the combination of Co and Ni, this phenomenon has been previously observed in other *Alyssum* species (Homer et al., 1991). Thus, hyperaccumulation of Ni in *Alyssum* seems to be associated with a rather nonspecifically increased capacity for heavy metal uptake and a relatively low preference for uptake of Ni under combined metal exposure, whereas hyperaccumulation of Zn in calamine *Thlaspi* apparently involves

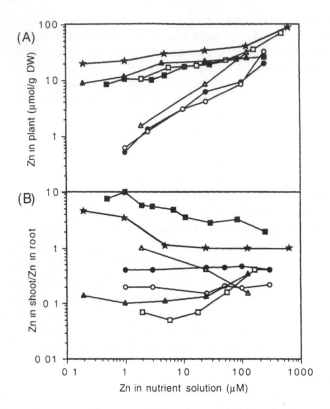

FIGURE 9.1 Total plant Zn contents (A) and shoot-to-root ratios of the plant-internal Zn concentrations (B) after 4 days of exposure to increasing Zn concentrations in an EDTA-free nutrient solution in *S. vulgaris* from: Am (closed circles) and Bl (open circles); *T. caerulescens* from Le (open squares), La (closed squares), and Mo (asterisks); *T. arvense* from Am (open triangles); and *A. bertolonii* from Pi (closed triangles). Roots were desorbed with ice-cold Pb(NO$_3$)$_2$ prior to harvest. (See Table 9.1 for site abbreviations.)

a highly specific increase in the capacity to take up Zn. The serpentine *Thlaspi* population (Mo), on the other hand, showed a strongly increased uptake of all the metals tested (Figures 9.1A to 9.4A), apart from Cu (data not shown). This population's metal preference patterns under combined metal exposure have not been established as of yet.

The metal preference patterns for root-to-shoot transport seem to be totally different from those for uptake. Compared to the nonhyperaccumulating species, calamine *Thlaspi* exhibited a similarly increased transport of Zn, Ni, Cd, and Co, but not of Cu. Also, the serpentine *Thlaspi* population showed a strongly increased transport of Ni, Co, and, to a lower degree, Zn and Cd (Figures 9.1B to 9.4B), but not of Cu (data not shown). *A. bertolonii* showed increased transport of Ni and, to a lower degree, Co, but not of Zn, Cd (Figures 9.1B to 9.4B), or Cu (data not shown). However, when under combined exposure, Zn transport in calamine *Thlaspi* was strongly inhibited by Ni and, though less effectively, Co, but not by Cd. Conversely, the transport of Ni, Co, and Cd was not inhibited by Zn, not even at extremely high

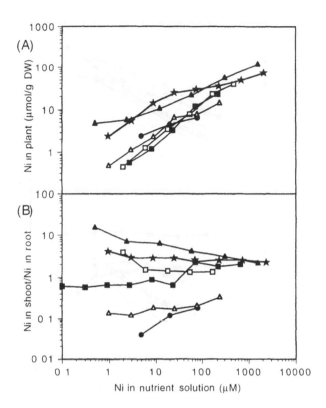

FIGURE 9.2 Total plant Ni contents (A) and shoot-to-root ratio's of the plant-internal Ni concentrations (B) after 4 days of exposure to increasing Ni concentrations in the nutrient solution. (Legends as in Figure 9.1.) Ni accumulation in Zn-tolerant *S. vulgaris* (Bl) was not different from that in nontolerant *S. vulgaris*. Ni transport was slightly lower, particularly at higher external concentrations (data not shown).

external Zn concentrations. In *Alyssum*, Ni transport did not respond to other metals and vice versa (data not shown). Thus, calamine *Thlaspi* and *Alyssum* may in fact possess similar metal preference patterns for transport as such (viz. Ni > Co > Zn > Cd) which is clearly different from the usual pattern in nonhyperaccumulators (viz. Zn > Cd > Co, Ni [Figures 9.1B to 9.4B]). In agreement with this hypothesis, the normal soil population of *Thlaspi* (Le) showed increased transport of Ni and, to a lower degree, Co, but not of Zn and Cd. Moreover, also the serpentine *Thlaspi* population seemed to exhibit a more strongly increased transport of Ni and Co than of Zn and, particularly, Cd (Figures 9.1B to 9.4B). Unfortunately, transport under combined metal exposure has not been studied in these populations yet. The suggested order of preference for transport (viz. Ni > Co > Zn > Cd) corresponds with that of the binding affinities for histidine (Smith and Martell, 1989), which has been shown to be involved in Ni transport in hyperaccumulating *Alyssum*. The xylem-histidine concentration in calamine *Thlaspi* was extremely high, indeed (10- to 20-fold higher than in nonhyperaccumulators and comparable to Ni-exposed *Alyssum*). Much lower xylem-histidine levels were found in the normal soil population (about

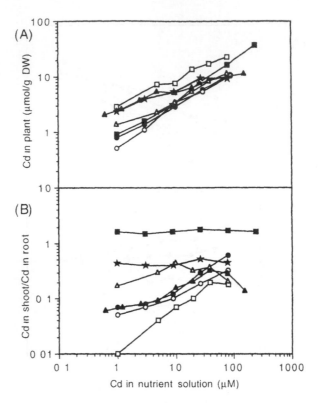

FIGURE 9.3 Total plant Cd contents (A) and shoot-to-root ratios of the plant internal Cd concentrations (B) after 4 days of exposure to increasing Cd concentrations in the nutrient solution. (Legends as in Figure 9.1.)

threefold higher than in nonhyperaccumulators), suggesting that intraspecific variation of metal transport in *Thlaspi* might be associated with differential xylem-histidine concentrations (H. Schat and M. Llugany, unpublished). To any extent, the high xylem-histidine levels in calamine *Thlaspi* were constitutive; they were also found in plants growing in control solutions and showed no response to metal exposure. Even prolonged Zn starvation had no effect. In hyperaccumulating *Alyssum*, on the other hand, histidine is only accumulated to high levels in the xylem under exposure to Ni or Co (Krämer et al., 1996), which might explain the low rates of Cd and Zn transport.

The above results clearly demonstrate independent variation of uptake and transport of Zn in *Thlaspi*. The calamine population and the normal soil population possessed equal rates of uptake, but strongly different rates of transport. The serpentine population showed a still higher uptake, but an intermediate transport of Zn. Also for the other metals, there was a largely independent variation in uptake and transport, particularly at lower exposure levels (Figures 9.1 to 9.4). It must be stressed, however, that this variation is not necessarily correlated with soil types. For example, the low rates of Zn and Cd transport in the normal soil population are probably highly unusual. Other populations from normal soil may show even higher

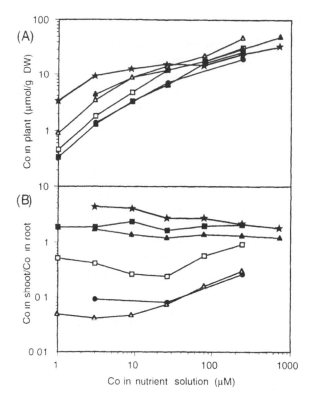

FIGURE 9.4 Total plant Co contents (A) and shoot-to-root ratios of the plant-internal Co concentrations (B) after 4 days of exposure to increasing Co concentrations in the nutrient solution. (Legends as in Figure 9.1.) Co accumulation in Zn-tolerant *S. vulgaris* (Bl) was no different from that in nontolerant *S. vulgaris*. Co transport was slightly lower (data not shown).

rates of uptake or transport of Zn, Pb, or Cd than those from calamine soil (H. Schat and R. Vooijs, unpublished). This is not surprising, because metal hyperaccumulation most likely represents a broad-spectrum defense against herbivores and pathogens (e.g., Boyd and Martens, 1992, 1994; Martens and Boyd, 1994; Boyd et al., 1994). Thus, in order to acquire effective defense, populations in normal soil might often require more high-affinity uptake capacity than populations from metalliferous soils, where metals are amply available. Unfortunately, studies on root herbivory in relation to root metal contents have not yet been performed. Evidently, there is a possibility that intraspecific variation in transport rates might reflect differential relative grazing pressures on roots and leaves. Tolerance of high external metal concentrations, uptake, and transport in *Thlaspi* seems to be largely unrelated and may exhibit strongly different metal preference patterns. This suggests that metal-tolerance mechanisms in hyperaccumulators, particularly those for Zn and Cd, might be fundamentally different from those in nonhyperaccumulating metallophytes, because the latter are often associated with increased metal retention in the roots. In view of the strongly differential metal preference patterns (see above), the root-to-shoot transport mechanisms seem to be different as well.

CONCLUSIONS

Heavy metal-adapted populations of metallophyte species exhibit a broad range of largely metal-specific, high-level tolerance mechanisms which seem to be under relatively simple genetic control. Heavy metal tolerance as such may be desirable for plants to be used in phytoextraction procedures, but the mechanisms operating in nonhyperaccumulating metallophytes seem to produce lower rather than higher rates of metal accumulation in the above-ground plant parts. Tolerance mechanisms in hyperaccumulators, on the other hand, do not appear to restrict the foliar accumulation of metals. In general, the tolerance, uptake, and root-to-shoot transport of metals in the hyperaccumulator *T. caerulescens* are subject to uncorrelated variation, even with respect to the patterns of metal specificity. This would allow for the selection and breeding of varieties with useful combinations of properties. Transferring the hyperaccumulation syndrome into highly productive crop species by genetic engineering techniques seems to be far away. First, hyperaccumulation may be a complex phenomenon, because metal uptake, transport, and tolerance are apparently under independent genetic control. Second, none of the genes involved have ever been identified and cloned. Molecular studies of artificial mutants or, when possible, intervarietal crosses are urgently required. Also, a better understanding of the underlying biochemical and physiological mechanisms could be helpful. It is possible, for example, that further studies on the role for free histidine and the plant-internal controls of its accumulation would provide tools to manipulate plant-internal metal translocation.

REFERENCES

Allen, W.R. and P.M. Sheppard. Copper Tolerance in some Californian Populations of the Monkey Flower *Mimulus guttatus. Proc. R. Soc. London. Ser. B* 177, 177-196, 1971.

Baker, A.J.M. Ecophysiological aspects of zinc tolerance in *Silene maritima* With. *New Phytol.* 80, 635-642, 1978.

Baker, A.J.M. and R.R. Brooks. Terrestrial higher plants which hyperaccumulate metallic elements — a review of their distribution, ecology and phytochemistry. *Biorecovery* 1, 81-126, 1989.

Baker A.J.M. and J. Proctor. The influence of cadmium, copper, lead and zinc on the distribution and evolution of metallophytes in the British Isles. *Plant System. Evol.* 173, 91-108, 1990.

Baker, A.J.M., R.D. Reeves, and A.S.M. Hajar. Heavy metal accumulation and tolerance in British populations of the metallophyte *Thlaspi caerulescens* J. and C. Presl. (Brassicaceae). *New Phytol.* 127, 155-177, 1994.

Bernal, M.P. and S.P. McGrath. Effects of pH and heavy metal concentrations in solution culture on the proton release, growth and elemental composition of *Alyssum murale* and *Raphanus sativus* L. *Plant Soil* 166, 83-92, 1994.

Boyd, R.S. and S.N. Martens. The raison d'étre for metal hyperaccumulation by plants. In *The Vegetation of Ultramafic (Serpentine) Soils.* Baker, A.J.M., J. Proctor, and R.D. Reeves, Eds, Intercept, Andover, U.K., 279-290. 1992.

Boyd, R.S. and S.N. Martens. Nickel hyperaccumulated by *Thlaspi montanum* var. *montanum* is acutely toxic to an insect herbivore. *Oikos* 70, 21-25, 1994.

Boyd, R.S., J.J. Shaw, and S.N. Martens. Nickel hyperaccumulation defends *Streptanthus polygaloides* (Brassicaceae) against pathogens. *Am. J Bot.* 81, 294-300, 1994.

Brooks, R.R. *Serpentine and Its Vegetation: A Multidisciplinary Approach.* Dioscorides Press, Portland, OR, 1987.

Brooks, R.R., F. Malaisse, and A. Empain. *The Heavy Metal-Tolerant Flora of Southcentral Africa.* Balkema, Rotterdam, The Netherlands, 1985.

Brown, G. and K. Brinkman. Heavy metal tolerance in *Festuca ovina* L. from contaminated sites in the Eifel Mountains, Germany. *Plant Soil* 143, 239-247, 1992.

De Knecht, J.A., *Cadmium Tolerance and Phytochelatin Production in Silene vulgaris.* Ph.D. thesis, Vrije Universiteit, Amsterdam, The Netherlands, 1994.

De Knecht, J.A., P.L.M. Koevoets, J.A.C. Verkleij, and W.H.O. Ernst. Evidence against a role for phytochelatins in naturally selected increased cadmium tolerance in *Silene vulgaris* (Moench) Garcke. *New Phytol.* 122, 681-688, 1992.

De Knecht, J.A., N. Van Baren, W.M. Ten Bookum, H.W. Wong Fong Sang, P.L.M. Koevoets, H. Schat, and J.A.C. Verkleij. Synthesis and degradation of phytochelatins in cadmium-sensitive and cadmium-tolerant *Silene vulgaris*. *Plant Sci.* 106, 9-18, 1995.

De Knecht, J.A., M. Van Dillen, P.L.M. Koevoets, H. Schat, J.A.C. Verkleij, and W.H.O. Ernst. Phytochelatins in cadmium-sensitive and cadmium-tolerant *Silene vulgaris*: chain length distribution and sulfide incorporation. *Plant Physiol.* 104, 255-261, 1994.

De Vos, C.H.R., H. Schat, M.A.M. De Waal, R. Vooijs, and W.H.O. Ernst. Increased resistance to copper-induced damage of the root cell plasmalemma in copper tolerant *Silene cucubalus*. *Physiol. Plant.* 82, 523-528, 1991.

De Vos, C.H.R., M.J. Vonk, R. Vooijs, and H. Schat. Glutathione depletion due to copper-induced phytochelatin synthesis causes oxidative stress in *Silene cucubalus*. *Plant Physiol.* 98, 853-858, 1992.

Ernst, W.H.O. Ecophysiological studies on heavy metal plants in south central Africa. *Kirkia* 8, 125-145, 1972.

Ernst, W.H.O. *Schwermetallvegetation der Erde.* Gustav Fischer Verlag, Stuttgart, Germany, 1974.

Ernst, W.H.O. Physiology of heavy metal resistance in plants. In *Proc. Int. Conf. Heavy Metals in the Environ., Toronto*, Vol. 2, Hutchinson, T.C., S. Epstein, A.L. Page, J. Van Loon, and T. Davey, Eds, CEP Consultants, Edinburgh, 121-136, 1975.

Gabbrielli, R., C. Mattioni, and O. Vergnano. Accumulation mechanisms and heavy metal tolerance in a hyperaccumulator. *J. Plant Nutr.* 14, 1067-1080, 1991.

Gabbrielli, R., T. Pandolfini, O. Vergnano, and M.R. Palandri. Comparison of two serpentine species with different nickel tolerance strategies. *Plant Soil* 122, 271-277, 1990.

Gartside, D.W. and T. McNeilly. The potential for evolution of heavy metal tolerance in plants. II. Copper tolerance in normal populations of different plant species. *Heredity* 32, 335-348, 1974.

Gilissen, L.W.J. and M.J. Van Staveren. Zinc resistant cell lines of *Haplopappus gracilis*. *J. Plant Physiol.* 125, 95-103, 1986.

Gregory, R.P.G. and A.D. Bradshaw. Heavy metal tolerance in populations of *Agrostis tenuis* Sibth. and other grasses. *New Phytol.* 64, 131-143, 1965.

Grill, E., E.L. Winnacker, and M.H. Zenk. Phytochelatins, a class of heavy metal-binding peptides from plants are functionally analogous to metallothioneins. *Proc. Nat. Acad. Sci. USA* 84, 439-443, 1987.

Harmens, H., E. Cornelisse, P.R. Den Hartog, W.M. Ten Bookum, and J.A.C. Verkleij. Phytochelatins do not play a key role in naturally selected zinc tolerance in *Silene vulgaris* (Moench) Garcke. *Plant Physiol.* 103, 1305-1309, 1993a.

Harmens, H., N.G.C.P.B. Gusmão, P.R. Den Hartog, J.A.C. Verkleij, and W.H.O. Ernst. Uptake and transport of zinc in zinc-sensitive and zinc-tolerant *Silene vulgaris. J. Plant Physiol.* 141, 309-315, 1993b.

Harmens, H., P.L.M. Koevoets, J.A.C. Verkleij, and W.H.O. Ernst. The role of low molecular weight organic acids in the mechanism of increased zinc tolerance in *Silene vulgaris* (Moench) Garcke. *New Phytol.* 126, 615-621, 1994.

Harper, F.A., S.E. Smith, and M.R. Macnair. Can an increased copper requirement in copper-tolerant *Mimulus guttatus* explain the cost of tolerance? I. vegetative growth. *New Phytol.* 136, 455-467, 1997.

Hogan, G.D. and W.E. Rauser. Tolerance and toxicity of cobalt, copper, nickel and zinc in clones of *Agrostis gigantea. New Phytol.* 83, 665-670, 1979.

Homer, F.A., R.S. Morrison, R.R. Brooks, J. Clemens, and R.D. Reeves. Comparative studies of nickel, cobalt, and copper uptake by some nickel hyperaccumulators of the genus *Alyssum. Plant Soil* 138, 195-205, 1991.

Howden, R., C.R. Anderson, P.B. Goldsbrough, and C.S. Cobbett. A cadmium-sensitive glutathione-deficient mutant of *Arabidopsis thaliana. Plant Physiol.* 107, 1067-1073, 1995a.

Howden, R., P.B. Goldsbrough, C.R. Andersen, and C.S. Cobbett. Cadmium-sensitive cad1 mutants of *Arabidopsis thaliana* are phytochelatin-deficient. *Plant Physiol.* 107, 1059-1066, 1995b.

Howden, R. and C.S. Cobbett. Cadmium-sensitive mutants of *Arabidopsis thaliana. Plant Physiol.* 100, 100-107, 1992.

Ingrouille, M.J. and N. Smirnoff. *Thlaspi caerulescens* J. and C. Presl. (*T. alpestre* L.) in Britain. *New Phytol.* 102, 219-233, 1986.

Jackson, P.J., E.J. Roth, P.R. McClure, and C.M. Naranjo. Selection, isolation and characterization of cadmium-resistant *Datura innoxia* suspension cultures. *Plant Physiol.* 75, 914-918, 1984.

Köhl, K.I., J.A.C. Smith, and A.J.M. Baker. Metal-allocation patterns in cuprophytes from the zairean copper belt contrast with those of nickel hyperaccumulators. *J. Exp. Bot.* 48 (Suppl.), 103, 1997.

Krämer, U., J.D. Cotter-Howells, J.M. Charnock, A.J.M. Baker, and J.A.C. Smith. Free histidine as a metal chelator in plants that accumulate nickel. *Nature* 379, 635-638, 1996.

Lasat, M.M., A.J.M. Baker, and L.V. Kochian. Physiological characterisation of root Zn^{2+} absorption and translocation to shoots in Zn hyperaccumulator and nonhyperaccumulator species of *Thlaspi. Plant Physiol.* 112, 1755-1722, 1996.

Leita, L., M. De Nobili, C. Mondini, and M.T. Baca Garcia. Response of Leguminosae to cadmium exposure. *J. Plant Nutr.* 16, 2001-2012, 1993.

Levitt, J. *Responses of Plants to Environmental Stresses.* Academic Press, New York, 1980.

Lin, S.L. and L.Wu. Effects of copper concentration on mineral nutrient uptake and copper accumulation in protein of copper-tolerant and nontolerant *Lotus purshianus* L. *Ecotoxicol. Environ. Safety* 29, 214-228, 1994.

Lolkema, P.C., M.H. Donker, A.J. Schouten, and W.H.O. Ernst. The possible role of metallothioneins in copper tolerance of *Silene cucubalus. Planta* 167, 174-179, 1984.

Macnair, M.R. The genetic control of copper tolerance in the yellow monkey flower, *Mimulus guttatus. Heredity* 50, 283-293, 1983.

Macnair, M.R. The genetics of metal tolerance in vascular plants. *New Phytol.* 124, 541-559, 1993.

Macnair, M.R., Q.J. Cumbes, and A.A. Meharg. The genetics of arsenate tolerance in yorkshire fog *Holcus lanatus. Heredity* 50, 283-293, 1992.

Macnair, M.R., S.E. Smith, and Q.J. Cumbes. The heritability and distribution of variation in degree of copper tolerance in *Mimulus guttatus* at Copperopolis, California. *Heredity* 71, 455-455, 1993.

Martens, S.N. and R.S. Boyd. The ecological significance of nickel hyperaccumulation: a plant chemical defense. *Oecologia (Berlin)* 98, 379-384, 1994.

Mathys, W. The role of malate, oxalate, and mustard oil glucosides in the evolution of zinc-resistance in herbage plants. *Physiol. Plant.* 40, 130-136, 1977.

Meharg, A.A. and M.R. Macnair. The mechanisms of arsenate tolerance in *Deschampsia cespitosa* (L.) Beauv. and *Agrostis capillaris* L. adaptation of the arsenate uptake system. *New Phytol.* 199, 291-297, 1991.

Meharg, A.A. and M.R. Macnair. Suppression of the phosphate uptake system: a mechanism of arsenate tolerance in *Holcus lanatus. J. Exp. Bot.* 43, 519-524, 1992a.

Meharg, A.A. and M.R. Macnair. Genetic correlation between arsenate tolerance and the rate of influx of arsenate and phosphate in *Holcus lanatus. Heredity* 69, 336-341, 1992b.

Murphy, A. and L.Taiz. Comparison of metallothionein expression and nonprotein thiols in ten *Arabidopsis* ecotypes. *Plant Physiol.* 109, 945-954, 1995.

Murphy, A., J. Zhou, P.B. Goldsbrough, and L.Taiz. Purification and immunological identification of metallothioneins 1 and 2 from *Arabidopsis thaliana. Plant Physiol.* 113, 1293-1301, 1997.

Ortiz, D.F., L. Kreppel, D.M. Speiser, G. Scheel, G. McDonald, and D.W. Ow. Heavy metal tolerance in the fission yeast requires an atp binding cassette-type vacuolar membrane transporter. *EMBO J.* 11, 3491-3499, 1992.

Ouzounidou, G., L. Symeonidis, D. Babalonas, and S. Karataglis. Comparative responses of a copper-tolerant and a copper-sensitive population of *Minuartia hirsuta* to copper toxicity. *J. Plant Physiol.* 144, 109-115, 1994.

Robinson, N.J., J.R. Wilson, and J.S. Turner. Expression of the type 2 metallothioneinlike gene MT2 from *Arabidopsis thaliana* in Zn(2+)-metallothionein-deficient *Synechococcus* PCC 7942: putative role for MT2 in Zn (2+) metabolism. *Plant Mol. Biol.* 30, 1169-1179, 1996.

Schat, H. and M.M.A. Kalff. Are phytochelatins involved in differential metal tolerance, or do they merely reflect metal-imposed strain? *Plant Physiol.* 99, 1475-1480, 1992.

Schat, H., E. Kuiper, W.M. Ten Bookum, and R. Vooijs. A general model for the genetic control of copper tolerance in *Silene vulgaris. Heredity* 69, 325-335, 1993.

Schat, H. and W.M. Ten Bookum. Genetic control of copper tolerance in *Silene vulgaris. Heredity* 68, 219-229, 1992a.

Schat, H. and W.M. Ten Bookum. Metal-specificity of heavy metal tolerance syndromes in higher plants. In *The Vegetation of Ultramaffic (Serpentine) Soils.* Baker, A.J.M., J. Proctor, and R.D. Reeves, Eds, Intercept, Andover, U.K., 337-352, 1992b.

Schat, H. and R. Vooijs. Multiple tolerance and co-tolerance to heavy metals in *Silene vulgaris*: a co-segregation analysis. *New Phytol.* 136, 489-496, 1997a.

Schat, H. and R. Vooijs. Effects of decreased cellular glutathione levels on growth, membrane integrity and lipid peroxidation in roots of copper-stressed *Silene vulgaris.* In *Proc. Third Int. Conf. Biogeochem. Trace Elements, Paris,* 1995. Prost, R., Ed., INRA, Paris, CD-ROM, 057.PDF. 1997b.

Schat, H., R. Vooijs, and E. Kuiper. Identical major genes for heavy metal tolerances that have independently evolved in different local populations and subspecies of *Silene vulgaris. Evolution* 50, 1888-1895, 1996.

Smith, R.M. and A.E. Martell. *Critical Stability Constants.* Plenum, New York, 1989.

Speiser, D.M., S.L. Abrahamson, and G. Bañuelos. *Brassica juncea* produces a phytochelatin-cadmium-sulfide complex. *Plant Physiol.* 99, 817-821, 1992a.

Speiser, D.M., D.F. Ortiz, L. Kreppel, G. Scheel, G. McDonald, and D.W. Ow. Purine biosynthetic genes are required for cadmium tolerance in *Schizosaccharomyces pombe*. *Mol. Cell. Biol.* 12, 5301-5310, 1992b.

Strange, J. and M.R. Macnair. Evidence for a role for the cell membrane in copper tolerance of *Mimulus guttatus* Fisher ex D.C. *New Phytol.* 119, 383-388, 1991.

Tilstone, G.H. and M.R. Macnair. Nickel tolerance and copper-nickel co-tolerance in *Mimulus guttatus* copper mine and serpentine habitats. *Plant Soil* 191, 173-180, 1997.

Verkleij, J.A.C. and W.B. Bast-Kramer. Co-tolerance and multiple heavy metal tolerance in *Silene vulgaris* from different heavy metal sites. In *Heavy Metal in the Environment, Athens, 1985*. Lekkas, T.D., Ed., CEP Consultants, Edinburgh, 174-176, 1985.

Verkleij, J.A.C., P.L.M. Koevoets, M.M.A. Blake-Kalff, and A.N. Chardonnens. Evidence for an important role of the tonoplast in the mechanism of naturally selected zinc tolerance in *Silene vulgaris*. *J. Plant Physiol.* 153, 188-191, 1998.

Verkleij, J.A.C., P.L.M. Koevoets, J. Van't Riet, J.A. De Knecht, and W.H.O. Ernst. The role of metal-binding compounds in the copper tolerance mechanism of *Silene cucubalus*. In *Metal Ion Homeostasis: Molecular Biology and Chemistry*. Winge, D. and D. Hamer, Eds, Alan R. Liss, New York, 347-357, 1989.

Verkleij, J.A.C. and H.E. Prast. Cadmium tolerance and co-tolerance in *Silene vulgaris* (Moench) Garcke (=*S. cucubalus* (L.) Wib.). *New Phytol.* 111, 637-645, 1989.

Vögeli-Lange, R. and G.J. Wagner. Subcellular localization of cadmium and cadmium-binding peptides in tobacco leaves. *Plant Physiol.* 92, 1086-1093, 1989.

Von Frenkell-Insam, B.A.K. and T.C. Hutchinson. Occurrence of heavy metal tolerance and co-tolerance in *Deschampsia cespitosa* (L.) Beauv. from European and Canadian populations. *New Phytol.* 125, 555-564, 1993.

Yang, X., V.C. Baligar, D.C. Martens, and R.B. Clark. Influx, transport, and accumulation of cadmium in plant species grown at different Cd (2+) activities. *J. Environ. Sci. Health Part B Pesticides Food Contam. Agric. Wastes* 30, 569-583, 1995.

Yang, X., V.C. Baligar, D.C. Martens, and R.B. Clark. Plant tolerance to nickel toxicity. I. influx, transport and accumulation of nickel in four species. *J. Plant Nutr.* 19, 73-85, 1996.

10 The Role of Root Exudates in Nickel Hyperaccumulation and Tolerance in Accumulator and Nonaccumulator Species of *Thlaspi*

David E. Salt, N. Kato, U. Krämer, R. D. Smith, and I. Raskin

CONTENTS

INTRODUCTION

The association of plant species with soils rich in various heavy metals has long been recognized (Antonovics, 1971). Examples of such distinct communities include serpentine (i.e., growing on Mg-, Ni-, Cr-, and Co-rich soils), seleniferous (i.e., growing on Se-rich soils), uraniferous (i.e., growing on U-rich soils), calamine (i.e., growing on Zn- and Cd-rich soils), and Cr/Co floras. Metal hyperaccumulator plant species are associated with these specialized metal floras, and they can concentrate metals in their above-ground tissues to levels far exceeding the concentration of metals present in the soil or in the nonaccumulating species growing nearby (Baker and Brooks, 1989). Because of their enhanced ability to accumulate metals, a recent U.S. Department of Energy (DOE) report concluded that "the genetic traits present in hyperaccumulator plants offer potential for the development of practical phytore-

1-56670-450-2/00/$0 00+$ 50
© 2000 by CRC Press LLC

mediation processes" (DOE/EM-0224, 1994). The genus *Thlaspi* (Brassicaceae) contains several species which can hyperaccumulate Ni. *T. goesingense* Hálácsy found growing on Ni-rich serpentinitic soils in Redschlag, Austria, can contain Ni at concentrations up to 15,000 μg g^{-1} in its shoot dry biomass (Reeves and Brooks, 1983). While the ecological role of metal hyperaccumulation is still unclear, recent evidence suggests that it may protect plants against herbivory and attack by fungal and bacterial pathogens (Boyd and Martens, 1994; Boyd et al., 1994; Pollard and Baker, 1997).

It has been shown for a number of metal-hyperaccumulating species that shoot metal concentrations reach very high levels after short-, medium-, or long-term exposure to various, even low concentrations of metals in soil or hydroponic culture (Homer et al., 1991; Lloyd-Thomas, 1995; Krämer et al., 1996, 1997). One possible explanation for this enhanced Ni accumulation is that hyperaccumulator species may release root exudates containing Ni-chelators with the potential to enhance Ni uptake, translocation, and resistance. Among the compounds that have been proposed to participate in Ni chelation in hyperaccumulators are citrate (Lee et al., 1978) and free histidine (Krämer et al., 1996).

By comparing the composition of Ni-binding compounds in root exudates from *T. goesingense* and the nonaccumulator species *T. arvense*, we have started to define the role of root exudates in Ni hyperaccumulation and tolerance in the *Thlaspi* genus.

EXPERIMENTAL APPROACH

Seeds of *T. arvense* L. were obtained from the Crucifer Genetics Cooperative (University of Wisconsin, Madison). Seeds of *T. goesingense* Hálácsy were collected from plants growing on a Ni-rich serpentinitic soil in Redschlag, Austria.

If not stated otherwise, seeds were germinated on filter paper moistened with 2.8 mmol L^{-1} Ca(NO$_3$)$_2$ for 1 week. Subsequently, 30 seedlings were transferred into 12 liter of hydroponic solution. Seedlings were initially supported by moist vermiculite, and later by cotton wool. Solutions were continuously aerated and exchanged in intervals of 7 to 18 days, according to plant size. Since growth rates of *T. goesingense* were substantially lower than those of *T. arvense*, experiments were performed using plants with equivalent numbers of leaves. The composition of all hydroponic solutions was as follows: 0.28 mmol L^{-1} Ca^{2+}, 0.6 mmol L^{-1} K$^+$, 0.2 mmol L^{-1} Mg^{2+}, 0.1 mmol L^{-1} NH$_4^+$, 1.16 mmol L^{-1} NO$_3^-$, 0.1 mmol L^{-1} H$_2$PO$_4^-$, 0.2 mmol L^{-1} SO$_4^{2-}$, 4.75 μmol L^{-1} ferric tartrate, 0.03 μmol L^{-1} Cu^{2+}, 0.08 μmol L^{-1} Zn^{2+}, 0.5 μmol L^{-1} Mn^{2+}, 4.6 μmol L^{-1} H$_3$BO$_3$ and 0.01 μmol L^{-1} MoO$_3$ (pH between 5.5 and 5.8). Plants were cultivated in the growth chamber with 10-h light periods. Light was provided by fluorescent and incandescent lamps. All plants were maintained at day/night temperatures of 22°C and a constant humidity of 50%.

In these experiments, plant roots were grown axenically. Seeds were surface sterilized by rinsing in 75% ethanol for 5 min, followed by incubation in 2.6% (w/v) sodium hypochlorite containing 0.01% (v/v) Triton X-100 as a surfactant for 15 min. After 4 washes in sterile deionized water, seeds were transferred onto agar plates containing 1 mmol L^{-1} Ca(NO$_3$)$_2$, 2 mmol L^{-1} KH$_2$PO$_4$, 4 mmol L^{-1} KNO$_3$, 0.3 mmol L^{-1} MgSO$_4$, 0.18 mmol L^{-1} FeCl$_3$, 42.26 μmol L^{-1} H$_3$BO$_3$, 0.312 μmol L^{-1}

$CuSO_4$, 9.10 µmol L^{-1} $MnCl_2$, 0.106 µmol L^{-1} MoO_3, 0.765 µmol L^{-1} $ZnSO_4$, 3% (w/v) sucrose, and 1.2% (w/v) agarose (pH 6.0 to 6.5) and germinated in the dark for 1 week. The etiolated seedlings were transferred individually into small glass vials containing 10 ml of sterile hydroponic solution. Soft styrofoam stoppers used to cap the vials were incised radially to provide support for the hypocotyls. Solutions were exchanged weekly and the vials were gently agitated on a rotary shaker (Lab-line Instruments, Inc., Melrose Park, IL) at 60 rpm to provide aeration and mixing. After various treatments outlined below, the hydroponic culture solution was collected and treated as dilute root exudate.

Four-week-old *T. goesingense* and two-week-old *T. arvense* of uniform sizes (three replicate plants for each time point), grown axenically as described above, were exposed to 10 µm $Ni(NO_3)_2$ in hydroponic culture solution. At 24-h intervals over a 7 day period, one set of plants was harvested. Roots were dried at 80°C and weighed. Hydroponic culture solution was aseptically transferred into centrifuge tubes, reduced in volume in a rotary evaporator, and analyzed for amino acids.

Concentrated sterile root'exudates were mixed with $NiCl_2$ traced with ^{63}Ni (59 µCi µmol^{-1} Ni) and applied to a silica thin layer chromatography (TLC) plate. The plate was developed in a solvent system containing water, 2-methoxyethanol, 1-butanol, acetone, and ammonium hydroxide (20%) (45:20:25:10:0.15 v/v), dried and placed face down onto a phosphoimager screen to visualize the ^{63}Ni.

Four-week-old *T. goesingense* and two-week-old *T. arvense* of uniform sizes (three replicate plants for each treatment), grown axenically as described above, were exposed to 25 µm $Ni(NO_3)_2$, $Co(NO_3)_2$, $Zn(NO_3)_2$, or $Cd(NO_3)_2$ added to the hydroponic culture solution. After 48 h, hydroponic culture solution was exchanged, and the plants were exposed to Ni, Co, Zn, or Cd for a further 48 h. Plants were harvested, separated into shoots and roots, dried at 80°C, and weighed. The hydroponic culture solution was aseptically transferred into centrifuge tubes, reduced in volume in a rotary evaporator, and analyzed for amino acids or other unknown Ni-chelating compounds using TLC.

Four-week-old *T. goesingense* and two-week-old *T. arvense* of uniform size, grown axenically as described above, were aseptically transferred into glass vials, containing 10 µmol L^{-1} $Ni(NO_3)_2$, in 10 ml of hydroponic culture solution. Transfers were made so that each vial contained either a pair of *T. goesingense*, a pair of *T. arvense*, or both *T. goesingense* and *T. arvense*. Each treatment was replicated four times. The hydroponic culture solution was changed daily, and plants were harvested after 7 days. Plants were separated into shoots and roots, dried at 80°C, and weighed. Dried plant material was wet ashed and analyzed for Ni by inductively coupled plasma spectrometry (ICP; Fisons Accuris, Fisons Instruments, Inc., Beverly, MA).

Amino acids in sterile root exudates were analyzed after derivatization with *o*-phthalaldehyde (OPA) by HPLC (Jones and Gilligan, 1983).

The compound preliminarily identified as citrate from the TLC Ni-binding assay was collected from a preparative TLC plate, eluted from the silica with water, and derivatized using *N*-methyl-*N*-trimethylsilyl-trifluoroacetimide (MSTFA) with 1% trimethylchlorosilane (TMCS). The sample was dissolved in pyridine, MSTFA + 1% TMCS was added, and the reaction mixture was heated to 60°C in a sealed reaction vial. After heating, the sample was dried under nitrogen, dissolved in pyridine, and injected into a Hewlett Packard gas chromatograph mass spectrometer (5890 GC/5971A MS). Compounds were identified from their fragmentation patterns.

Dried plant material was wet ashed using nitric acid and perchloric acid according to standard methods (Jones and Case, 1990). The resulting solution was analyzed for metal content by ICP. Certified National Institute of Standards and Technology plant (peach leaf) standards were carried through the digestions and analyzed as part of the QA/QC protocol. Reagent blanks and spikes were used where appropriate to ensure accuracy and precision.

RECENT EXPERIMENTAL RESULTS

In the absence of Ni-chelating compounds, ^{63}Ni was found to remain localized at the origin on the TLC plate (Figure 10.1, lane B). However, in the presence of the Ni-chelates histidine and citrate, ^{63}Ni moved from the origin as discrete spots with Rf values of 0.18 and 0.8, respectively (Figure 10.1, lanes A, C). By mixing various unknown compounds with ^{63}Ni and applying the mixture to a TLC plate, it was possible to determine if these unknown compounds were able to chelate Ni by their ability to enhance the mobility of ^{63}Ni on the TLC plate. Using this assay it was determined that sterile root exudate from the hyperaccumulator *T. goesingense* contained no major Ni-chelating compounds (Figure 10.1, lanes D, E). In contrast, root exudate from the nonhyperaccumulator *T. arvense* appeared to contain two major compounds capable of chelating Ni (Figure 10.1, lane F). These two com-

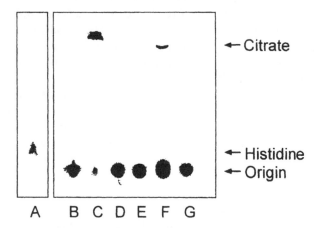

FIGURE 10.1 Silica TLC plate containing Ni-binding compounds. All samples were spiked with NiCl₂ containing ^{63}Ni (59 μCi μmol⁻¹ Ni), loaded at the origin, and the plate developed in a solvent system containing water, 2-methoxyethanol, 1-butanol, acetone, and ammonium hydroxide (20%; 45:20:25:10:0.15 v/v). The plate was air-dried and the ^{63}Ni imaged using a phosphoimager. Lane (A) histidine (7 nmol); (B) water; (C) citrate (0.1 μmol); (D) Root exudate from *T. goesingense* exposed to 25 μm Ni(NO₃)₂ for 48 h; (E) Root exudate from *T. goesingense* not exposed to Ni; (F) Root exudate from *T. arvense* exposed to 25 μm Ni(NO₃)₂ for 48 h; (G) Root exudate from *T. arvense* not exposed to Ni.

pounds had the same Rf values as histidine and citrate (Figure 10.1, lanes A, C) and appeared to accumulate mainly in root exudates produced by *T. arvense* exposed to Ni (Figure 10.1, lanes F, G). Further analysis, by GCMS, of the compound with a similar Rf value to citrate confirmed the presence of citrate in this material.

Histidine exudation from roots of the hyperaccumulator *T. goesingense* was 3.7 ± 0.6 nmol g⁻¹ root dry biomass over a 24-h period (Figure 10.2), and did not change significantly after exposure to 10 µm $Ni(NO_3)_2$ (Figure 10.2). In contrast, root exudation of histidine in the nonhyperaccumulator *T. arvense* increased from 30.3 ± 15.9 nmol g⁻¹ root dry biomass to a mean production of 127 ± 48 nmol His g⁻¹ root dry biomass after exposure to 10 µm Ni nitrate (Figure 10.2).

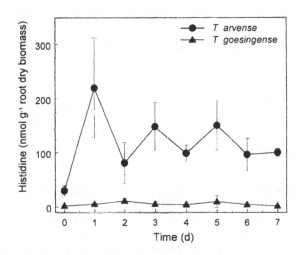

FIGURE 10.2 Histidine production in root exudates of *T. goesingense* and *T. arvense*. Sterile root exudates were collected over a 24-h period from plants exposed to 10 µm $Ni(NO_3)_2$ in hydroponic culture for 0 to 7 days. Root exudates were concentrated in a rotary evaporator and analyzed for amino acids by HPLC after derivatization with *o*-phthaladehyde (OPA). Data points represent mean values ±SD where n = 3.

Root exudation of histidine, serine, glycine, and glutamine in *T. goesingense* remained relatively constant after exposure to 25 µm $Ni(NO_3)_2$ (Table 10.1, Figure 10.3A). However, exposure to 25 µm $Co(NO_3)_2$, $Zn(NO_3)_2$, or $Cd(NO_3)_2$ suppressed amino acid release (Table 10.1, Figure 10.3A). In contrast, roots of the nonhyperaccumulator *T. arvense* responded to Ni exposure by specifically increasing exudation of histidine from 12 ± 5 to 286 ± 162 nmol g⁻¹ root dry biomass (Table 10.2, Figure 10.3B). Smaller increases in histidine were also observed on exposure to Co, Zn, and Cd (Table 10.2, Figure 10.3B). Glutamine concentration also increased in root exudates of *T. arvense* exposed to Cd from 21 ± 1 to 193 ± 48 nmol g⁻¹ root dry biomass (Table 10.2, Figure 10.3B).

Nickel concentrations in shoots of the nonhyperaccumulator *T. arvense* were 4.2 ± 0.3 µmol g⁻¹ dry biomass and decreased slightly to 2.0 ± 0.35 µmol g⁻¹ dry biomass after growth in the presence of the hyperaccumulator *T. goesingense* (Table 10.3). Root Ni concentrations of 15 µmol g⁻¹ dry biomass were measured in *T. arvense*

TABLE 10.1

Amino Acids in Sterile Root Exudates of *T. goesingense* Exposed to Various Metals

	His	Ser	Gly	Gln
Control	21 ± 7	156 ± 55	174 ± 42	20 ± 10
Ni	25 ± 6	185 ± 100	138 ± 68	12 ± 4
Co	24 ± 3	35 ± 4	31 ± 5	4 ± 1
Zn	12 ± 4	88 ± 73	94 ± 68	17 ± 22
Cd	8 ± 1	47 ± 17	42 ± 10	7 ± 2

Note: nmol g^{-1} root dry biomass (SD; n = 3).

growing separately or together with *T. goesingense* (Table 10.3). Nickel concentrations in shoots and roots of *T. goesingense* increased slightly after co-culture with *T. arvense* and shoot Ni concentrations increased from 14 ± 2.7 μmol g^{-1} dry biomass to 18 ± 1.7 μmol g^{-1} dry biomass, and root concentrations increased from 3.7 ± 1.3 μmol g^{-1} dry biomass to 6.6 ± 1.4 μmol g^{-1} dry biomass (Table 10.3).

IMPLICATIONS OF RECENT RESEARCH FINDINGS

We have shown previously that *T. goesingense* grown in hydroponic culture accumulates high concentrations of Ni in shoots and also tolerates high Ni concentrations in both shoots and the hydroponic culture solution (Krämer et al., 1997). However, in the closely related species *T. arvense*, shoot Ni concentrations were found to be low, and a significant reduction in shoot biomass was detected after Ni exposure (Krämer et al., 1997). Intriguingly, the Ni content of roots displayed the opposite trend, with root Ni concentrations being higher in *T. arvense* than in *T. goesingense*. The existence of a more efficient root-to-shoot Ni translocation mechanism in *T. goesingense* would be one possible explanation for the enhanced shoot Ni accumulation observed in the hyperaccumulator species *T. goesingense*. However, it was recently demonstrated that at nontoxic Ni concentrations both *T. arvense* and *T. goesingense* translocate Ni to shoots at equivalent rates (Krämer et al., 1997).

The release of specific metal-chelating compounds into the rhizosphere by plant roots is a well established phenomenon involved in the solubilization and uptake of Fe (Ma and Nomoto, 1996) and possibly Zn (Cakmak et al., 1996a,b). Therefore, it is appealing to suggest that similar Ni-specific chelators may be involved in the Ni hyperaccumulation phenotype observed in *T. goesingense*. However, using a TLC-based assay system, we were unable to identify any high-affinity Ni-chelating compounds in the root exudate of the Ni hyperaccumulator *T. goesingense* (Figure 10.1). In contrast, we observed that the Ni-chelators histidine and citrate accumulate in the root exudate of the nonhyperaccumulator *T. arvense* during exposure to 25 μm Ni for 48 h (Figure 10.1). Further analysis of root exudates from *T. goesingense* and *T. arvense* confirmed that histidine levels in root exudates remain unchanged in the

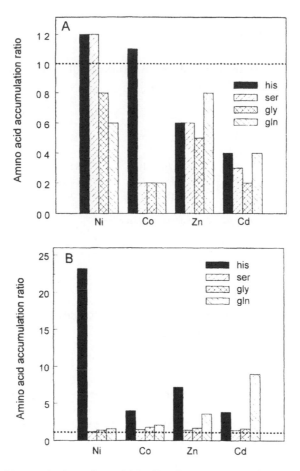

FIGURE 10.3 Changes in the amino acid profile of root exudates of *T. goesingense* (A) and *T. arvense* (B) after exposure to Ni, Co, Zn, and Cd. Sterile root exudates were collected over a 48-h period from plants exposed to either 25 μm $Ni(NO_3)_2$, $Co(NO_3)_2$, $Zn(NO_3)_2$, or $Cd(NO_3)_2$. Root exudates were concentrated in a rotary evaporator and analyzed for amino acids by HPLC after derivatization with *o*-phthaladehyde (OPA). Data is presented as a ratio of the amino acid exudation in the presence of metal, divided by production in the absence of metal. The dashed line represents an amino acid accumulation ratio of 1 signifying that amino acid production remained constant. Data used in these calculations are presented in Tables 10.1 and 10.2.

hyperaccumulator *T. goesingense* and increase in the nonhyperaccumulator *T. arvense* on exposure to Ni (Tables 10.1 and 10.2, Figures 10.2 and 10.3). Histidine concentrations in root exudates of *T. arvense* after Ni exposure are also at least 10-fold higher than in *T. goesingense* (Tables 10.1 and 10.2, Figure 10.2). These data suggest that histidine, exuded by roots, does not play a significant role in the enhanced foliar Ni accumulation observed in the Ni hyperaccumulator *T. goesingense*. It is possible that Ni-chelating compounds exuded by roots, but not detected using the TLC Ni-binding assay described here, are involved in Ni hyperaccumula-

TABLE 10.2

Amino Acids in Sterile Root Exudates of *T. arvense* Exposed to Various Metals

	His	Ser	Gly	Gln
Control	12 ± 5	133 ± 52	89 ± 28	21 ± 1
Ni	286 ± 162	157	129	34
Co	50 ± 23	201 ± 108	164 ± 82	44 ± 15
Zn	89 ± 123	184 ± 33	154 ± 14	76 ± 24
Cd	47 ± 3	183 ± 61	144 ± 47	193 ± 48

Note: nmol g⁻¹ root dry biomass (SD; n = 3)

TABLE 10.3

Nickel Accumulation During Hydroponic Co-Culture of *Thlaspi* Species

Species		Shoots	Roots
Thlaspi goesingense[a]	GG	14.0 ± 2.0	3.7 ± 0.9
	GA	18.0 ± 1.5	6.6 ± 1.2
T. arvense[a]	AA	4.2 ± 0.3	14.6 ± 1.1
	AG	2.0 ± 0.3	14.9 ± 0.7

Note: μmol Ni g⁻¹ dry biomass (SD; n = 4)

[a] Plants cultured as follows: GG, *T. goesingense*; GA, *T. goesingense* and *T. arvense*; AA, *T. arvense*; AG, *T. arvense* and *T. goesingense*.

tion. However, experiments in which *T. goesingense* was co-cultured with *T. arvense* (Table 10.3) suggest that if these compounds exist they cannot be utilized by closely related nonhyperaccumulator species to enhance Ni uptake.

Even though root-exudate histidine does not appear to be involved in Ni hyperaccumulation in *T. goesingense*, the question remains as to its function in the nonhyperaccumulator *T. arvense*. Release of Al chelators, including citric and malic acids, into the rhizosphere reduces the chemical activity of the toxic Al^{3+} ion in solution and is known to be involved in Al resistance in certain Al-tolerant plant cultivars (Kochian, 1995). This raises the intriguing possibility that the specific release of histidine by *T. arvense* roots may be part of a Ni detoxification strategy. Histidine forms stable complexes with Ni^{2+} (Sundberg and Martin, 1974) and may be expected to chelate Ni in the rhizosphere, reducing the solution activity of the Ni^{2+} ion and thereby lowering its toxicity. The release of citrate by *T. arvense* into the rhizosphere would also be expected to lower Ni toxicity by chelation of Ni^{2+} (Dawson et. al., 1986). It is therefore possible that the enhanced release of both histidine and citrate by *T. arvense* roots may be a plausible strategy to reduce Ni uptake and toxicity (Cumming and Taylor, 1990). However, it is clear that this mechanism is rapidly overwhelmed at elevated Ni concentrations in the Ni nontolerant *T. arvense* used in this study (Krämer et al., 1997). It is interesting to speculate that the enhancement of this type of mechanism may be involved in the adaptation of certain Ni-tolerant plant ecotypes to growth on Ni-enriched soils (Cox and Hutchinson, 1979).

Therefore, release of histidine and citrate into the rhizosphere does not appear to play an important role in Ni hyperaccumulation in *T. goesingense*. However, root exudation of the Ni chelators histidine and citrate may play a role in the reduction of Ni uptake and toxicity in nonhyperaccumulator plants such as *T. arvense*.

FUTURE DIRECTIONS

Recent data from our laboratory suggest that Ni tolerance is sufficient to explain the Ni hyperaccumulation phenotype observed in hydroponically cultured *T. goesingense* when compared with the Ni-sensitive nonaccumulator *T. arvense* (Krämer et al., 1997). The fact that protoplasts isolated from *T. goesingense* were more Nitolerant than those isolated from *T. arvense* suggests the existence of a cellular mechanism of Ni tolerance in the leaves of the hyperaccumulator. Our future research efforts therefore will be to address the biochemical and molecular basis of Ni detoxification in *T. goesingense*.

By isolating protoplasts and intact vacuoles from leaves of *T. goesingense*, and using ^{63}Ni as a sensitive tracer for Ni, we are working toward gaining a better understanding of the distribution of Ni in the leaves of the hyperaccumulator. Our preliminary data suggest that cellular Ni is localized within the vacuole of the hyperaccumulator, as has been found for Cd (Volgeli-Lange and Wagner, 1990) and Zn (Brune et al., 1994). If this is confirmed, we will proceed to investigate the mechanism(s) of Ni transport across the tonoplast using the techniques applied to investigating the tonoplast transport of Cd (Salt and Wagner, 1993; Salt and Rauser, 1995).

In support of these localization studies, we have also been using x-ray spectroscopy as a noninvasive tool to establish the speciation of Ni within the Ni hyperaccumulator and nonaccumulator plants. Our data suggest that the majority of cellular Ni in *T. goesingense* is associated with organic acids. As the major pool of organic acids within plant cells is known to be vacuolar, this observation supports the vacuolar localization for Ni in the hyperaccumulator. However, in the Ni non-tolerant nonaccumulator *T. arvense*, up to 45% of the foliar Ni appears to be associated with histidine-like ligands (containing N and O-Ni coordination sites). This represents a sevenfold higher concentration of Ni in this form than in the hyperaccumulator, suggesting that in the Ni nontolerant plants, Ni accumulates within the cytoplasm, where it is coordinated by mixed N-O ligands within proteins and possibly nucleic acids. It is this uncontrolled binding of Ni within the cytoplasm of the cells that leads to the toxic symptoms observed in these Ni nontolerant nonaccumulator plants after Ni treatment.

Recently, free histidine has been proposed to play a key role in Ni hyperaccumulation in several species of *Alyssum* (Krämer et al., 1996). Exposure of these plants to Ni caused a pronounced increase in the xylem sap concentration of free histidine, and the concentration of histidine was directly correlated with the Ni concentration in the sap. Because of these interesting findings, we have also been investigating the possible involvement of histidine in Ni hyperaccumulation in *T. goesingense*. Our preliminary observations have revealed that there are no significant increases in free histidine concentrations in either xylem sap or bulk extracts of shoot or root tissue of *T. goesingense* after Ni exposure. However, in order to investigate the possible involvement of free histidine in Ni hyperaccumulation in more detail at the tissue and cellular level, we have cloned, by functional complementation in *E. coli* mutants, three key genes involved in the biosynthesis of histidine: ATP phosphoribosyl transferase (ATP PRT), imidazoleglycerol phosphate dehydratase (IGPD), and histidinol dehydrogenase (HD). We are now studying the regulation of these enzymes by Ni and the regulation of gene expression during Ni exposure.

By investigating the basic physiology, biochemistry, and molecular biology of Ni hyperaccumulation in *T. goesingense* we hope to be able to define, at the genetic level, the key processes that determine the Ni hyperaccumulation phenotype. This knowledge will then be used to create plants more ideally suited to the phytoremediation of metal-contaminated soils and waters.

ACKNOWLEDGMENTS

This research was supported by the U.S. Department of Energy Grant No. DE-FG07-96ER20251 to D.E.S, a NATO fellowship awarded to U.K. by the German Academic Exchange Service (DAAD), and by Phytotech, Inc. to I.R. We would like to thank Dr. David M. Ribnicky for his assistance in the derivatization and analysis of samples by GCMS, and Dr. Angus Murphy for his help in the analysis of the GCMS mass profiles. We would also like to thank Dr. Xianghe Yan for cloning of the histidine biosynthetic genes, and Dr. Ingrid Pickering and Dr. Roger Prince for assistance with the collection and analysis of the x-ray spectroscopy data at the Stanford Synchrotron Radiation Laboratory, Stanford, CA.

REFERENCES

Antonovics, J., Bradshaw, A.D., and Turner, R.G. (1971) Heavy metal tolerance in plants. *Adv. Ecol. Res.* 7: 1-85.

Baker, A.J.M. and Brooks, R.R. (1989) Terrestrial higher plants which hyperaccumulate metallic elements — a review of their distribution, ecology and phytochemistry. *Biorecovery* 1: 81-126.

Boyd, R.S. and Martens, S.N. (1994) Nickel hyperaccumulated by *Thlaspi montanum* var. *montanum* is acutely toxic to an insect herbivore. *Oikos* 70: 21-25.

Boyd, R.S., Shaw, J.J., and Martens, S.N. (1994) Nickel hyperaccumulation defends *Streptanthus polygaloides* (Brassicaceae) against pathogens. *Am. J. Bot.* 81: 294-300.

Brune, A., Urbach, W., and Dietz, K.-J. (1994) Compartmentation and transport of zinc in barley primary leaves as basic mechanisms involved in zinc tolerance. *Plant Cell. Environ.* 17: 153-62

Cakmak, I., Sari, N., Marschner, H., Ekiz, H., Kalayci, M., Yilmaz, A., and Braun, H.J. (1996a) Phytosiderophore release in bread and durum wheat genotypes differing in zinc efficiency. *Plant Soil* 180: 183-189.

Cakmak, I., Ozturk, L., Karanlik, S., Marschner, H., and Ekiz, H. (1996b) Zinc-efficient wild grasses enhance release of phytosiderophores under zinc deficiency. *J. Plant Nutr.* 19: 551-563.

Cox, R.M. and Hutchinson, T.C. (1979) Metal co-tolerance in the grass *Deschampsia caespitosa*. *Nature* 279: 231-233.

Cumming, J.R. and Taylor, G.J. (1990) Mechanisms of metal tolerance in plants: physiological adaptations for exclusion of metal ions from the cytoplasm. 329-359. In *Stress Responses in Plants: Adaptation and Acclimation Mechanisms*. Alscher, R.G. and Cumming, J.R. (Eds.). Wiley-Liss.

Dawson, R.M.C., Elliott, D.C., Elliott, W.H., and Jones, K.M., Eds. (1986) *Data for Biochemical Research*, 3rd ed., Clarendon Press, Oxford, U.K.

DOE/EM-0224, (1994) Summary Report of a Workshop on Phytoremediation Research Needs, July 24-26, Santa Rosa, CA.

Homer, F.A., Morrison, R.S., Brooks, R.R., Clemens, J., and Reeves, R.D. (1991) Comparative studies of nickel, cobalt and copper uptake by some nickel hyperaccumulators of the genus *Alyssum*. *Plant Soil* 138: 195-205.

Jones, J.B., Jr. and Case, V.V. (1990) *Soil Testing and Plant Analysis*, 3rd ed.; Westerman, R.L. (Ed.); SSSA Book Series, No. 3. Soil Science Society of America, Madison, WI.

Jones, B.N. and Gilligan, J.P. (1983) *o*-Phthalaldehyde precolumn derivatization and reversed-phase high-performance liquid chromatography of polypeptide hydrolysates and physiological fluids. *J. Chromatogr.* 266: 471-482.

Kochian, L.V. (1995) Cellular mechanisms of aluminum toxicity and resistance in plants. *Ann. Rev. Plant Physiol. Plant Mol. Biol.* 46: 237-260.

Krämer, U., Cotter-Howells, J.D., Charnock, J.M., Baker, A.J.M., and Smith, J.A.C. (1996) Free histidine as a metal chelator in plants that accumulate nickel. *Nature* 379: 635-638.

Krämer, U., Smith, R.D., Wenzel, W.W., Raskin, I., and Salt, D.E. (1997) The role of metal transport and tolerance in nickel hyperaccumulation by *Thlaspi goesingense* Hálácsy. *Plant Physiol.* 115: 1641-1650.

Lee, J., Reeves, R.D., Brooks, R.R., and Jaffré, T. (1978) The relation between nickel and citric acid in some nickel-accumulating plants. *Phytochemistry* 17: 1033-1035.

Lloyd-Thomas, D.H. (1995) Heavy metal hyperaccumulation by *Thlaspi caerulescens* J. & C. Presl. Ph.D. thesis. University of Sheffield, Sheffield, U.K.

Ma, J.F. and Nomoto, K. (1996) Effective regulation of iron acquisition in graminaceous plants — the role of mugineic acids as phytosiderophores. *Physiol. Plant* 97: 609-617.

Pollard, A.J. and Baker, A.J.M. (1997) Deterrence of herbivory by zinc hyperaccumulation in *Thlaspi caerulescens* (Brassicaceae). *New Phytol.* 135, 655-658.

Reeves, R.D. and Brooks, R.R. (1983) European species of *Thlaspi* L. (Cruciferae) as indicators of nickel and zinc. *J. Geochem. Explor.* 18: 275-283.

Salt, D.E. and Rauser, W.E. (1995) MgATP-dependent transport of phytochelatins across the tonoplast of oat roots. *Plant Physiol.* 107: 1293-301.

Salt, D.E. and Wagner, G.J. (1993) Cadmium transport across tonoplast of vesicles from oat roots. Evidence for a Cd^{+2}/H^+ antiport activity. *J. Biol. Chem.* 268: 12297-302.

Sundberg, R.J. and Martin, R.B. (1974) Interactions of histidine and other imidazole derivatives with transition metal ions in chemical and biological systems. *Chem. Rev.* 74: 471-517.

Vogeli-Lange, R. and Wagner, G.J. (1990) Subcellular localization of cadmium-binding peptides in tobacco leaves. Implications of a transport function for cadmium-binding peptides. *Plant Physiol.* 92: 1086-93.

11 ENGINEERED PHYTOREMEDIATION OF MERCURY POLLUTION IN SOIL AND WATER USING BACTERIAL GENES

R. B. Meagher, C. L. Rugh, M. K. Kandasamy, G. Gragson, and N. J. Wang

CONTENTS

Key words: toxic metals, heavy metals, pollution, sequestration, reduction, phytoremediation

1-56670-450-2/00/$0 00+$.50
© 2000 by CRC Press LLC

INTRODUCTION

Many industrial, defense, and agricultural processes have contaminated large areas of soil, sediment, and water with mercury. Divalent mercury (Hg[II]) forms hydroxides, thiol salts, and organomercury compounds that are extremely toxic. In particular, methylmercury is biomagnified up the food chain in many contaminated environments. Our present research focuses on using plants to extract, sequester, and/or eliminate mercury from contaminated sites.[1] We have successfully engineered a modified bacterial mercuric ion reductase gene (merA) to be expressed in several diverse plant species. The MerA enzyme reduces ionic mercury (Hg[II]) to less toxic and volatile metallic mercury (Hg[0]). When the small model plant *Arabidopsis thaliana* is engineered to express a modified bacterial gene, merA9, the seeds germinate and the plants root, grow, and flower on medium with Hg(II) levels lethal to wildtype plants (25-250 μm).[2] *Arabidopsis* seedlings suspended in 5 μm Hg(II) volatilize 10 ng of Hg(0)/min/mg of plant tissue. We have extended this system from a small model plant to larger macrophytes for the remediation of soil and water. Transgenic *Brassica napus* (canola) and *Nicotiana tabacum* (tobacco) plant lines were constructed expressing modified bacterial merA genes. Their transgenic seeds germinate and plantlets grow at normal rates on medium with levels of Hg(II) that kill the parent plant line. When hydroponically grown transgenic merA tobacco lines are exposed to as little as 5 μm Hg(II) in their medium, they can reduce the mercury concentration to 1 μm in less than 1 week. Control plants have little impact on the Hg(II) content of their media. The advantages of phytoremediation of trace element contamination are discussed. An initial assessment of the risks to the local and global environments suggest that this phytoremediation strategy is quite sound.

In this chapter, we present the general advantages of phytoremediation over other existing technologies for cleaning up environmental toxic waste, particularly mercury. We describe our work on the engineered phytoremediation of mercury pollution to illustrate some of these advantages. The potential risks and global impact of the large-scale remediation of mercury pollution are also discussed.

SOURCES AND CONSEQUENCES OF MERCURY CONTAMINATION

Man-made mercury contamination occurs throughout the globe and has increased steadily with the industrial revolution.[3] In the U.S., mercury is a common pollutant at government production sites, where it is used in energy- and defense-related activities (e.g., as a coolant in reactors, as ballast in submarines). The largest industrially contaminated sites have resulted from its use in the bleaching industry (e.g., chlorine and NaOH production, paper, textiles). Mercury's use in agricultural pesticides directed against bacteria and fungi is also a major source of mercury pollution. Other man-made sources of mercury pollution include mining operations, coal-burning power plants, urban waste disposal such as the incineration of medical waste, the use of mercury as a catalyst, and as a pigment in paints. Once in the environment, mercury and its various species enter complex global and local cycles, moving between the atmosphere, sediment, and water, as summarized in Figure 11.1. The

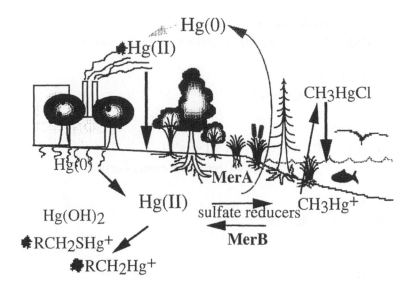

FIGURE 11.1 Biogeochemical cycle for mercury. This abbreviated diagram of the global mercury cycle emphasizes the rapid cycling of Hg(II) species between soil, water, and the atmosphere. In contrast metallic mercury, Hg(0) vapor is slowly and inefficiently returned to the Earth's surface.

numerous bio- and geochemical forces acting on mercury result in its rapid movement away from point sources.

In the U.S. alone, industrial and agricultural sources account for thousands of square miles of land, rivers, lakes, and estuaries contaminated with millions of kilograms of mercury. Many contaminated sites worldwide are the result of accidents during the production, use, or storage of mercury. For example, the production of mercury-based fungicides was responsible for the notorious contamination of Minamata Bay, off Kyushu Island, Japan. Under today's stricter guidelines for industry, relatively few enormous mercury spills occur within the U.S. or most other developed countries (e.g., most of the Superfund sites are more than 20 years old). However, the available literature indicates that *few of the large mercury-contaminated sites in the U.S. or anywhere in the world have been completely reclaimed.* The contamination at these sites continues to spread because there are few, if any, affordable and effective means of containing the mercury.

We have focused our research on methylmercury because it is extremely toxic[4] and has a devastating impact on humans and other animals.[5] The neurological diseases methylmercury produces are, thus far, untreatable. It is true that much of the ionic mercury at polluted sites is bound in high molecular weight compounds in the soil and sediment, and most mercury species move slowly and inefficiently into the food chain. However, methylmercury, which is organic, is a major threat to wildlife and humans (Figure 11.1) because it moves rapidly into the food chain.[6]

The prevailing theory, supported by mounting evidence, is that most methylmercury is catalytically produced from Hg(II) in aquatic or marine environments[7,8] by anaerobes like the sulfate-reducing bacteria at these sites (Figure 11.2, Equation 1) living at the aerobic/anaerobic interface (Figure 11.1).[9,10] Methylmercury is very spontaneous & catalyzed

$$methyl - R_1 + Hg(II)TH + \cdots \rightarrow CH_3 - Hg^+ + R_1H \; (\text{sulfate reducing bacteria})$$ 1.

spontaneous

$$CH_3 - Hg^+ + R_1SH \cdots \rightarrow CH_3 - Hg - SR_1 + H^+$$ 2.

MerB

$$R_2 - CH_2 - Hg - SR_1 + 2H^+ \cdots \rightarrow R_2CH_3 + R_1SH + Hg(II)$$ 3.

$$(\text{broad spectrum resistance } mer \text{ operons})$$

spontaneous

$$Hg(II) + R_1SH \cdots \rightarrow R_1S - Hg^+ + H^+$$ 4.

MerA

$$R_1S - Hg^+ + NADPH \cdots \rightarrow Hg(0) + R_1SH + NADP^+ \; (\text{all } mer \text{ operons})$$ 5.

FIGURE 11.2 Biological reactions involving ionic and organic mercury. Mono-methylmercury (CH_3Hg^+) is formed catalytically in sulfate-reducing bacteria (Equation 1) and degraded by two enzymes encoded in the *mer* operon found in gram-negative bacterial plasmids. Methylmercury may react with sulfhydryl compounds (Equation 2) to form a substrate for its decay in broad-spectrum, mercury-resistant bacteria. *In vitro* R1-SH can be almost any thiol-containing compound but *in vivo* it is probably small molecular weight compounds like glutathione. In a coupled reaction, organomercury lyase (MerB) catalyzes the protonolysis of organomercury compounds (Equation 3) like methylmercury ($R_2 = H$) or phenylmercury (R_2-CH_2 = benzene) to release free mercury ions (Hg(II)). MerB protonolyzes the carbon-mercury bond of a wide range of other organic mercury compounds (e.g., ethylmercury, vinylmercury, propenylmercury), freeing the mercuric ion, Hg(II).[49,50] Ionic mercury reacts rapidly with the sulfhydryl groups on any available thio-organic compound, forming very stable thiol salts (e.g., $(SR)_2Hg$, $RSHg^+$, $Kd \sim 10^{-45}$) (Equation 4). Mercuric ion reductase (MerA, Equation 5) removes mercury from stable thiol salts by electrochemically reducing it to the less toxic, relatively volatile metallic mercury Hg(0) in an NADPH-coupled redox reaction.[51]

mobile and is quickly biomagnified up the food chain from sulfate bacteria to fungi, crustaceans, and annelids living on detritus at the sediment–water interface, and then to fish, birds, and mammals.[11] As a result, even if the sediment has less than 1 ppm methylmercury, the fish, birds, and mammals may end up with 100 ppm methylmercury after its biomagnification through the food chain. Methylmercury concentrates primarily because it and some of its salts, such as methylmercury chloride, are nonpolar and thus move efficiently into the membrane systems of eukaryotic cells and organisms (discussed below).

Mercury pollution enters the environment in a number of ways (Figure 11.1). For example, most of the mercury released into the atmosphere by incineration is ionic mercury, Hg(II), bound to high molecular weight compounds and particulates.[12] The chloride salt of methylmercury is also volatilized by evaporation into the atmosphere from aquatic sites. Some mercury from spills of Hg(0) is volatilized directly into the atmosphere, but most of it is sequestered in soil and sediments where it is later oxidized to ionic mercury, $Hg(II)$. $Hg(II)$ reacts rapidly with many constituents in soil and sediment to produce organic, sulfhydryl, and hydroxide derivatives of mercury. For example, humic acids form a number of related and relatively stable high molecular weight compounds and produce the increased mercury-binding capacity of rich soils. In soil, as in aquatic environments, only limited supplies of free reactive ionic mercury are available because methylation, reduction, and particulate scavenging consume ionic Hg.[13]

As noted in a recent EPA report on mercury pollution: "Mercury species are subject to much faster atmospheric removal than elemental mercury."[12] All known species of $Hg(II)$ that make it into the atmosphere (i.e., large particulates bound to mercury and made airborne by incineration, gaseous organics of mercury such as phenylmercury chloride or methylmercury chloride) are rapidly returned to Earth by dry deposition or washed from the atmosphere to the surface by rainfall.[13] These compounds have a half-life in the atmosphere of a few weeks. "In contrast, elemental mercury ($Hg[0]$) vapor has a strong tendency to remain airborne and is not susceptible to any major process resulting in direct deposition to the Earth's surface."[12] $Hg(0)$ has a half-life in the atmosphere on the order of years.[14,15] Thus, the metallic mercury volatilized into the atmosphere at contaminated sites is rapidly diluted into a large global atmospheric pool which ends up dispersed over the Earth's surface,[6] whereas the mercury species released into the atmosphere are redeposited near their release sites.

STRATEGY FOR DEVELOPING GENETICALLY ENGINEERED, MERCURY-RESISTANT TRANSGENIC PLANTS

Bacteria having the *mer* operon of genes can detoxify and reduce various mercury species to the much less harmful $Hg(0)$ (Figure 11.2) and these genes are proving highly useful to phytoremediation research. We have extended the metal ion reduction portion of the bacterial system for mercury detoxification to transgenic plants in an effort to harness these natural, bacterial remediation capabilities to the tremendous extraction and reduction capacity of plants. Plants engineered to metabolize organic mercury species, reduce $Hg(II)$ to metallic $Hg(0)$, and transpire $Hg(0)$ into the atmosphere may be able to remediate mercury-contaminated sites. The subject of our research is the manipulation of plants to carry out this process more efficiently than bacteria.

Because plants are multicellular, most of their natural mechanisms of metal tolerance involve sequestering metals on root surfaces, in vascular tissues, and in vacuoles. Photosynthetic electron transport (PET) and oxidative electron transport (OET) systems are particularly sensitive to strong terminal electron acceptors (e.g., cyanide) and thio-binding reagents, such as some heavy metals cations (e.g., $Hg[II]$, $Cd[II]$) and oxyanions (e.g., arsenite). In plants and other eukaryotes, these sensitive electron transport systems are contained in chloroplasts and mitochondria that are protected from toxins by a layer of cytoplasm (Figure 11.3).

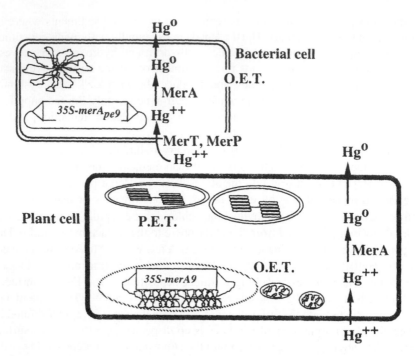

FIGURE 11.3 Plant electron transport systems are protected from toxins by a layer of cytoplasm in contrast to the exposed systems in bacteria. Essential energy transfer systems like the photosynthetic electron transport chain (PET) or oxidative electron transport chain (OET) are easily poisoned by a number of toxins. In plants and other eukaryotes, toxins, such as terminal electron acceptors and sulfhydryl reagents, must pass through a layer of cytoplasm where they either react with normal cellular metabolites or can be detoxified by specific enzymes before poisoning these systems. Bacteria have these essential energy systems in their outer membranes where they are more exposed to the environment and thus have evolved more immediate and assertive solutions to these toxins. This is well illustrated by the bacterial mercury resistance operon. In plants, only a functional *merA* gene is needed to protect cells from Hg(II).

In contrast, bacterial electron transport systems are more exposed to environmental toxins because they are protected only by their cell walls. As a result, bacteria have developed more assertive solutions to toxin exposure. Bacteria are able to degrade a variety of organic pollutants, including chlorinated hydrocarbons and aromatics, toxic nitroaromatics, and a wide variety of mutagens.[16-18] To cope with heavy metals, which are immutable, bacteria have evolved two complementary mechanisms: (1) pumps to eliminate the toxic metal ions from their cytoplasm[19] and (2) the ability to reduce, oxidize, or modify metal ions to less toxic metal species. For example, there are bacteria that can catalyze the reduction of chromate, ferric ion, selenate, tellurate, and uranate[20-22] to less toxic cations or oxyanions.

Only a few bacterial genes with activities directed toward toxic heavy metals have been sufficiently characterized at the molecular level to be of immediate use in the phytoremediation of heavy metals. The best studied system, that for mercury resistance, is encoded by the *mer* operon.[23] Microbial *mer* operon-based processing of mercury occurs at all contaminated sites.[24,25] The *mer* operon encodes genes for sensing mercury (*merR*, a mercury-responsive regulatory protein), mercury transport through the membrane (*merT*), sequestration in the periplasmic space (*merP*), and metal ion reduction (*merA*). Of particular interest to our research, the MerA enzyme catalyzes the reduction of Hg(II) to Hg(0), the least toxic and most volatile form of mercury (Figure 11.2). As a group, these genes provide bacteria with what is called "narrow-spectrum resistance" specifically to mercury ions. In addition to these genes, a subset of the *mer* operons have been characterized that contain a methylmercury or organomercurial lyase gene (*merB*). The MerB enzyme catalyzes the breakdown of various forms of organic mercury to Hg(II) and a reduced carbon compound. Together, the *merA* and *merB* operons provide bacteria with "broad-spectrum resistance" to a variety of organomercurial compounds (see Figure 11.2).

ENGINEERED *Arabidopsis* PLANTS EXPRESSING A MODIFIED *merA* SEQUENCE DETOXIFY HG(II)

The original *merA* gene with which we initiated our studies came from the bacterial transposon, *Tn21*. It had a very high GC composition and used codons that are rare in plants and *E. coli*. After moderate flanking sequence reengineering for plant expression, this native bacterial *merA* protein coding region was not stably expressed in transgenic plants.[26,27] Therefore, we constructed a modified *merA* sequence, *merA9*, to make a gene more compatible with *E. coli* and plant expression. This was done using PCR and large, mutagenic oligonucleotide primers to modify the *merA* sequence. In the flanking regions, we added plant and bacterial translation signals and appropriate restriction endonuclease cloning sites. In a central region covering 9% of the codons (53 codons), we lowered the GC composition and replaced rare codons with more common ones.[2] The resulting *merA9* gene was placed under control of the bacterial *lac o/p* region (i.e., the bacterial lactose operon's operator/promotor sequence, which directs transcription) to make the pNS2 plasmid. To test that *merA9* encoded a functional mercuric ion reductase protein, pNS2 was transformed into an *E. coli* strain containing the *Tn21 mer* operon with a dysfunctional *merA* gene (a mercury-supersensitive strain, Hg[ss]). pNS2 conferred high-level mercury resistance to this strain (Hg[r]), demonstrating that *merA9* produced a functional mercuric ion reductase enzyme. Sequence analysis revealed that the modified *merA9* gene encoded the expected MerA protein with only a few minor changes in the amino acid sequence.

This *merA9* sequence was excised from the bacterial plasmid and placed under control of plant regulatory elements, the CaMV 35S promoter, and NOS 3′ polyadenylation signals on the plant binary vector, pVST1. This construct was inserted into the genome of the small model plant, *Arabidopsis thaliana*, via *Agrobacterium tumefaciens*-mediated transformation of embryos. These embryos and the subsequent transgenic shoots were induced to develop from root tissue through the use of phytohormones. Transgenic plants were regenerated (T0) and T1 generation seeds were collected. T2 and later generation seeds, seedlings, and plants were assayed for their ability to metabolize Hg(II). Transgenic *A. thaliana* seeds expressing *merA9* germinated and these seedlings grew, flowered, and set seed on medium containing HgCl$_2$ concentrations of 25-100 μm, levels toxic to most plant species. The parent plant line and transgenic controls died rapidly when grown with these concentrations of mercury. Transgenic MerA *Arabidopsis* seedlings suspended in 5 μm of Hg(II) volatilized 10 ng of Hg(0)/min/mg plant tissue. The rate of mercury evolution and the level of resistance were proportional to the steady-state *merA9* mRNA levels in the various independent plant lines, confirming that resistance was due to expression of the *merA9* enzyme. Plants and bacteria expressing *merA9* were also resistant to toxic levels of Au(III), suggesting that, in the future, related enzymes or modified *merA* sequences might be used to confer reduction of and resistance to a wide variety of toxic metal ions.

ENGINEERING MACROPHYTES FOR *merA* EXPRESSION AND RESISTANCE TO HG(II): *Nicotiana tabacum* AND *Brassica napus*

The *merA9* gene is functional in *Arabidopsis* plants and efficiently reduces Hg(II) to Hg(0). The goal of our recent research has been to extend this result from *Arabidopsis* to larger macrophytes in which we can better assay the potential of engineered plants to remediate mercury-contaminated soil and water. Extending this technology into appropriate conservation species (i.e., native or naturalized plant species that grow in a variety of ecosystems) should enable the cleanup of mercury-polluted sites that are prohibitively expensive to clean with existing technologies. Constructed ecosystems of mercury-processing herbs, grasses, shrubs, and trees (Figure 11.1) could be used to clean waste streams from ongoing industrial operations, waste sediment from population centers, or large contaminated sites. To this end, we have engineered tobacco and *Brassica* plants to express the *merA* gene. Because *Arabidopsis* has not had a common ancestor with *B. napus* for 40 to 60 million years (MY), nor with tobacco for approximately 110 MY,[28] success with these two plant species would demonstrate that related gene constructs should work in all dicots and possibly in the more-distant monocots.

Agrobacterium tumefaciens-mediated transformation methods similar to those described previously[2] were used to insert the *merA* gene constructs into explants of aseptically grown tobacco (*N. tabacum*) and *Brassica* (*B. napus* var. Westar and Oscar). *Brassica* was transformed with the previously described *merA9* gene construct, and tobacco was transformed with *merA9* and a second-generation construct encoding *merA18*.[29] *merA18* encodes a protein identical to *merA9*, but the *merA18* coding sequence has had 18% of the protein coding sequence reconstructed from synthetic DNA to reflect the preferred codon usage and GC composition of *E. coli* and higher plants. Both *merA9* and *merA18* were carried in the plant binary vector, pVST1, described above. Shoots of transformed tobacco carrying *merA18* and *Brassica* carrying *merA9* were selected for the linked *nptII* (a kanamycin resistance) gene on the pVST1 vector on medium containing kanamycin. The T0 plants were allowed to set seed (T1), and only T2 or later generation seeds or plants were assayed in this study. Seeds were germinated on normal Gamborg's B5 medium containing 25-250 μm $HgCl_2$. A portion of these data are shown in Figure 11.4 and described in the legend.

Transgenic tobacco lines expressing *merA18* (Figure 11.4A) and transgenic *Brassica* lines expressing *merA9* (Figure 11.4B) were resistant to Hg(II) levels that

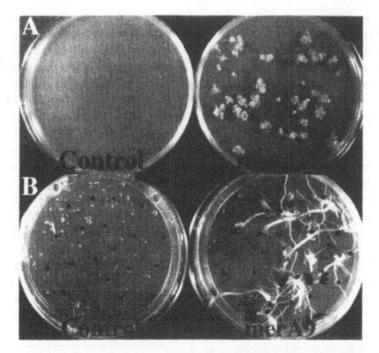

FIGURE 11.4 Transgenic tobacco and *Brassica* plants expressing *merA* constructs were resistant to toxic levels of Hg(II) in their media. (A) Tobacco seeds from wildtype controls (left) and from transgenic lines expressing the *merA18* gene (right) were germinated on medium containing 50 μm mercury. (B) *Brassica napus* seeds from wildtype controls (left) and from transgenic lines expressing the *merA9* gene (right) were germinated on medium containing 250 μm mercury.

killed wildtype control plants. Transgenic tobacco and *Brassica* seeds germinated normally on mercury-containing medium, while the controls did not. Both plant species matured normally and root growth was vigorous on mercury-containing medium that killed the controls. Cuttings of *merA*-expressing tobacco plants could be rooted on medium containing levels of Hg(II) that killed control tissue. These transgenic lines of tobacco and *Brassica* are currently being used in a variety of experiments designed to test the efficacy of our mercury remediation strategy. Parallel work has been carried out successfully in an even more distant tree species, yellow poplar.[27]

PHYSIOLOGICAL RESPONSES OF MERCURY-REMEDIATING TRANSGENIC PLANTS

The ability of MerA plants to process mercury in a liquid waste stream was simulated with a hydroponic medium. Tobacco plants for hydroponic assays were prepared from seeds screened for Hg(II) resistance by germination on plates containing 50 μm Hg(II) and grown in soil. When the plants were about 5 weeks old (12 to 15 in. tall), they were suspended with their roots in 1 l of Kent's hydroponic medium[30] in a Ball canning jar. Their stems were held in place with cotton in a 1-cm hole in a #13 rubber stopper. A 1 x 2 cm air stone was used to slowly bubble air through the hydroponic medium. Plants were allowed 1 week to adapt to hydroponic growth. At time zero, when Hg(II) was added to the medium, most of the plants had about 5 g of roots. After the appropriate time of exposure to Hg(II), tissue samples were ground in liquid nitrogen, lyophilized for 24 h, and acid hydrolyzed for 48 h. The tissue hydrolysate was analyzed against known standards by cold vapor atomic adsorption spectroscopy at the University of Georgia Chemical Analysis Laboratory.

The 12- to 15-in. tall *merA18*-expressing tobacco plants and wildtype control plants were adapted to growth in hydroponic medium for 1 week. At time zero, 5 μm Hg(II) were added to the medium as $HgCl_2$. At the appropriate time the plants were harvested (roots, tops) and samples of the medium were taken. These samples were assayed for total mercury content. Most of the mercury had adsorbed to roots of both control and transgenic plants after 24 h of exposure (not shown). After 7 days, the *merA9* plants had eliminated 75 to 80% of the Hg(II) from the system, reducing it to about 1.25 μm (Figure 11.5, day 7). Thus, when these transgenic tobacco plants had about 5 g of roots exposed to 5 μm mercury (1 mg in a liter) in hydroponic medium, they vaporized approximately 0.75 mg of Hg(0) in a week (Figure 11.5). This is about 1.5 ng Hg(0) evolved per milligrams of root tissue per

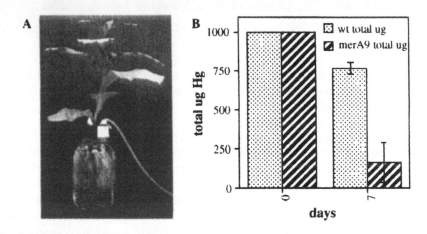

FIGURE 11.5 Transgenic tobacco processes significant amounts of Hg(II) from hydroponic growth medium. (A) Six-week-old *merA9* expressing tobacco plants (shown above) and wildtype control plants were adapted to growth in hydroponic medium and were treated 5 μm mercury in their medium. (B) At time zero, 1000 μg of Hg(II) were added to the 1 l of hydroponic growth media surrounding the roots of *merA9* transgenic and wildtype control plants (i.e., media = 5 μm $HgCl_2$). Within 24 h, most of the Hg(II) in the medium was bound to roots of either the wildtype or *merA9* transgenic plants (not shown). After 7 days, the *merA9* transgenic plants had reduced the total amount of Hg(II) in the system to less than 20 to 25% of the starting levels (roots + leaves + medium; all of remaining mercury was still bound to roots). The wildtype controls left 75 to 80% of the mercury in the system, with all of this still bound to roots. These data points represent the average from three wildtype control and three *merA9* transgenic plants; the range of raw values are represented by the error bars.

minute, only a fewfold below the levels of volatilization obtained with whole *Arabidopsis* seedlings submerged in a solution of ionic mercury.

This preliminary experiment is being repeated and expanded to include many more samples and controls, but the result appears quite clear. The *merA18* transgenic tobacco eliminated three- to fourfold more mercury from hydroponic medium than nontransgenic, wildtype controls. It seems likely that this technology can be extended to remediate mercury from constructed wetlands, previously contaminated wetlands, marine sites, and industrial waste streams. The appropriate transgenic plants are being constructed to test this technology in the field.[27] For this technology to have its full impact, we need plants that can carry out the extraction, reduction, and volatilization of mercury from soil. Some of our current research focuses on determining the transpiration rates of Hg(0) from these plants (Heaton, Rugh, Wang, and Meagher, unpublished results). Rates of extraction from soil will undoubtedly be significantly slower than what we have demonstrated in hydroponic medium. However, our initial data on soil are promising (not shown).

We have demonstrated that transgenic *Arabidopsis* and tobacco plants expressing a modified *merA* gene, release large amounts of Hg(0) when suspended in a solution with as little as 5 μm Hg(II).[2] In fact, it is somewhat surprising that the mercury was volatilized at the efficiencies shown for these plants when one considers that 5 μm is below the 12 μm Km (Mechalis Constant) of mercuric ion reductase for Hg(II), and thus only a small fraction of the MerA enzyme present may actually be functioning in plant cells. We have demonstrated that the level of MerA expression may be limiting for volatilization,[2] but it is certainly sufficient to make these plant species resistant to the toxic effects of Hg(II). Future work will have to look more closely at the kinetics of this reaction in whole plants. If plants can be used to eliminate such low levels of mercury from soil and water, this technology may prove even more powerful than anticipated. The existence of a robust new technology that can be used on very low but toxic levels of metals may encourage regulatory agencies to take tougher stands to protect the environment and the public from heavy metal contamination.

AN INITIAL ASSESSMENT OF THE ENVIRONMENTAL RISKS OF THIS PHYTOREMEDIATION TECHNOLOGY

Methylmercury is so dangerous that transgenic plants which even moderately reduce the biomagnification of methylmercury will be advantageous. Data from our laboratory on engineered model plants expressing the modified bacterial organomercury lyase gene, *merB*, demonstrate that they can detoxify a wide range of organic mercury compounds, including phenyl mercury and methylmercury.[31] These plants trap the Hg(II) product of the MerB reaction in their tissues. Plants expressing both *merA* and *merB* should be able to block the flow of methylmercury into the food chain and permanently clean sites of all mercury species.

Given that in the near future the appropriate conservation plant species may be engineered to efficiently clean contaminated sites by extracting mercury and evolving metallic mercury gas, an important question becomes: What rate of Hg(0) release into the environment will be acceptable?[1] Current, state-of-the art technology (e.g., steam-cleaning soil, soil roasting) also releases mercury as vapor into the environment, although these technologies are so expensive that they are not as widely used as we believe will be the less-expensive phytoremediation technology. The safety of such thermal treatments and the planned phytovolatilization of mercury from large sites needs to be carefully considered. If genetically engineered plants make the cleanup of mercury-contaminated sites efficacious, what will be the local and global environmental impact of increased phytovolatilization of mercury as Hg(0)?

At present, natural bacterial reduction, photochemical, and chemical reduction are the major sources of Hg(0) released from contaminated sites.[25] For example, one site in Oak Ridge, TN contains 80,000 kg of Hg in a 250 ha area. The levels of atmospheric Hg(0) in the first 1 to 1.5 m above the soil[32,33] at this site are 0.004-0.01 $\mu g/m^3$. This level of atmospheric Hg(0) is typical of what is found over most of the Earth's land mass and is about four orders of magnitude below the federally regulated levels (OSHA, 50 $\mu g/m^3$) or the levels in the breath of a person with dental amalgams.[34,35] However, the mercury levels in soil, water, and wildlife at this and most other mercury-contaminated sites exceed acceptable levels, suggesting serious efforts are needed to clean up these sites. The entire Oak Ridge site has a maximum efflux rate of only 10 kg/year. At this rate, even if all the forms of mercury became available to natural reduction mechanisms, it would take 8000 years for the site to volatilize most of its mercury. If the endogenous mercury-resistant bacteria resident at these sites are producing significant amounts of Hg(0), it is evidently recaptured by the soil and does not efficiently escape. In contrast, we believe that Hg(0) gas will be transpired up from roots and out the leaves of our engineered plants, just like other waste gases (e.g., O_2, CO_2, H_2O). In a scenario where *mer*-expressing transgenic plants increased the volatilization rate of Hg(0) 400-fold, the site would take only 20 years to clean up. The efflux rate and dilution of Hg(0) into the environment also should be expected to increase 400-fold. The 400-fold increase in the level of Hg(0) in the air over the site would still be 25-fold below most government guidelines. Thus, it is unlikely that the increase in vaporization of mercury will endanger local animals and plants. Furthermore, the lack of any significant mechanism for the rapid return of Hg(0) to the Earth's surface ensures that there will not be a plume of mercury deposited near the site.

But what about the impact on the global levels of Hg(0)? As mentioned above, Hg(0) is not efficiently redeposited to the Earth's surface.[12] Although mercuric species emitted from an incineration facility, for example, are deposited locally near their site of origin, Hg(0) should be rapidly diluted into the global atmospheric pool of Hg(0). This pool has be conservatively estimated to be about 1.0 x 10^6 kg,[6] and it has increased moderately since industrialization. 1.0 x 10^6 kg is a conservative estimate of the mercury contaminating all the surface soil, marsh land, and estuarian sediment that might be available to phytoremediation in the U.S. If we examine the unlikely and extreme scenario that phytoremediation technology is applied to all the available sites in the U.S. immediately and all of this mercury is released as Hg(0) over the next 20 years, say 5% (0.05 x 10^6 kg) per year, the global Hg(0) pool would increase by 5% the first year. This pool is also cycling, albeit slowly, and by the end of the year, most of this Hg(0) would be reoxidized and returned to Earth diluted over the entire Earth's surface. The natural pool of mercury in surface soil and water is several orders of magnitude greater than that in the atmosphere.[6] The 1.0×10^6 kg of mercury redeposited over 20 years will have an insignificant impact on this surface pool. Thus, based on this initial assessment, even if this technology were to be applied to its limits in the U.S., there should be little global impact. However, this scenario does raise important questions, such as, what limits the size of the atmospheric Hg(0) pool? Why has the Hg(0) pool not increased more significantly since the industrial revolution even though global contamination with Hg(II) species has increased by 50 to 70%? Future discussions about the potential impact of engineered phytoremediation of mercury among ecologists, atmospheric chemists, geologists, and biologists should provide guidance on the scale at which this exciting new technology can be applied.

THE ADVANTAGES PLANTS OFFER AS REMEDIATION AGENTS

Technological remediation agents currently being used to clean up waste sites — such as moving vast areas of contaminated earth and containing it within imperme able plastic liners, electrolytic extraction, chemical leaching, or burying contaminants — are all prohibitively expensive and questionably effective in the long run. Chemical stabilization is gaining favor as the least expensive of physical remediation methods. Of these techniques, only electrolytic extraction and chemical leaching actually remediate the contaminants; the other methods merely postpone cleanup for a later date. In contrast, higher plants can actually take pollutants out of the soil or water through their root systems (extraction), store and concentrate pollutants in their cells (sequestration), and/or convert harmful substances to less-toxic forms (detoxification). Furthermore, they can do this for a fraction of the cost of most of the chemical/physical techniques.

The following are some of the most important reasons why we believe phytoremediation has tremendous potential as a tool for combating heavy-metal pollution. First, a number of species in diverse plant families already tolerate or hyperaccumulate heavy metals, suggesting great flexibility in these plants' ability to adapt to metals in their environment.[36,37] Second, plants can break down many organic pollutants into harmless constituents like CO_2, NH_4^+, and Cl_2.[38] This is important because many sites are contaminated with organic as well as heavy-metal pollutants. Third, the release of genetically engineered higher plants into the environment can be carefully controlled, relative to microorganisms. Fourth, plants may produce as much as 100×10^6 miles of roots per acre, which means that plants are in contact with vast amounts of soil surface area.[39] Plant root systems also can reach reasonable depths into the soil for extraction.[40] Roots are the means by which plants exercise their genetic capacity (using hundreds if not thousands of genes) to extract at least 16 nutrients from the soil and ground water. This capacity can be chemically and genetically manipulated to extract environmental pollutants. As a bonus, while extracting pollutants from the environment, plants do not damage the ecosystem as do chemical leaching or steam cleaning or any technique involving soil removal. Instead, plant root systems buffer the soil chemically and stabilize it to erosion. This prevents secondary pollution of areas adjacent to barren, polluted sites.

Fifth, plants are photosynthetic, governing as much as 80% of the energy at any time in most ecosystems.[41,42] Thus, plants have the greatest amount of energy to devote to detoxification and/or sequestration of pollutants. Nonphotosynthetic organisms can tap only a small part of this energy. Sixth, through photosystem I (a system not found in photosynthetic bacteria), plants and blue-green algae use light energy to (1) generate vast amounts of reducing power that can be used to efficiently reduce toxic metal ions, (2) fix and reduce CO_2 to make their own carbon/energy source which supports the heterotrophic growth of roots, and (3) produce large amounts of biomass[43] (e.g., above-ground portions can be harvested to remove metal ions; root systems enrich the soil). Seventh, it is estimated that phytoremediation will cost orders of magnitude less than most existing remediation technologies. The high cost of current remediation efforts is preventing most existing sites from being cleaned up. Finally, plants are esthetically pleasing, particularly in contrast to the barren landscape they replace at some heavily contaminated sites. This should result in badly needed public support for large cleanup projects.

COMMON QUESTIONS AND CONCERNS ABOUT THE USE OF GENETICALLY ENGINEERED PLANTS IN PHYTOREMEDIATION

There have been concerns expressed by the public about phytoremediation and the use of genetically engineered plants. Most of these concerns are based on misunderstandings and misconceptions about these and related technologies. The following are some of the major concerns that have been expressed.

A primary concern is that mutant engineered plants will spread uncontrollably, destroying the environment. However, our results indicate that some of the plant lines we engineered to detoxify Hg(II) cannot, in fact, grow well without mercury.[2] Thus, they are self-limiting in their ability to spread in an ecosystem. In fact, there are few significant concerns in most academic groups and U.S. government regulatory agencies about the release of most genetically engineered plants.[44,45] Most genetically engineered plants have little or no advantage in either pristine or contaminated environments. Genetically engineered plants are already in use to lower pesticide[46] and herbicide use in agriculture and will soon produce edible vaccines against cholera and other serious childhood diseases.[47,48]

A second concern is that genetic engineering will lower genetic variability. It is true that standard agricultural plant breeding techniques have had genetic uniformity and yield as their goal for hundreds of years. However, in contrast, genetic uniformity is not a goal of engineering conservation plant species to clean the environment. We can engineer numerous genetically diverse individuals within a species. For example, forest and tree species that might be engineered for phytoremediation are already propagated by methods (i.e., recurrent selection) that ensure maximum genetic variability. This is done because many tree species are inviable when inbred for even one generation.

Another concern is that engineered plants will strip the soil and ecosystem of vital nutrients. On the contrary, most plants enrich the soil and increase both its fertility and tolerance (binding capacity) for contaminants. Plants have often been used to stabilize marginal lands against erosion and to restore the ecology. Plants engineered to clean the environment are no different.

Fears have also been expressed that engineered plants are full of unnatural chemicals and are hazardous to humans. However, although engineered plants do contain one or two new proteins, they are neither toxic nor unnatural. In fact, 99.9% of the proteins expressed by an engineered plant remain unchanged.

From an economic standpoint, a concern is that the investment will not pay off; that is, plants alone will not achieve the level of decontamination required by the EPA. It is true that the full potential of phytoremediation technology is yet to be seen. However, plants offer a low-cost, low-impact solution that should be the first line of attack at a polluted site. Even when it is not known if plants can achieve complete remediation, they will at a minimum help prevent further environmental damage by controlling erosion and runoff and contamination of adjacent environments. Our research and that of others gives every reason to assume that plant phytoremediation technology will be highly successful. Our initial data suggest that even starting with extremely low toxic mercuric ion concentrations, our engineered plants quickly reduce contaminant concentrations. More research on the mechanisms by which plants make metals available will, we are confident, lead to even greater improvements.

Another concern is that in hyperaccumulating toxic wastes, the problem has not really been solved because the plants are now contaminated with metals. True metals are immutable (i.e., they cannot be broken down, short of nuclear fusion or fission), and this makes metals a particularly difficult pollution problem. However, once the metals are efficiently concentrated up from the soil and water into plants, these plants can be harvested and the metals safely processed out.

A companion fear is that plants engineered to harvest heavy metals will be passed on to consumers and into the food chain. There are a variety of phytoremediation strategies, only some of which result in plant leaves being contaminated with heavy metals. Responsible phytoremediation strategies would not utilize food-source plants to hyperaccumulate toxic metals. Insects generally will not feed on plants with high levels of toxic metals and thus may pose no danger to the food chain. Clearly, food-source animals, such as cattle or pigs, should be kept away from these plants. In the case of phytovolatilization of mercury, the hazards to herbivores are avoided. For example, our initial data on *merA*-expressing transgenic plants indicate that when they are grown on highly contaminated soil, they contain 10 to 100 times less mercury than control plants (unpublished data, Rugh and Meagher) because they are so efficiently vaporizing it from plant tissues.

Finally, there are those who would recommend merely leaving these sites alone and fencing them off from the public and wildlife. However, there are even greater, more immediate risks to not taking action. The leaching and mobility of mercury are major problems. Many of the mercury-contaminated sites that were localized, concentrated spills of mercury a few decades ago now cover tens to hundreds of square miles. The resulting spread of mercury pollution and contamination of the food chain continues to be a serious treat to our environment, to wildlife, and to human health.

CONCLUSIONS

Mercury contamination of the environment has been a serious problem for hundreds of years, but the risks it poses to humans have only recently been fully realized. Engineered phytoremediation of Hg(II) using modified bacterial mercuric ion reductase gene (*merA*) sequences in plants has great potential to be an effective new technology for remediating mercury pollution. These transgenic plants extract, reduce, and volatilize large amounts of mercury from their growth media. This technology appears to work for a wide variety of dicot species and future work should focus on extending this technology to aquatic monocots. A similar genetic engineering approach using the bacterial organomercury lyase gene (*merB*) blocks the flow of methylmercury into the environment. Using both the *merA* and *merB* genes together should prevent biomagnification of mercury while cleaning mercury from these sites.

An initial assessment suggests that applying this technology to cleaning polluted soils, sediments, and water has few risks to the local environment. Plants are the perfect remediation agents because they have extensive and efficient root systems to extract metals from the soil, supply their own energy, stabilize damaged soil and improve the environment, and are esthetically pleasing. Many of the fears initially expressed about the use of genetically engineered plants appear to be unfounded. At a minimum, the risks of using engineered plants certainly appear lower than taking no action at all, and plants will be far cheaper to use than most existing technologies. The global impact of applying this technology to all contaminated sites where the mercury is available to plants should be the subject of lively and informative discussion for all those involved in this research area.

ACKNOWLEDGMENTS

We want to thank Sidney Kushner, Debra Devedjian, George Boyajian, and Anne Summers for their input and helpful suggestions. Scott Bizily and Andrew Heaton generously allowed us to make reference to their as yet published research. This work has been supported by recent grants from the U.S. National Science Foundation and the Department of Energy's Environmental Management Science Program and past grants from the University of Georgia Research Foundation's Biotechnology Development Program.

REFERENCES

1. Meagher, R.B. and Rugh, C.L. Phytoremediation of heavy metal pollution: Ionic and methyl mercury. In *OECD Biotechnology for Water Use and Conservation Workshop.* 1996. Cocoyoc, Mexico: Organization for Econonmic Co-Operation and Development.

2. Rugh, C., Wilde, D., Stack, N.M., Thompson, D.M., Summers, A.O., and Meagher, R.B. 1996. Mercuric ion reduction and resistance in transgenic *Arabidopsis thaliana* plants expressing a modified bacterial *merA* gene. *Proc. Nat. Acad. Sci. USA* 93: 3182-3187.

3. Adriano, D.C. 1986. *Trace Elements in the Terrestrial Environment.* Springer-Verlag: New York, 298-328.

4. Clarkson, T.W. 1994. The toxicology of mercury and its compounds, In *Mercury Pollution Integration and Syntehesis,* C.J. Watras and J.W. Huckabee, Eds. Lewis Publishers: Ann Arbor, MI, 631-642.

5. D'Itri, P.A. and D'Itri, F.M. 1978. Mercury contamination: a human tragedy. *Environ. Manage.* 2: 3-16.

6. Nriagu, J.O. 1979. *The Biogeochemistry of Mercury in the Environment.* Elsevier: New York, 23-41.

7. Compeau, G.C. and Bartha, R. 1985. Sulfate-reducing bacteria: Principal methylators of mercury in anoxic estuarine sediment. *Appl. Environ. Microbiol.* 50: 498-502.

8. Gilmour, C.C., Henry, E.A., and Mitchell, R. 1992. Sulfate stimulation of mercury methylation in freshwater sediments. *Environ. Sci. Technol.* 26: 2281-2287.

9. Choi, S.C., Chase, J.T., and Bartha, R. 1994. Metabolic pathways leading to mercury methylation in *Desulfovibrio desulfuricans* LS. *Appl. Environ. Microbiol.* 60: 4072-4077.

10. Choi, S.-C., Chase, T., Jr., and Bartha, R. 1994. Enzymatic catalysis of mercury methylation by *Desufovibrio desulfuricans* LS. *Appl. Environ. Microbiol.* 60: 1342-1346.

11. Gardner, W.S., Kendall, D.R., Odom, R.R., Windom, H.L., and Stephens, J.A. 1978. The distribution of methyl mercury in a contaminated salt marsh ecosystem. *Environ. Pollut.* 15: 243-251.

12. Keating, M.H., Mahaffey, K.R., and Schoeny, R., 1997. *Mercury Study Report to Congress.* EPA-4521, H-97-003, U.S. Environmental Protection Agency: Washington, D.C., 1, 2-11.

13. Mason, R.P., Rolfhus, K.R., and Fitzgerald, W.F. 1995. Methylated and elemental mercury cycling in surface and deep ocean waters of the north atlantic, in *Water, Air, and Soil Pollution.* Kluwer: Amsterdam, 665-677.

14. EPRI, Mercury Atmospheric Processes: A Synthesis Report. 1994, Electric Power Research Institution: Palo Alto, CA.

15. Petersen, G., Iverfeldt, A., and Munthe, J. 1995. Atmospheric mercury species over Central and Northern Europe. Model calculations and comparison with observations from the Nordic Air and Precipitation Network for 1987 and 1988. *Atmos. Environ.* 29: 47-68.

16. Roberts, L. 1987. Discovering microbes with a taste for PCBs. *Science* 237: 975-977.

17. Timmis, K.N., Steffean, R.J., and Unterman, R. 1994. Designing microorganisms for the treatment of toxic wastes. *Annu. Rev. Microbiol.* 48: 528-557.

18. Janssen, D.B., Pries, F., and van der Ploeg, J.R. 1994. Genetics and biochemistry of dehalogenating enzymes. *Annu. Rev. Microbiol.* 48: 163-191.

19. Tsai, K.-J., Hsu, C.-M., and Rosen, B.P. 1997. Efflux mechanisms of resistance to cadmium, arsenic, and antimony in prokaryotes and eukaryotes. *Zool. Stud.* 36: 1-16.
20. Ji, G. and Silver, S. 1995. Bacterial resistance mechanisms for heavy metals of environmental concern. *J. Ind. Microbiol.* 14: 61-75.
21. Moore, M.D. and Kaplan, S. 1992. Identification of intrinsic high-level resistance to rare earth oxides and oxyanions in members of the class *Proteiobacteria*: characterization of tellurite, selenite, and rhodium sesquioxide reduction in *Rhodobacter sphaeroides. J. Bacteriol.* 74: 1505-1514.
22. Moore, M.D. and Kaplan, S. 1994. Members of the family Rhodospirillaceae reduce heavy-metal oxyanions to maintain redox pois during photosynthetic growth. *ASM News* 60: 17-23.
23. Summers, A.O. 1986. Organization, expression, and evolution of genes for mercury resistance. *Annu. Rev. Microbiol.* 40: 607-634.
24. Nakamura, K. and Silver, S. 1994. Molecular analysis of mercury-resistant *Bacillus* isolates from sediment of Minamata Bay, Japan. *Appl. Environ. Microbiol.* 60: 4596-4599.
25. Barkay, T., Turner, R., Saouter, E., and Horn, J. 1992. Mercury biotransformations and their potential for remediation of mercury contamination. *Biodegradation* 3: 147-159.
26. Thompson, D.M. 1990. Transcriptional and post-transcriptional regulation of the genes encoding the small subunit of ribulose-1,5-bisphosphate carboxylase. Ph.D. thesis, University of Georgia, 209.
27. Rugh, C.L., Senecoff, J.F., Meagher, R.B., and Merkle, S.A. 1998. Development of transgenic yellow-poplar for mercury phytoremediation. *Nature Biotech.* 33: 616-621.
28. Cronquist, A. 1981. *An Integrated System of Classification of Flowering Plants.* Columbia Unversity Press: New York, 1262.
29. Rugh, C.L. 1997. Transgenic plants engineered for remediation of mercury pollution using a modified bacterial gene., Ph.D. thesis, University of Georgia, Athens, 121.
30. Clark, R.B. 1982. Nutrient solution growth of sorghum and corn in mineral nutrition studies. *J. Plant Nutr.* 5: 1039-1057.
31. Bizily, S., Rugh, C.L., Summers, A.O., and Meagher, R.B. 1999. Engineering the phytoremediation of methyl mercury pollution with the bacterial organo mercury lyase gene. *Proc. Natl. Acad. Sci. USA.* In press.
32. Lindberg, S.E., Kim, K.-H., and Munthe, J. 1995. The precise measurement of concentration gradients of mercury in air over soils: a review of past and recent measurements. *Water Air Soil Pollut.* 80: 383-392.
33. Lindberg, S.E., Kim, K.H., Meyers, T.P., and Owens, J.G. 1995. Micrometerological gradient approach for quantifying air-surface exchange of mercury vapor: tests over contaminated soils. *Environ. Sci. Technol.* 29: 126-135.
34. Vimy, M.J. and Lorscheider, F.L. 1990. Dental amalgam mercury daily dose estimated from intra-oral vapor measurements: a predictor of mercury accumulation in human tissues. *J. Trace Elements Exp. Med.* 3: 111-123.
35. Skare, I. and Engqvist, A. 1994. Human exposure to mercury and silver released from dental amalgam restorations. *Arch. Environ. Health* 49: 384-394.
36. Baker, A.J.M., McGrath, S.P., Sidoli, C.M.D., and Reeves, R.D. 1994. The possibility of *in situ* heavy metal decontamination of polluted soils using crops of metal accumulating plants. *Res. Conserv. Recyc.* 11: 41-49.
37. Baker, A.J.M. 1989. Terrestrial higher plants which hyperaccumulate metallic elements — a review of their distribution, ecology and phytochemistry. *Biorecovery* 1: 81-126.

38. Schnoor, J.L., Light, L.A., McCutcheon, S.C., Wolfe, N.L., and Carreira, L.H. 1995. Phytoremediation of organic and nutrient contaminants. *Environ. Sci. Technol.* 29: 318-324.

39. Dittmer, H.J. 1937. A quantitative study of the roots and root hairs of a winter rye plant (*Secale cereale*). *Am. J. Bot.* 24: 417-420.

40. Stone, E.L. and Kalisz, P.J. 1991. On the maximum extent of tree roots, In *Forest Ecology and Management.* Elsevier Science Publishers, Amsterdam, 59-102.

41. Pomeroy, L.R., Darley, W.M., Dunn, E.L., Gallagher, J.L., Haines, E.B., and Whitney, D.M. 1981. Salt marsh populations: primary production, In *The Ecology of the Salt Marsh,* L.R. Pomeroy and R.G. Wiegert, Eds. Springer-Verlag: New York, 39-67.

42. Wiegert, R.G. and Freeman, B.J. 1990. Tidal salt marshes of the southeast Atlantic coast: a community profile. U.S. Dept. Interior, Fish and Wildlife Service, *Biol. Rep.* 85(7.29): 1-70.

43. Puri, S., Singh, V., Bhushan, B., and Singh, S. 1994. Biomass production and distribution of roots in three stands of *Populus deltoides*. *Forest Ecol. Manage.* 65: 135-147.

44. Conner, A.J. and Dale, P.J. 1996. Reconsideration of pollen dispersal data from field trials of transgenic potatoes. *Theor. Appl. Genet.* 92: 505-508.

45. Beachy, R.N., Cantrell, R.P., Chitwood, D.J., Cook, R.J., Cregan, P.B., Day, P.R., Devine, T.E., Gantt, E., Gilcrist, D.G., Kennedy, G.G., Ng, T.J., Qualset, C.O., Thenell, J.S., Tolin, S.A., and Vidaver, A.K. 1996. On appropriate oversite for plants with inherited traits that show resistance to pests. *Genet. Eng. News* 16: 38-39.

46. Stewart, C.N., Adang, M.J., All, J.N., Boerma, H.R., Cardineau, G., Tucker, D., and Parrott, W.A. 1996. Genetic transformation, recovery, and characterization of fertile soybean transgenic for a synthetic *Bacillus thuringiensis cryIAc* gene. *Plant Physiol.* 112: 121-129.

47. Richter, L., Mason, S., and Arntzen, C.J. 1996. Transgenic plants created for oral immunization against diarrheal diseases. *J. Travel Med.* 3: 52-56.

48. Arntzen, C.A. Edible vaccines produced in transgenic plants., in The Jordan Report. 1996, National Institute of Allergy and Infectious Diseases, Bethesda, MD, 43-48.

49. Begley, T.P., Walts, A.E., and Walsh, C.T. 1986. Bacterial organomercurial lyase: overproduction, isolation, and characterization. *Biochemistry* 25: 7186-7192.

50. Begley, T.P., Walts, A.E., and Walsh, C.T. 1986. Mechanistic studies of protonolytic organomercurial cleaving enzyme: bacterial organomercurial lyase. *Biochemistry* 25: 7192-7200.

51. Walsh, C.T., Moore, M.J., and Distefano, M.D. 1987. Conserved cysteine pairs of mercuric ion reductase: an investigation of function via site-directed mutagenesis. In *Flavins and Flavoproteins: Proc. Ninth Int. Symp.* Berlin. (New York: de Gruyter).

12 METAL TOLERANCE IN PLANTS: THE ROLE OF PHYTOCHELATINS AND METALLOTHIONEINS

Peter Goldsbrough

CONTENTS

INTRODUCTION

The successful development of phytoremediation as a method for treatment of contaminated sites depends in part on identifying plant material that is well adapted to the environmental conditions that prevail at such sites. There will be a great deal of variation in these sites, from those that are severely degraded and need significant modification and amendment before any plants will grow, to others that have relatively good conditions apart from the presence of toxic contaminants. Developing both cultivation practices and plant varieties for these environments are major challenges that must be addressed if phytoremediation is to develop into a widely adopted technology for the restoration of polluted environments. As with virtually all agronomic and horticultural practices, plant varieties will have to be selected that are well adapted and able to perform under specific conditions. Selection of suitable varieties may also be complemented by biotechnological approaches, including gene transfer, to further improve the capacity of plants to function under these conditions.

1-56670-450-2/00/$0.00+$.50
© 2000 by CRC Press LLC

Remediation of metal-contaminated sites by plants depends on metal uptake by roots and transport of toxic metals to shoots for subsequent harvest and removal. One trait that is of great significance to these physiological processes is the ability of plants to tolerate the metals that are being extracted from the soil. This is complicated by the fact that several metals are essential for normal plant growth but are toxic at excessive concentrations, and that many nonessential metals have chemical properties similar to those of essential metals. Plants are not unique in having to protect themselves against the toxic effects of metals. Thus, a variety of tolerance and resistance mechanisms have evolved, including exclusion or active efflux systems to minimize the cellular accumulation of metals. While these are effective protective strategies, they result in low concentrations of metal ions in the organism, precisely the opposite outcome of that desired for phytoremediation, where the goal is to maximize metal accumulation in plant material. Therefore, physiological mechanisms that are based on tolerance rather than avoidance of metals are likely to be important for phytoremediation, as these will allow plants to survive (and hopefully thrive) while accumulating high concentrations of metals. This chapter will address the role of two types of metal ligands, phytochelatins and metallothioneins, in tolerance of plants to heavy metals.

PHYTOCHELATINS

Phytochelatins (PCs) were first identified as Cd-binding peptides in *Schizosaccharomyces pombe* and subsequently shown to perform a similar function in a number of plant species (for recent reviews, see Rauser, 1995; Fordham-Skelton et al., 1997b). Phytochelatins are comprised of a family of peptides with the general structure $(\gamma\text{-GluCys})_n\text{-Gly}$, where n = 2 to 11. Similar $(\gamma\text{-GluCys})_n$ peptides with carboxy-terminal amino acids other than Gly have been identified in a number of plant species, but it is likely that these serve the same function as PCs. PCs appear to be ubiquitous in the plant kingdom, having been shown to accumulate in a wide variety of species (Gekeler et al., 1989). PCs are also found in fungi other than *S. pombe*, including *Candida glabrata*, an opportunistic pathogen of humans that infects immunocompromised patients (Mehra et al., 1988).

The γ-carboxamide linkage between glutamate and cysteine indicates that PCs are not synthesized by translation of a mRNA, but are instead the product of an enzymatic reaction. A number of studies demonstrated that glutathione (γ-GluCys-Gly, GSH) is the substrate for synthesis of PCs (the pathway of PC synthesis is illustrated in Figure 12.1). An enzyme activity that catalyzes the formation of PCs from GSH has been described in cell-free extracts from a number of plant species (Grill et al., 1989; Klapheck et al., 1995; Chen et al., 1997). This enzyme, PC synthase, transfers γ-GluCys from GSH to an acceptor GSH to produce $(\gamma\text{-GluCys})_2$-Gly (PC_2); the same enzyme can add additional γ-GluCys moieties, derived from either GSH or PCs, to PC_2 to produce larger PC peptides. PC synthase activity is dependent on the presence of one of a number of free metal ions, e.g., Cd^{2+}, Zn^{2+}, Ag^+. Chelation of metal ions, for example, by newly synthesized PCs, inactivates PC synthase, thereby providing a simple method to regulate the synthesis of PCs. The enzyme is constitutively expressed in plant roots and stems (Chen et al., 1997)

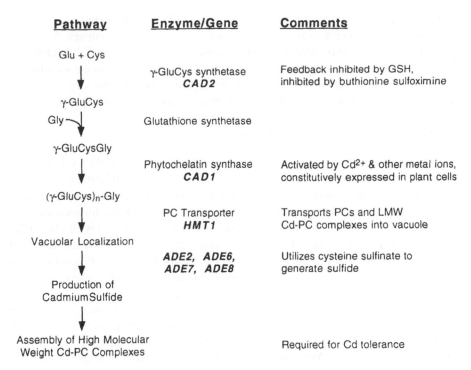

Pathway	Enzyme/Gene	Comments
Glu + Cys	γ-GluCys synthetase *CAD2*	Feedback inhibited by GSH, inhibited by buthionine sulfoximine
γ-GluCys / Gly	Glutathione synthetase	
γ-GluCysGly	Phytochelatin synthase *CAD1*	Activated by Cd^{2+} & other metal ions, constitutively expressed in plant cells
(γ-GluCys)$_n$-Gly	PC Transporter *HMT1*	Transports PCs and LMW Cd-PC complexes into vacuole
Vacuolar Localization	*ADE2, ADE6, ADE7, ADE8*	Utilizes cysteine sulfinate to generate sulfide
Production of Cadmium Sulfide		
Assembly of High Molecular Weight Cd-PC Complexes		Required for Cd tolerance

FIGURE 12.1 Synthesis of phytochelatins and formation of HMW Cd-PC complexes. The left column shows defined steps in the PC detoxification system used by plants and some fungi. The middle column shows the proteins and genes that are responsible for these activities. *CAD1* and *CAD2* are genes identified in *Arabidopsis*, whereas *HMT1*, *ADE2*, *ADE6*, *ADE7*, and *ADE8* have been identified in *Schizosaccharomyces pombe*. Comments on these steps are given in the right column.

and in plant cells growing in culture (Grill et al., 1989), perhaps providing a constant protective mechanism against heavy metal toxicity. In spite of the fact that PC synthase was reported to have been purified from plant cells several years ago (Grill et al., 1989), the gene encoding this protein has proved elusive. It is likely that this gene will finally be cloned either by complementation (in fungi) or positional cloning (in *Arabidopsis*), using Cd-sensitive mutants that lack PC synthase activity.

Phytochelatin synthesis in plant cells and PC synthase activity can be induced by a wide variety of metal ions, and PCs are able to bind a number of metal ions *in vitro* through thiolate bonds. However, the only metal-PC complexes that have been isolated from plants contain ions of Cd, Cu, or Ag (Maitani et al., 1996). Cd-PC complexes have been extensively studied and are classified as either high or low molecular weight (HMW or LMW) complexes. Two important differences distinguish these complexes: cellular location and incorporation of cadmium sulfide. The HMW complexes accumulate in the vacuole and contain CdS, perhaps in the form of a microcrystalline structure (Dameron et al., 1989). Partition of HMW complexes in the vacuole provides an effective method to separate Cd from the majority of metabolic processes. Incorporation of CdS in HMW complexes increases the amount of Cd that is sequestered per molecule of PC. HMW Cd-PC complexes are also more stable, requiring a lower pH to dissociate than LMW complexes. *S. pombe* mutants that do not make HMW Cd-PC complexes are sensitive to Cd (Mutoh and Hayashi, 1988), demonstrating the importance of vacuolar compartmentation for Cd tolerance. Vacuolar HMW Cd-PC complexes appear to be the final step in cellular detoxification of Cd, with no evidence for further metabolism of PCs or export of Cd.

GENETIC ANALYSIS OF CADMIUM TOLERANCE

A large number of physiological studies indicate that PCs are critical for Cd tolerance in plants. Indirect inhibition of PC synthesis with buthionine sulfoximine, an inhibitor of GSH synthesis, reduces Cd tolerance, whereas an exogenous supply of GSH increases both PC synthesis and Cd tolerance. Cell lines selected for increased Cd tolerance do not exclude Cd from cells but instead accumulate Cd with essentially all of the Cd present in the form of Cd-PC complexes. However, the identification and characterization of Cd-sensitive mutants has clarified the role of PCs in tolerance of plants not only to Cd but also to other metals. Genetic analysis has also contributed to our understanding of cellular partitioning of Cd-PC complexes.

Cadmium-sensitive mutants of *Arabidopsis thaliana* were identified by Howden and Cobbett (1992), initially using a screen for root growth inhibition on Cd-containing medium. Two loci were identified that are essential for normal Cd tolerance, *CAD1* and *CAD2*. Mutants at either locus had a number of similar characteristics, including reduced uptake of Cd and lower accumulation of PCs. *cad1* mutants had normal levels of GSH but were deficient in PC synthase activity (Howden et al., 1995b), whereas the single *cad2* mutant had a reduced level of GSH but normal PC synthase activity (Howden et al., 1995a). These mutants have confirmed the importance of PCs for Cd tolerance in plants. It was initially proposed that PCs provided tolerance to all heavy metals in plants based on the ability of many metals to induce PC synthesis (Grill et al., 1987). However, PC-deficient *cad* mutants are hypersensitive to cadmium, mercury, and lead (Howden and Cobbett, 1992; Chen and Cunningham, personal communication), but have essentially normal levels of tolerance to other metals, including copper and zinc. This demonstrates that PCs are not required for tolerance to all heavy metals and may be restricted to detoxification of nonessential metals. A similar spectrum of metal tolerance has been described for GSH-deficient (and, therefore, PC-deficient) mutants of *S. pombe* (Glaeser et al., 1991).

Genetic complementation of Cd-sensitive mutants of *S. pombe* has identified genes that are involved in the accumulation of HMW vacuolar Cd-PC complexes (Figure 12.1). *HMT1* encodes a vacuolar membrane ABC-type transporter that can transport both PCs and LMW Cd-PC complexes from the cytoplasm into the vacuole (Ortiz et al., 1992, 1995). Mutants that lack this transporter do not accumulate HMW complexes and are Cd-sensitive. A similar vacuolar PC transport activity has been identified in plants (Salt and Rauser, 1995), indicating that not only PC synthesis but also cellular compartmentation of the PC detoxification system is conserved between plants and fungi. The second group of genes identified by complementation of *S. pombe* mutants was in the adenine synthesis pathway (Speiser et al., 1992).

This pathway is believed to generate the sulfide that is incorporated into the HMW complexes, starting from cysteine sulfinate (Juang et al., 1993). While plants clearly accumulate similar HMW complexes (Reese et al., 1992), the pathway for sulfide production is unknown and may be derived in the same manner as in *S. pombe*.

Analysis of Cd-sensitive mutants has made a central contribution to our understanding of Cd tolerance in plants and fungi. There are likely to be several other genes that could be identified using the same approach, and this should be an objective for future research.

METALLOTHIONEINS

The discovery of PCs in plants led to the proposal that plants do not possess metallothionein (MT) proteins, i.e., gene-encoded, cysteine-rich proteins translated from mRNAs, but instead utilized PCs to fulfill the functions of metal homeostasis and detoxification (Grill et al., 1987). However, only 2 years after the first reports of PCs in plants, Lane et al. (1987) purified the E_c protein from wheat embryos and demonstrated that the amino acid sequence of this protein was consistent with that of an MT and that this protein bound Zn^{2+}. This was followed by the cloning of genes that encoded MT-like proteins from several plant species. While the functions of these MT genes are still unknown, it is clear that plants are equipped with at least two ligands that use cysteine coordination of metals, namely PCs and MTs. Because this is the most extensively documented gene family, the *Arabidopsis* MT gene family will be used as a model to discuss the structure, expression, and possible function of MTs in plants. Studies on MTs from other species will be discussed where they add to the overall view of the function of plant MTs.

THE *ARABIDOPSIS* MT GENE FAMILY

The first *Arabidopsis* MT gene that was cloned in this laboratory (MT1a) was identified while screening a library for cDNAs representing transcripts that were induced by ethylene (Zhou and Goldsbrough, 1994). A number of other plant MT genes had already been cloned using differential screening procedures to identify genes expressed in particular tissues or under specific environmental conditions. The frequent identification of MT genes in this type of screen indicates that at least some plant MT genes are expressed at relatively high levels in terms of RNA abundance. Using the cDNA for MT1a, and the sequence of another *Arabidopsis* MT gene in the Genbank database (now called MT2a), homologous genomic DNA sequences were cloned and characterized. This revealed the presence of at least five MT genes in the *Arabidopsis* genome (Zhou and Goldsbrough, 1995). More recently, the *Arabidopsis* EST database and genome sequencing project have revealed the presence of three additional MT genes. The predicted amino acid sequences of these genes are shown in Figure 12.2. *Arabidopsis* MT genes are placed in four categories based on sequence similarity and relationship with MT genes from other plant species. With the exception of MT3, each of the other classes contains two active genes. Additionally, there is at least one pseudogene, MT1b (Zhou and Goldsbrough, 1995).

MT Class	Amino Acid Sequence
MT1a	MADSNCGCGS SCKCGDSCSC EKNYNKECDN CSCGSNCSCG SNCNC
MT1c	MAGSNCGCGS SCKCGDSCSC EKNYNKECDN CSCGSNCSCG SSCNC
MT2a	MSCCGGNCGC GSGCKCGNGC GGCKMYPDLG FSGETTTTET FVLGVAPAMK NQYEASGESN NAESDACKCG SDCKCDPCTC k
MT2b	MSCCGGSCGC GSACKCGNGC GGCKRYPDL. ...ENTATET LVLGVAPAMN SQYEASGETF VAENDACKCG SDCKCNPCTC k
MT3	MSSNCGSCDC ADKTQCVKKG TSYTFDIVET QESYKEAMIM DVGAEENNAN CKCKCGSSCS CVNCTCCPN
MT4a	AGCNDSCGCP SPCPGGNSCR CRMREASAGE .QGHMVCPCG EHCGCNPCNC PKTQTQTSDK GCTCGEGCTC ASCDT
MT4b	ASCNDRCGCP SPCPGGESCR CKMMSEASGG DQEHNTCPCG EHCGCNPCNC PKTQTQTSAK GCTCGEGCTC ATCAA

FIGURE 12.2 Amino acid sequences predicted for *Arabidopsis* MTs. These are predicted from the DNA sequences of *Arabidopsis* MT genes that are known to be expressed. Cysteine residues are in bold. The four classes of MT genes are based on similarity to each other and to MT genes identified in other plant species. Note that the protein sequences for members of the MT4 class do not initiate with a methionine because the cDNAs encoding these proteins are not full length.

The *Arabidopsis* genome is normally regarded as a model of simplicity, but there are other examples of large gene families in this species, including those for β-tubulins and chlorophyll a/b binding proteins. Metallothioneins in animals are typically encoded by a gene family of varying complexity. Is the extensive MT gene family in *Arabidopsis* representative of other plant species? Examples of each of the classes present in *Arabidopsis* have been found in at least one other species. The *Arabidopsis* MT1 class is homologous to Type 1 plant MTs in the classification proposed by Robinson et al. (1993). Twelve cysteine residues in Type 1 MTs are present as Cys-X-Cys motifs in two distinct domains at the amino- and carboxy-termini of these proteins. *Arabidopsis* and *Brassica napus* MT1 proteins are distinguished from other Type 1 MTs by having a "spacer" of only 10 amino acids separating the two cysteine domains (Buchanan-Wollaston, 1994), compared to approximately 45 amino acids in other Type 1 MTs. However, it is likely that these MT genes have a common progenitor given the conservation of both the cysteine residues and the position of the single intron in Type 1 MT genes. *Arabidopsis* MT2 genes are similar to Type 2 plant MTs, where the first pair of cysteines are arranged as CysCys.

Arabidopsis MT3 was found in a search of the *Arabidopsis* EST database (Murphy et al., 1997). This gene is present as a single copy in the *Arabidopsis* genome (Bundithya and Goldsbrough, unpublished observations). Homologous genes have been described from kiwi fruit and rice. The final class of MT genes in *Arabidopsis*, MT4, is related to the wheat E_c genes that are expressed during embryo development. cDNAs for two genes with homology to wheat E_c MTs were sequenced from a library prepared from RNA from dry seeds.

There is now evidence that other species contain more than one class of MT gene. For example, maize has genes encoding a Type 1 MT, expressed primarily in roots (de Framond, 1991), and a homolog of the E_c MT that is expressed in seeds (White and Rivin, 1995). Gene families encoding a single class of MT protein have been characterized in tomato (Whitelaw et al., 1997) and cotton (Hudspeth et al., 1996). Therefore, it is likely that the size of the *Arabidopsis* MT gene family is not unusual but merely a consequence of the effort put into understanding the structure and content of this species' genome.

RNA Expression of *Arabidopsis* MT Genes

The frequency with which MT genes have been isolated from plant cDNA libraries in various differential screening experiments indicates that many MT mRNAs are expressed at relatively high levels. Most of the *Arabidopsis* MT genes that are expressed in vegetative tissues have been sequenced several times in assembling the *Arabidopsis* EST database. For example, more than 40 cDNAs corresponding to MT3 have been identified. The overall pattern of RNA expression of *Arabidopsis* MT genes is shown in Table 12.1. MT1 RNA is more abundant in roots than in leaves, whereas RNAs for MT2 and MT3 are expressed at higher levels in leaves than roots (Zhou and Goldsbrough, 1994, 1995; Bundithya and Goldsbrough, unpublished observations). RNA hybridization experiments indicate that the *Arabidopsis* MT4 genes are only expressed during seed development (Dandelet, Bundithya, and Goldsbrough, unpublished observations). This is supported by the lack of any ESTs corresponding to MT4 in cDNA libraries prepared from vegetative tissues.

TABLE 12.1

Summary of RNA Expression of *Arabidopsis* MT Genes

| | RNA Expression | | | | | Copper Induction |
Gene	Seedling	Roots	Leaves	Flowers	Seeds	(Tissue)
MT1a/c	+++	+++	+			++	(leaves)
MT2a	+	+	++	+		+++	(seedlings)
MT2b	++	+	++	+		+	(seedlings)
MT3	++	+	++	nd	nd	+	(leaves)
MT4					++	nd	

Note: The relative level of expression of RNAs from each MT gene is indicated. The tissues in which copper induction of MT RNAs have been observed are also indicated. Gene-specific probes have not been used to examine specifically the expression of MT1a and MT1c.

nd = not determined

Expression of some *Arabidopsis* MT genes can be induced by metals, notably copper. In the MT2 family, the level of MT2a RNA increases when seedlings are exposed to copper ions (Zhou and Goldsbrough, 1995; Murphy and Taiz, 1995). Copper induction of MT RNA expression has been demonstrated for other MT genes, both in *Arabidopsis* and other species, suggesting a role for MTs in an adaptive response to copper (Hsieh et al., 1995; Robinson et al., 1993). However, two observations suggest caution is warranted with this interpretation. First, many other MT genes have been shown not to be induced by copper or other metals. This may be the result of examining the expression of MT genes that are not metal regulated in these species, or using conditions where the MT genes are already expressed at a high level and are refractory to further induction. Second, MT RNA expression can be induced by a variety of other environmental and developmental conditions, including heat shock, aluminum stress, nutrient starvation, senescence, and abscission (reviewed by Fordham-Skelton et al., 1997b). Therefore, while *Arabidopsis* MT genes have been shown to be regulated by copper, it is not yet clear if copper induction can be separated from a general stress response. Answers to these questions will come from a detailed analysis of the transcriptional regulation and promoter activities of a number of MT genes. One approach is to study reporter gene expression driven by MT gene promoters. Fordham-Skelton et al. (1997a) have shown that the pea $PsMT_A$ promoter is active in many tissues in transgenic *Arabidopsis*, including leaves, cotyledons, and floral organs, but is maximally expressed in roots, in agreement with RNA hybridization results. A promoter from a cotton MT gene is also highly expressed in roots, notably the root apex (Hudspeth et al., 1996). Comprehensive analysis of the tissues where individual MT genes are expressed and of the conditions that modulate this expression should provide some insight into the functions of MT genes in plants.

EXPRESSION OF PLANT MT PROTEINS

The first evidence that plants synthesized MT proteins, in addition to PCs, came from the work of Lane et al. (1987), who demonstrated not only that the wheat E_c protein bound Zn^{2+}, but that its amino acid sequence was consistent with that of an MT. E_c proteins can bind approximately 5% of the zinc in a seed, but they are not expressed in vegetative tissues. In spite of the large number of genes encoding MTs that have been cloned, there has, until recently, been no information on the expression of these "nonseed" MT genes at the protein level. Results of Murphy et al. (1997) may help explain some of the difficulties encountered in trying to identify MT proteins in plants. Low molecular weight, copper-binding proteins were purified from various *Arabidopsis* tissues. Amino acid sequences of tryptic fragments obtained from some of these proteins corresponded perfectly with those predicted from the sequences of MT1a/c, MT2a, MT2b, and MT3, providing a categorical demonstration that these MT genes are indeed expressed as proteins. If the protein extracts were exposed to oxygen during the first steps of the isolation procedure, MT proteins could not be recovered. The sensitivity of these proteins to oxygen likely accounts for the difficulty in isolating MTs from plants.

In addition to demonstrating the presence of MTs in vegetative tissues, Murphy et al. (1997) used antibodies raised against MT-GST fusion proteins to show that expression of MT1 and MT2 proteins reflected the RNA expression of these genes in terms of tissue specificity and copper induction. This correspondence between RNA and protein expression does not rule out the possibility of more complex regulation of the expression of these genes through a number of post-transcriptional mechanisms.

METAL BINDING PROPERTIES OF MTS

The wheat E_c protein was identified as a Zn-binding protein, and *Arabidopsis* MTs were purified using copper-affinity chromatography. However, because of the difficulties in purifying MTs from plants, there is a lack of information about the metals that are bound to MTs *in vivo*. An alternative approach to address this question has been to express plant MT genes in a number of microbial hosts and either directly assess the metal binding properties of these proteins or examine the ability of plant MTs to confer metal tolerance.

When expressed in *E. coli*, either as the native protein or as a fusion protein, the pea MT was shown to bind Cu, Cd, and Zn ions. In its native form, i.e., not a fusion protein, the MT was cleaved within the spacer region, giving rise to two cysteine-rich peptides which could function as independent metal ligands (Kille et al., 1991). Similar processing of MTs in plants might contribute to the difficulties encountered in trying to purify these proteins from plants. The affinity of various metals for the MT fusion proteins was assessed by examining metal dissociation at low pH. The pea MT had the highest affinity for Cu and the lowest for Zn (Tommey et al., 1991).

A number of microbes contain MTs that are required for metal tolerance. Mutant strains that lack MTs and, therefore, have reduced tolerance to metals have been used as transformation hosts to examine the functional properties of plant MTs. *Arabidopsis* MT1a and MT2a proteins were expressed in a yeast strain in which one of its endogenous MT genes, *CUP1*, had been deleted. Constitutive expression of the *Arabidopsis* MTs restored copper tolerance and increased cadmium tolerance (Zhou and Goldsbrough, 1994). Similarly, the MT2a protein was able to restore some degree of zinc tolerance to a *Synechococcus* mutant that lacked its own zinc MT (Robinson et al. 1996). These experiments have established that *Arabidopsis* MTs indeed function *in vivo* as metal of binding proteins.

Differences in metal tolerance between yeast transformants expressing *Arabidopsis* MT1 or MT2 may reflect differences in the affinity of these MTs for metals, raising the possibility that the complexity of the *Arabidopsis* MT gene family is necessary to deal with a variety of metals. The arrangement of cysteine residues in different plant MTs may affect the metal-binding specificity of these proteins. Important objectives for future research are to identify the metal ions that bind to MTs *in vivo* and to determine the intracellular localization of these proteins. Answers to these questions should contribute to an understanding of the functions of MTs in plants.

ARE MTS REQUIRED FOR METAL TOLERANCE IN PLANTS?

Phytochelatin-deficient mutants of *Arabidopsis* have essentially normal tolerance to Cu and Zn, indicating that PCs are not required for tolerance to these metals and that there must be other mechanisms to provide tolerance to these metals in plants. MTs are one candidate to fill this role, and a number of observations support this hypothesis. As discussed above, plant MTs can bind Cu and Zn and function *in vivo* to provide tolerance to these metals in other organisms. Expression of MT genes can be induced by Cu, and expression of MT2 RNA is elevated in a Cu-sensitive mutant of *Arabidopsis*, *cup1*, which accumulates higher concentrations of Cu (van Vliet et al., 1995). In a survey of *Arabidopsis* ecotypes for differences in metal tolerance, Murphy and Taiz (1995) demonstrated a positive correlation between Cu tolerance of seedlings, measured as root growth after transfer to a Cu-supplemented medium, and expression of MT2 RNA. This suggests that expression of at least some MTs is important for Cu tolerance. More direct evidence will have to await studies on transgenic plants with altered expression of MT genes and detailed analysis of mutants with altered tolerance to Cu, Zn, and other metals.

MANIPULATION OF METAL LIGANDS FOR PHYTOREMEDIATION

Plants that are selected and developed for phytoremediation will need to have a number of advantageous physiological traits, including tolerance of metals and other environmental conditions at the contaminated site, enhanced uptake and transport of metals, and sequestration of metals in shoot tissues. Manipulating the expression of PCs and MTs might play a part in one or more of these traits. However, there is only limited information available about the best targets for this approach or the likely outcome of such efforts.

While cloning a gene for PC synthase has not yet been accomplished, altering the expression of this gene in plants may not have a significant impact on metal tolerance. The enzyme is constitutively expressed in many tissues, and its activity is regulated by free metal ions. However, genes for enzymes of GSH synthesis may hold more promise. Increased activity of γ-GluCys synthetase in selected Cd-tolerant tomato cells could increase GSH and PC synthesis and contribute to Cd tolerance (Chen and Goldsbrough, 1994). Genes encoding γ-GluCys synthetase and GSH synthetase have been isolated from tomato. Surprisingly, neither of these genes shows any change in RNA expression in plants or cells that are exposed to Cd (Kovari and Goldsbrough, unpublished observations). Regulation of these genes may occur at a post-translational level, but this remains to be demonstrated. The potential application of gene transfer to manipulating metal tolerance has been indicated by showing that expression of tomato γ-GluCys synthetase could restore some degree of Cd tolerance to the *cad2 Arabidopsis* mutant (Kovari, Cobbett, and Goldsbrough, unpublished observations). However, this gene did not increase Cd tolerance of wildtype plants, perhaps due to an inadequate level of expression or other regulatory problems. The vacuolar sequestration pathway may provide another target to increase metal tolerance. Increasing the expression of the HMT1 transporter in *S. pombe* resulted in Cd hypertolerance (Ortiz et al., 1992). A plant homolog of this transporter has not yet been cloned. There are a number of possible approaches to altering PC metabolism that might contribute to increased metal tolerance, and in general there is a positive correlation between PC synthesis and Cd accumulation. However, the outcome of such experiments is uncertain because of our lack of understanding of how GSH synthesis, PC synthesis, and vacuolar compartmentation are regulated.

The possibility of altering the expression of plant MT genes to increase metal tolerance or accumulation has not yet been addressed. One potential difficulty with this approach is that many of the plant MT RNAs are already expressed at relatively high levels, and increasing the expression of a single MT gene may have only a small effect on metal tolerance. However, a number of attempts have been made to express animal MTs in plants, either to affect tolerance or to modify the distribution of metals within the plant. These studies have been reviewed by Wagner (1993). While it is possible to express these genes in a variety of plant tissues, they have little effect on metal tolerance and accumulation overall. It is possible that plant MTs have different affinities for specific metals than the animal MTs, and most of these studies have examined effects on Cd, where PCs are likely to play the dominant role in plants. There is still a need to manipulate the expression of plant MT genes in plants and examine the effects of either reduced or increased expression on tolerance to different metals.

To date, there have been few studies on the role of PCs and MTs in species that are being considered for phytoremediation. However, Salt et al. (1995) have shown that PCs are likely involved in binding Cd in the roots of *Brassica juncea* plants exposed to Cd, but not in the transport of Cd from root to shoot. Expression of genes encoding MT2 and γ-GluCys synthetase did not change in *B. juncea* seedlings exposed to Cd, but MT2 RNA decreased and γ-GluCys synthetase RNA increased in response to Cu (Schäfer et al., 1997). Further investigation of both MT gene expression and PC accumulation in plants that accumulate high concentrations of metals is warranted. The use of cysteine-rich ligands for detoxification of metals in these plants may be limited by the availability of reduced sulfur for cysteine synthesis.

In conclusion, our understanding of the role of PCs in metal tolerance is based on sound physiological and genetic data. By comparison, the function of MTs in plants is not yet clear. Information on gene structure and expression has accumulated but has not been matched by a similar level of insight into the role of these proteins in metal tolerance or other aspects of metal homeostasis. However, it is likely that both of these ligands, which are widely distributed in plants, will be of importance in the development of plant varieties for phytoremediation.

REFERENCES

Buchanan-Wollaston, V. Isolation of cDNA clones for genes that are expressed during leaf senescence in *Brassica napus* — Identification of a gene encoding a senescence-specific metallothionein-like protein. *Plant Physiol.* 105, 839-846, 1994.

Chen, J. and P.B. Goldsbrough. Increased activity of γ-glutamylcysteine synthetase in tomato cells selected for cadmium tolerance. *Plant Physiol.* 106, 233-239, 1994.

Chen, J., J. Zhou, and P.B. Goldsbrough. Characterization of phytochelatin synthase from tomato. *Physiol. Plant* 101, 165-172, 1997.

Dameron, C.T., R.N. Reese, R.K. Mehra, A.R. Kortan, P.J. Carroll, M.L. Steigerwald, L.E. Brus, and D.R. Winge. Biosynthesis of cadmium sulfide quantum semiconductor crystallites. *Nature* 338, 596-597, 1989.

de Framond, A.J. A metallothionein-like gene from maize (*Zea mays*). *FEBS Lett.* 290, 103-106, 1991.

Fordham-Skelton, A.P., C. Lilley, P.E. Urwin, and N.J. Robinson. GUS expression in *Arabidopsis* directed by 5' regions of a pea metallothionein-like gene, *PsMT$_A$*. *Plant Mol. Biol.* 34, 659-669, 1997a.

Fordham-Skelton, A.P., N.J. Robinson, and P.B. Goldsbrough. Methallothionein-like Genes and Phytochelatins in Higher Plants, in *Metal Ions in Gene Regulation*, Silver. S. and Walden, W., Eds. Chapman and Hall, New York, 398-430, 1997b.

Gekeler, W., E. Grill, E.-L. Winnacker, and M.H. Zenk. Survey of the plant kingdom for the ability to bind heavy metals through phytochelatins. *Z. Naturforsch. Sec. C Biosci.* 44, 361-369, 1989.

Glaeser, H., A. Coblenz, R. Kruczek, I. Ruttke, A. Ebert-Jung, and K. Wolf. Glutathione metabolism and heavy metal detoxification, in *Schizosaccharomyces pombe. Curr. Genet.* 19, 207-213, 1991.

Grill, E., S. Loffler, E.-L. Winnacker, and M.H. Zenk. Phytochelatins, the heavy-metal-binding peptides of plants, are synthesized from glutathione by a specific γ-glutamylcysteine dipeptidyl transpeptidase (phytochelatin synthase). *Proc. Natl. Acad. Sci. USA* 86, 6838-6842, 1989.

Grill, E., E.-L. Winnacker, and M.H. Zenk. Phytochelatins, a class of heavy-metal-binding peptides from plants are functionally analogous to metallothioneins. *Proc. Natl. Acad. Sci. USA* 84, 439-443, 1987.

Howden, R. and C.S. Cobbett. Cadmium-sensitive mutants of *Arabidopsis thaliana. Plant Physiol.* 100, 100-107, 1992.

Howden, R., C.R. Andersen, P.B. Goldsbrough, and C.S. Cobbett.. A cadmium-sensitive, glutathione-deficient mutant of *Arabidopsis thaliana. Plant Physiol.* 107, 1067-1073, 1995a.

Howden, R., P.B. Goldsbrough, C.R. Andersen, and C.S. Cobbett. Cadmium-sensitive, *cad1*, mutants of *Arabidopsis thaliana* are phytochelatin deficient. *Plant Physiol.* 107, 1059-1066, 1995b.

Hsieh, H.-M., W.-K. Liu, and P.C. Huang. A novel stress-inducible metallothionein-like gene from rice. *Plant Mol. Biol.* 28, 381-389, 1995.

Hudspeth, R.L., S.L. Hobbs, D.M. Anderson, K. Rajasekaran, and J. Grula. Characterization and expression of metallothionein-like genes in cotton. *Plant Mol. Biol.* 31, 701-705, 1996.

Juang, R.-H., K.F. MacCue, and D.W. Ow. Two purine biosynthetic enzymes that are required for cadmium tolerance in *Schizosaccharomyces pombe* utilize cysteine sulfinate *in vitro. Arch. Biochem. Biophys.* 304, 392-401, 1993.

Kille, P., D.R. Winge, J.L. Harwood, and J. Kay. A plant metallothionein produced in *E. coli. FEBS Lett.* 295, 171-175, 1991.

Klapheck, S., S. Schlunz, and L. Bergmann. Synthesis of phytochelatins and homophytochelatins in *Pisum sativum* L. *Plant Physiol.* 107, 515-521, 1995.

Lane, B., R. Kajioka, and T. Kennedy. The wheat germ E$_c$ protein is a zinc-containing metallothionein. *Biochem. Cell. Biol.* 65, 1001-1005, 1987.

Maitani, T., H. Kubota, K. Sato, and T. Yamada. The composition of metals bound to class III metallothionein (phytochelatin and its desglycyl peptide) induced by various metals in root cultures of *Rubia tinctorum. Plant Physiol.* 110, 1145-1150, 1996.

Mehra, R.K., E.B. Tarbet, W.R. Gray, and D.R. Winge. Metal-specific synthesis of two metallothioneins and γ-glutamyl peptides in *Candida glabrata. Proc. Natl. Acad. Sci. USA* 85, 8815-8819, 1988.

Murphy, A. and L. Taiz. Comparison of metallothionein gene expression and nonprotein thiols in ten *Arabidopsis* ecotypes. *Plant Physiol.* 109, 945-954, 1995.

Murphy, A., J. Zhou, P.B. Goldsbrough, and L. Taiz. Purification and immunological identification of metallothioneins 1 and 2 from *Arabidopsis thaliana*. *Plant Physiol*. 113, 1293-1301, 1997.

Mutoh, N. and Y. Hayashi. Isolation of mutants of *Schizosaccharomyces pombe* unable to synthesize cadystin, small cadmium-binding peptides. *Biochem. Biophys. Res. Commun*. 151, 32-39, 1988.

Ortiz, D.F., L. Kreppel, D.M. Speiser, G. Scheel, G. McDonald, and D.W. Ow. Heavy-metal tolerance in the fission yeast requires an ATP-binding cassette-type vacuolar membrane transporter. *EMBO J*. 11, 3491-3499, 1992.

Ortiz, D.F., T. Ruscitti, K.F. McCue, and D.W. Ow. Transport of metal-binding peptides by HMT1, a fission yeast ABC-type vacuolar membrane protein. *J. Biol. Chem*. 270, 4721-4728, 1995.

Rauser, W.E. Phytochelatins and related peptides. *Plant Physiol*. 109, 1141-1149, 1995.

Reese, R.N., C.A. White, and D.R. Winge. Cadmium sulfide crystallites in Cd-(γ-EC)$_n$G peptide complexes from tomato. *Plant Physiol*. 98, 225-229, 1992.

Robinson, N.J., A.M. Tommey, C. Kuske, and P.J. Jackson. Plant metallothioneins. *Biochem. J*. 295, 1-10, 1993.

Robinson, N.J., J.R. Wilson, and J.S. Turner. Expression of the type 2 metallothionein-like gene *MT2* from *Arabidopsis thaliana* in Zn^{2+}-metallothionein deficient *Synechococcus* PCC 7942: Putative role for MT2 in Zn^{2+}-metabolism. *Plant Mol. Biol*. 30, 1169-1179, 1996.

Salt, D.E., R. C. Prince, I.J. Pickering, and I. Raskin. Mechanisms of cadmium mobility and accumulation in Indian mustard. *Plant Physiol*. 109, 1427-1433, 1995.

Salt, D.E. and W.E. Rauser. MgATP-dependent transport of phytochelatins across the tonoplast of oat roots. *Plant Physiol*. 107, 1293-1301, 1995.

Speiser, D.M., D.F. Ortiz, L. Kreppel, and D.W. Ow. Purine biosynthetic genes are required for cadmium tolerance in *Schizosaccharomyces pombe*. *Mol. Cell. Biol*. 12, 5301-5310, 1992.

Schäfer, H.J., S. Greiner, T. Rausch, and A. Haag-Kerwer. In seedlings of the heavy metal accumulator *Brassica juncea* Cu^{2+} differentially affects transcript amounts for γ-glutamycysteine synthetase (γ-ECS) and metallothionein (MT2). *FEBS Lett*. 404, 216-220, 1997.

Tommey, A.M., J. Shi, W.P. Lindsay, P.E. Urwin, and N.J. Robinson. Expression of the pea gene *PsMT$_A$* in *E. coli*. *FEBS Lett*. 292, 48-52, 1991.

van Vliet, C., C.R. Andersen, and C.S. Cobbett. Copper-sensitive mutant of *Arabidopsis thaliana*. *Plant Physiol*. 109, 871-878, 1995.

Wagner, G.J. Accumulation of cadmium in crop plants and its consequences to human health. *Adv. Agron*. 51, 173-212, 1993.

White, C.N. and C.J. Rivin. Characterization and expression of a cDNA encoding a seed-specific metallothionein in maize. *Plant Physiol*. 108, 831-832, 1995.

Whitelaw, C.A., J.A. LeHuquet, D.A. Thurman, and A.B. Tomsett. The isolation and characterisation of type II metallothionein-like genes from tomato (*Lycopersicon esculentum* L.). *Plant. Mol. Biol*. 33, 503-511, 1997.

Zhou, J. and P.B. Goldsbrough. Functional homologs of fungal metallothionein genes from *Arabidopsis*. *Plant Cell* 6, 875-884, 1994.

Zhou, J. and P.B. Goldsbrough. Structure, organization and expression of the metallothionein gene family in *Arabidopsis*. *Mol. Gen. Genet*. 248, 318-328, 1995.

13 The Genetics of Metal Tolerance and Accumulation in Higher Plants

Mark R. Macnair, Gavin H. Tilstone, and Susanne E. Smith

CONTENTS

INTRODUCTION

The activities of man have left many sites contaminated with one or more heavy metals. The affected land is toxic to plants and animals, including humans, and there is considerable public and legislative pressure to remediate the land, or at least alleviate many of the adverse properties of these soils. Existing technologies are *ex situ* and expensive, and often render the soil barren and unusable for agriculture or domestic use. Because of the considerable attraction in using plants to assist in the remediation of these sites, phytoremediation offers the prospect of a cheaper and "greener" way of dealing with these problems.

    The toxicity of the metals may cause the death of the very plants being used for phytoremediation. However, many plants have evolved tolerant races, varieties that can survive in soils that are toxic to "normal" plants, thus obviating this source of difficulty. Plants interact with the minerals in the soil, and will take many up into their roots actively or passively with the transpiration stream. Some of this metal will move from roots and be accumulated in the shoots. The accumulation of toxic

1-56670-450-2/00/$0 00+$.50

materials in the aerial parts of plants means that there has been a net movement of metal, and possibly a change in its chemical state. This may be of concern to the public and regulators, since it may be perceived to be more readily assimilated into the food chain, or of greater ecological significance, in plants than in the soil.

Plants can be used in a number of fundamentally different ways in order to assist in bioremediating metal contaminated soils or waters. These strategies are discussed in Chapter 18. But in the context of this chapter, it is important to recognize that different strategies will have differing requirements of plants in terms of tolerance and accumulation. Thus, Cunningham et al. (1995) recognize two major uses for plants.

1. Phytostabilization — Plants are used to stabilize the land and reduce or eliminate the movement of toxic elements from the contaminated soil to the general environment. The value of the land may also be increased if previously derelict land could be used for some economically or socially beneficial purpose. Examples include growing a biofuel or forestry on the land, or turning the land into a park or nature reserve. Plants used for phytostabilization will need to be *tolerant* of the metals present in the particular site, but the *accumulation* of metals in their aerial parts may be positively disadvantageous. If the objective of phytostabilization is purely to prevent erosion and improve the visual amenity of a derelict site, then the accumulation of metals in the plants may be irrelevant. If, on the other hand, the site is to be used by the public or cropped for biofuels or timber, then the presence of toxic metals in the plants or crop might render the site less valuable for its purpose.

2. Phytoextraction — In this case, plants are used to extract metals from the soil. The objective may be to clean up mildly contaminated soils, so that the metal is removed from the site *in situ*, or to biomine metals from heavily contaminated soils. It is important that plants are tolerant of the soils in which they are to grow, but in this case, it is also important that they translocate significant amounts of metal and accumulate in their aerial parts. In practice, only plants that accumulate more than 1% (dry weight) of metal are going to be useful in either context (Chaney et al., 1997). Plants which accumulate very high concentrations of metals are called hyperaccumulators.

Thus, the questions of tolerance and accumulation are central to the prospects of using plants for these technologies. This chapter will review briefly what is known about the genetics of these phenomena and discuss recent work in our laboratory on these topics. A knowledge of the genetics is important for at least two reasons. First, it is likely that existing plant varieties will not be suitable for some applications of the technologies, particularly phytoextraction. Phytoextraction, to be a truly useful technology, will require a deep-rooting, high biomass, hyperaccumulating species; all of the species currently investigated are slow growing or of low biomass. If we know the genetics of hyperaccumulation we will know the prospects for genetically engineering a suitable crop. If there is a comparatively simple genetic basis, then this should be readily achievable. Second, a genetic approach can greatly assist in investigating the mechanisms of these phenomena; at present we know distressingly little about either.

TOLERANCE

In plants, tolerance can be defined as the ability to survive in a soil that is toxic to other plants of the same or different species. The important point is that the toxicity of the soil is defined by reference to its effects on other plants. Technically, tolerance is manifested as a genotype x environment interaction. Tolerance was first recognised by Prat (1934) who showed that seeds of *Silene vulgaris* from a mine population were able to grow on copper-contaminated soil while those from a normal population could not. Since this initial observation, the tolerance of many species to many metals has been reported (e.g., see Antonovics et al., 1971). Tolerance in plants is most dramatically seen on heavily contaminated soils such as those found at abandoned mine sites, but has been demonstrated in such environments as lead-contaminated road-side verges and the soil beneath galvanized electricity pylons (Al-Hiyaly et al, 1988). Note that tolerance is not a universal phenomenon — there are far more plant species that are *not* found on mines than are found. One species that is not tolerant is *Arabidopsis thaliana* which is the best characterized model plant species and has never been found on contaminated soils. This is unfortunate, since it would have been much easier to investigate the genetics and mechanisms of tolerance in this species had it shown the phenomenon. There are a number of mutations in *A. thaliana* which make it hypersensitive to normal levels of metals, but these genes may well have little to do with tolerance. A number of researchers (e.g., Ow, 1996) have suggested that the understanding of normal metal homeostasis gleaned from the study of nontolerant model systems may enable us to genetically engineer tolerant or accumulating plants; however, we are skeptical.

There are a number of issues that have attracted recent interest concerning the genetics of tolerance. The first is the consideration of whether tolerance is determined by major genes, and what causes the variation in levels of tolerance seen within and between tolerant populations. Second, does tolerance to one metal confer tolerance to others? Many sites are multiply contaminated and it would be easier to breed crops able to remediate these sites if one gene was able to confer tolerance to a range of potential contaminants. Third, what are the other effects of tolerance genes? If tolerance is associated with considerable costs then multiply tolerant plants engineered for phytoremediation may be unacceptably noncompetitive.

THE ROLE OF MAJOR GENES

Early work on the genetics of tolerance suggested that the phenomenon was governed by many genes of small effect (Antonovics et al., 1971). However, the techniques used to measure and define tolerance almost guaranteed that major genes could not be detected, even if present (see Macnair, 1990, 1993). Macnair (1993) reviewed the genetic basis of tolerance in wild and agricultural species. In general, he showed that, where it had been investigated fully, metal tolerance was governed by a small number of major genes (see Table 13.1). A number of authors had also suggested that there were additional, minor "modifier" genes contributing to the phenomenon.

For instance, the systems studied in most detail are copper tolerance in *Mimulus guttatus* (Macnair, 1983; Smith and Macnair, 1998) and in *S. vulgaris* (Schat and Ten Bookum, 1992a, Schat et al., 1993). In both species, a single major gene has been found in most populations studied, though in *S. vulgaris* the most tolerant population also has an additional additive major gene. However, in both species individual tolerant plants vary in the degree of tolerance they manifest.

TABLE 13.1

Studies that Have Found One or More Major Genes Giving Metal Tolerance in Higher Plants

Metal	Species	Number of Genes	Ref.
Aluminium	Barley	1 Major gene	
	Zea mais	1 Major gene, with multiple alleles	Macnair, 1993
	Nicotiana plumbaginifolia	1 Major gene	Macnair, 1993
	Wheat (cv Chinese Spring)	1 Major gene + at least 2 modifiers	Macnair, 1993
	Wheat (cv BH1146)	3 Major genes, + some minor genes	Macnair, 1993
Arsenic	*Holcus lanatus*	1 Major gene + modifiers	Macnair, 1993
	Agrostis capillaris	1 Major gene + modifiers	Macnair, 1993
Boron	Wheat	3 Major genes	Macnair, 1993
Cadmium	*Chlamydomonas reinhardtii*	2 Major genes	Macnair, 1993
Copper	*Mimulus guttatus*	1 Major gene + modifiers	Macnair, 1993
	Silene vulgaris	1 Major gene + modifiers	Macnair, 1993
	S. vulgaris	2 Major genes + modifiers	Macnair, 1993
Zinc	*S. vulgaris*	2 Major genes	Schat et al., 1996
Manganese	Wheat	2 Major genes	Macnair, 1993

Note: In all cases, the major genes detected could be a number of tightly linked genes.

Source: Macnair, M.R. *New Phytol.* 124, 541–559, 1993. With permission.

Macnair et al. (1993) and Harper et al. (1997a) selected copper tolerant *M. guttatus* for increased and decreased tolerance, and within six generations had achieved great differences in tolerance between the two lines. Smith and Macnair (1998) investigated the difference in degree of copper tolerance between these two lines. Both lines were crossed to a single nontolerant individual (*tt*). The F_1s were selfed, and both F_2s segregated tolerants:nontolerants in a 3:1 ratio. This showed that both lines were homozygous for the major tolerance gene, *TT*, and the difference between them was not due to a major gene, or any gene that acted additively with *T*. The nontolerant F_2 progeny were crossed to a single tester plant, a plant of low tolerance also of genotype *TT*. Some typical results are shown in Figure 13.1. Plants 1/20 and 13/14 are nontolerant progeny from two different F_2 families. Figures 13.1a and b show the results of crossing these plants to the tester plant. Clearly 13/14 enhances the tolerance of the tester (as indicated by the rootlength of its progeny in copper solution) more than 1/20 does. The two nontolerant F_2 plants were selfed, and five offspring again crossed to the tester plant. The tolerance of the resulting progeny is given in Figures 13.1c and d. Again the progeny derived from 13/14 are more tolerant than those from 1/20, showing that the differences observed in Figures 13.1a and b are heritable. These results show that the difference in degree of tolerance is caused by modifier genes that only have any effect on tolerance when the gene for tolerance is present (technically, this means that the genes are *hypostatic* to tolerance). Schat and Ten Bookum (1992a) came to a similar conclusion in *S. vulgaris*.

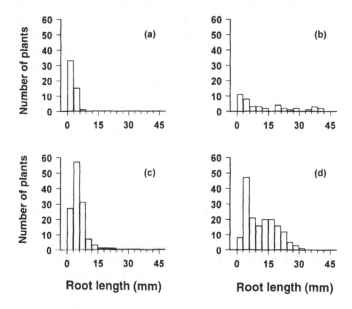

FIGURE 13.1 Histograms of rootlengths of families of plants produced by crossing non-tolerant F_2s (a,b) or F_3s (c,d) to a standard recurrent tolerant homozygote with low tolerance. (a,c) Families derived from plant 1/20 (from a line selected for low tolerance) and (b,d) families derived from plant 13/14 (from a line selected for high tolerance).

This genetic control of tolerance has significance for studying the mechanisms of tolerance and for breeding plants with increased tolerance for use in phytoreme-diation. If there is only a single major gene for tolerance, then it follows that there is a single change to the physiology or biochemistry of the plants that enables the plants to grow in contaminated soils. It is this we must identify in order to explain the mechanism of tolerance. If there was more than one independent mechanism, then the modifier genes would have acted additively, not hypostatically. Thus far, though many mechanisms have been postulated (Baker, 1987; Cumming and Tomsett, 1992), only that for arsenic tolerance in *Holcus lanatus* (Meharg and Macnair,

1990, 1992) and aluminium tolerance in wheat (Delhaize et al., 1993) have been shown to be correlated with the genes for tolerance. The ubiquity of major genes for tolerance suggests that it should, in principle, be possible to identify and clone the genes for tolerance, and transform species with desired agronomic qualities with the tolerance genes needed to grow in particular contaminated environments. The ease with which we were able to select for enhanced tolerance in *M. guttatus* suggests that, once the major gene has been introduced to a particular species, standard selective breeding may be able to produce highly tolerant lines for phytoremediation.

Cotolerance

The artificial breeding of tolerant varieties for phytoremediation will be easier if the tolerance to a range of metals is achieved by a single genetic change. The early work on metal tolerance suggested that tolerance to different metals was achieved independently and that plants which showed tolerance to more than one metal were caused by multiple contaminations of the site (e.g., Gregory and Bradshaw, 1965; Antonovics et al., 1971). However, over the years there have been a number of studies suggesting that cotolerance might occur. One of the first was the study of Allen and Sheppard (1971) of copper tolerance in *M. guttatus* at the large mine at Copperopolis. They found that, while the levels of zinc, lead, and nickel were no greater at Copperopolis than in control sites, the level of tolerance to these metals shown by the copper-tolerant Copperopolis population was significantly greater. Schat and Ten Bookum (1992b) suggested that *S. vulgaris* might also show cotolerance, since many populations from northern Europe showed tolerance to metals not present at their site of origin.

We have been investigating whether copper tolerance in *M. guttatus* really does impart tolerance to other metals. If it does, then the cotolerance could be shown by either the major gene for copper tolerance, the modifers, or both. To test whether the major gene gave cotolerance, we obtained isogenic lines which differed essentially only in the major gene by repeated backcrossing to a nontolerant population (Strange and Macnair, 1991; Macnair et al., 1993). The tolerance of the two lines to copper, nickel and cadmium is given in Figure 13.2. Figure 13.2c shows that in copper, the tolerant line produces longer roots than the nontolerant. However, this line does not produce longer roots in cadmium (Figure 13.2b) or nickel (Figure 13.2a). Indeed, if anything the copper-tolerant line shows less nickel tolerance (though the difference is not significant). Thus, the major gene does not show cotolerance. The effect of the copper-tolerance modifiers was tested by screening five independent selection lines, each of which had been selected for both increased and decreased copper tolerance (Harper et al., 1997a). The difference between the lines is due to modifiers (see above), and cotolerance due to these modifier genes would be indicated if the high lines were consistently more tolerant to other metals than the low lines. Table 13.2 shows that this is not true for cadmium or nickel. Tilstone (1996) also tested for cotolerance between copper and a range of other heavy metals, and between nickel and zinc. He found no evidence that *M. guttatus* shows any case of cotolerance. Schat and coworkers have also failed to substantiate their earlier suggestion (Schat and Ten Bookum, 1992b) of cotolerance in *S. vulgaris* following

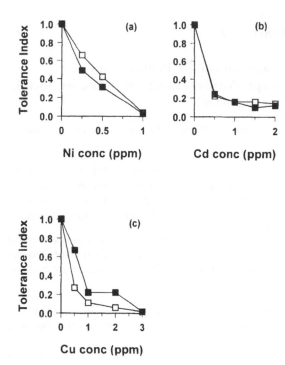

FIGURE 13.2 Dose response curves of isogenic lines of *Mimulus guttatus* homozygous for the tolerant allele (solid symbols) or the nontolerant allele (open symbols) against nickel (a), cadmium (b) and copper (c).

more detailed genetic analysis, though they have found evidence that one of the two genes implicated in zinc tolerance also confers nickel and cobalt tolerance (Schat and Vooijs, 1997).

The general lack of cotolerance has implications for the mechanisms of tolerance. It suggests that tolerance is unlikely to be produced by quantitative or qualitative differences in relatively nonspecific binding compounds, such as organic acids, phytochelatins or metallothioneins. The primary tolerance mechanism must reside in compounds that show much greater specificity for individual cations. The absence of cotolerance also suggests that breeding a plant showing general metal tolerance for widespread phytoremediation of metal-contaminated soils will require a large number of genetic changes.

THE COST OF TOLERANCE

It has been widely asserted that metal-tolerant plants are at a disadvantage compared with normal conspecifics in uncontaminated soil, the so-called cost of tolerance. If the cost is high, then it may prove difficult to produce vigorous and productive metal tolerant varieties for phytoremediation.

There is circumstantial evidence that tolerance is disadvantageous in the absence of the selective metal. First, the genes for tolerance are generally at very low

TABLE 13.2

Comparison of the Tolerance of Five
Independent Selection Lines to Copper, Nickel,
and Cadmium

Line	Mine of Origin	Copper	Nickel	Cadmium
1	Copperopolis	Higher	Equal	Equal
2	Copperopolis	Higher	Higher	(Higher)
3	Copperopolis	Higher	Higher	Equal
4	Quail	Higher	Equal	Lower
5	Penn	(Higher)	Higher	Lower

Note: Each line consisted of sublines selected for both increased
and decreased copper tolerance. The direction of the difference
(high line–low line) is given. Parentheses indicate that the difference is not statistically significant.

Source: Tilstone, G.H. and M.R. Macnair. *Plant Soil* 191, 173-
180, 1997a and Tilstone et al., *Heredity* 79, 445-452, 1997. With
permission.

frequency in populations growing on normal soils, and there are very steep clines
for tolerance at the edge of mines (Hickey and McNeilly, 1975). McNeilly (1968)
found that the seed of the wind-pollinated grass *Agrostis capillaris* growing downwind of a mine was more tolerant than its maternal parent, suggesting that selection
acted against the gene flow coming from the mine. In the case of *M. guttatus*, the
clines are steeper for the modifiers of tolerance than for tolerance itself, suggesting
that selection may act more on the degree of tolerance than on the possession of
tolerance per se in this species (Macnair et al., 1993). Second, tolerant plants seem
to often have slower growth rate or reduced competitiveness than nontolerant
ecotypes (Hickey and McNeilly,1975; Wilson, 1988; Ernst et al, 1990). Third, mine
plants often appear to have a greater requirement for the metal to which they are
tolerant. Many root-growth experiments have found an apparent stimulation of root-
growth at low levels of the metal in tolerant plants. There have also been some
reports that some tolerant plants require exogenous metal in order to grow them
successfully in normal potting compost (Antonovics et al., 1971). Schat and Ten
Bookum (1992b) found that highly copper-tolerant *S. vulgaris* showed symptoms
of copper deficiency when grown in soil from a lead/zinc mine.

   However, while these observations show that mine plants are often at a disadvantage when compared to nonmine plants, they do not show that tolerance per se
is disadvatageous, nor that it is impossible to produce a vigorous metal-tolerant
plant. Mine plants differ from normal plants in many ways apart from just being
tolerant. The whole ecotype may be disadvantageous, but that does not mean that

any one of the genes contributing to the whole phenotype need be so. In particular, many mines are free draining and have low nutrient status, and it is not unlikely that adaptations to these features of mines may include slow growth rate and reduced competitiveness in a richer environment (Chapin, 1991).

In this laboratory we have recently been examining the cost of copper tolerance in *M. guttatus* in more detail. We have explicitly tested the two principal hypotheses for how the cost could arise: (1) that the mechanism of tolerance requires energy/resources for its maintenance, that are diverted from other activities, reducing growth rate and/or competitiveness and (2) that the mechanism of tolerance reduces the intracellular concentration of copper available to normal metabolism, thus creating a greater requirement for this essential micronutrient. We have looked at both the effects of the major gene (by comparing the isogenic lines, see above) and the effects of the modifiers of tolerance (by comparing the performance of lines selected for increased and decreased tolerance). No consistent effects of tolerance on growth and competitiveness could be found, and the only possible effects on fitness was a tendency for a change in resource allocation between male and female function (Harper et al., 1997a,b). Many experiments were carried out looking at the effect on the plants' growth and fitness in suboptimal copper concentrations. There is absolutely no evidence that highly tolerant plants require more copper than less tolerant ones. It seems that the evolution of tolerance has *increased* the range of environments in which the plant is buffered in terms of its copper nutrition.

Overall, this work has shown that if there is a cost to tolerance, it is either very small, or is produced by a mechanism other than those postulated hitherto. Thus, there should be no intrinsic reason why metal-tolerant plants produced for phytoremediation should be competitively inferior or slow growing.

ACCUMULATION

In this chapter, accumulation is taken to mean the concentration of metal in the aerial parts of the plant. This metal will result from the combined effects of uptake from external medium into the roots and the translocation from the roots to the shoots. Baker (1981) drew a distinction between three types of plants showing different patterns of accumulation: (hyper)accumulators, which show a very rapid increase in shoot concentration with rising external concentration; indicators which show a more or less linear relation between external and shoot concentration; and excluders, in which the shoot concentration rises only slowly as external concentration rises. In practice, the distinction between the latter two types may be difficult to make, since it essentially depends on a difference in slope of the graph connecting external and shoot concentrations and this may not be easy to determine. Hyperaccumulation may be easier to define since it produces plants that have unusually high concentrations of metals in their shoots. There have been a number of recent reviews (e.g., Brooks, 1994) defining hyperaccumulation on the basis of an (arbitrary) threshold value and listing species that have been found to exceed that value in the field. The threshold is different for different metals, e.g., Zn 1%, Ni and Cu 0.1%, and Cd 0.01%, to name just a few.

GENETIC VARIATION IN ACCUMULATION IN
NONHYPERACCUMULATORS

There are many reports in the literature that varieties or species differ in the levels
of metals in their shoots when grown under standard conditions (e.g., Clark, 1983).
These differences must reflect genetic differences, but in general little is known
about the nature of the genes causing this variation. It is probable that much is
polygenic, reflecting many different ways that the uptake or translocation of metals
could be affected by genetic variation. There are a number of major mutants of
plants affecting metal uptake and translocation (e.g., *brz* in peas [Cary et al., 1994];
man-1 in *Arabidopsis thaliana* [Delhaize, 1996]), but it is not clear whether these
loci are involved in natural variation in accumulation.

One class of genes that might be expected to affect accumulation is tolerance
genes. Does an increase in tolerance lead to an increase or decrease in accumulation
of that metal in the aerial parts? There have been a large number of studies in which
the accumulation by tolerant and nontolerant clones of a species has been compared.
No consistent pattern emerges — sometimes more tolerant plants having higher
concentrations in their roots than nontolerants, and sometimes lower. Relative shoot
concentrations are equally variable. No pattern arises when only a particular metal
is considered. Thus, if we consider only copper, several authors have found that
tolerance is negatively correlated with shoot copper concentrations (Morishima and
Oka, 1977; Baker et al., 1983; Ouzounidou et al. 1994). Lolkema et al. (1984) and
Lin and Wu (1994) found no difference in shoot concentrations between tolerant
and nontolerant clones, while Macnair (1981) and Wu and Antonovics (1975) found
tolerant plants accumulated more copper than nontolerants. All these studies compared tolerant and nontolerant plants, i.e., they looked at the effects of major genes
for tolerance. Tilstone and Macnair (1997b) compared the accumulation of copper
over 4 weeks by tolerant plants differing in their degree of tolerance (Figure 13.3),
and thus investigated the effects of the modifier genes. Three lines, IT, LT, and HT

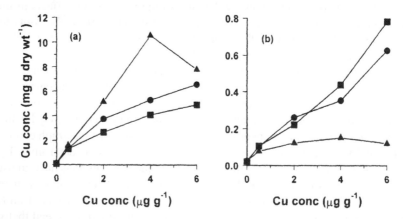

FIGURE 13.3 Copper concentration in the roots (a) and shoots (b) of three lines of *Mimulus
guttatus* after 30 days in nutrient solutions containing various levels of copper. IT (triangles),
LT (circles), or HT (squares). The lines differ in tolerance in the order IT<LT<HT.

were compared and these lines differ in tolerance in the order IT<LT<HT. The least tolerant strain has a higher concentration of copper in its roots, while the most tolerant translocates most to its shoots. Maximum concentrations obtained in the shoots are about 400 mg kg^{-1} at 4 mg L^{-1} solution concentration, rising to about 700 mg kg^{-1} at 6 mg L^{-1}, where all strains are severely stressed by the copper. Thus, by selecting for increased tolerance (the HT line was produced by six generations of directional selection for increased tolerance), we have produced an increase in accumulation, and the levels of copper achieved by the HT line (700 mg kg^{-1}) approach those often cited for copper hyperaccumulation (1000 mg kg^{-1} [Brooks, 1994]). However, we think this is unlikely to be a general conclusion, and that further studies on the relationship between variation in tolerance and variation in accumulation are required before any generalizations can be made (if, indeed, there will be any). It is not improbable that in other species selection for increased tolerance may lead to either a reduced accumulation or no change at all.

HYPERACCUMULATION

Little is known about the genetic basis of hyperaccumulation. One problem is that this phenomenon appears to vary only at the species level. We are not aware of any species that show the sort of ecotypic variation typically displayed by tolerant plants. The most extensively studied species is *Thlaspi caerulescens*, which in Britain is found primarily on lead/zinc mines, but also on serpentine and uncontaminated sites (Ingrouille and Smirnoff, 1986). Baker et al. (1994) found little variation in tolerance or accumulating ability for a range of metals between a range of populations collected from a variety of different habitats. Ingrouille and Smirnoff (1986), however, had found some difference in zinc tolerance between populations from zinc mine and uncontaminated sites. Little or no variation has been found between plants or populations in their ability to hyperaccumulate zinc (Ingrouille and Smirnoff, 1986; Pollard and Baker, 1996), though Pollard and Baker (1996) did find some evidence for heritable variation in degree of hyperaccumulation in one population. Because zinc tolerance was variable, but hyperaccumulation was not, Ingrouille and Smirnoff (1986) suggested that these characters might be independent. The lack of within-species variation for hyperaccumulation means that a standard genetic analysis is not possible. There is, however, intrageneric variation in this character in *Thlaspi*, *Alyssum*, and *Cardaminopsis*, and it may be possible to use interspecific crosses to generate the variation required to study the genetics of this phenomenon.

We have been studying the hyperaccumulation of zinc by *Cardaminopsis halleri*. *C. halleri* is closely related to *Arabidopsis thaliana*, and is a well-known zinc hyperaccumulator that is an obligate metallophyte in northern Europe (Ernst, 1974). Field samples of *C. halleri* regularly show concentrations of zinc of between 3 and 5% dry weight, at least 10 times greater than specimens of highly zinc tolerant species (e.g., *S. vulgaris* and *Armeria maritima*) growing in the same environment. Other species in *Cardaminopsis*, however, have not been reported to display this character. Figure 13.4 shows the pattern of zinc accumulation by this species. Initially roots show higher zinc concentrations than shoots, but by 14 days the shoots contain more zinc than the roots. This pattern is very unusual; most studies of plants in

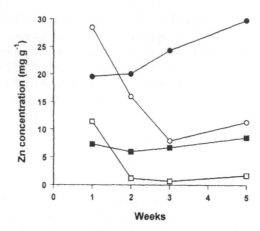

Weeks

FIGURE 13.4 Zinc concentrations (μg g dry wt^{-1}) of the roots (open symbols) or shoots (closed symbols) of *Cardaminopsis halleri* plants grown in nutrient solution containing 10 μm Zn (squares) or 100 μm Zn (circles) for differing times. At week zero all plants had less than 0.5 mg g^{-1} Zn in both roots and shoots.

metal solutions show much higher root concentrations than shoots (e.g., see Figure 13.3). This tendency is also found in other well-studied hyperaccumulators (e.g., for nickel in *A. lesbiacum* [Krämer et al.,1996] and for zinc in *T. caerulescens* [Vazquez et al., 1994]).

We have crossed *C. halleri* to *C. petraea*, and obtained the F_2. Figure 13.5 shows some preliminary results from an experiment in which we grew *C. halleri*, *C. petraea*, *C. neglecta*, and *Arabidopsis thaliana*, and the F_1 and F_2 of the *C. halleri* x *C. petraea* cross in zinc solution culture. Accumulation was determined by measuring leaf zinc concentrations, while tolerance to zinc was scored subjectively by visual symptoms of stress (severe chlorosis or death). The difference between *C. halleri* and the related nonaccumulating species is clear, and there is no overlap between them. All the F_1 plants have high zinc contents, indicating that hyperaccumulation is a dominant character. The plants also all appear to be zinc tolerant. The F_2 plants were separated, prior to the determination of their zinc content, into "stressed" (i.e., nontolerant) and tolerant classes. Both have very similar distributions of zinc content, though there is a suggestion that the tolerant plants may have a higher mean zinc content than the nontolerants. Even this may be misleading, since plants which were dead (and therefore nontolerant) are excluded from this graph, since the fresh weight of dead leaves is not comparable to the live leaves otherwise used. All the dead plants had high zinc contents, however. The distribution of zinc contents of the F_2 is highly variable, overlapping both parental classes.

This preliminary analysis of the genetics of zinc hyperaccumulation in *C. halleri* shows that it is possible to get both tolerance and hyperaccumulation to segregate. At this stage, speculation on the number of genes responsible for either character is not possible. However, the high variance may suggest that there may be one or a few major genes. Zinc tolerance is known to be determined by one or two major genes in *S. vulgaris* (Schat et al., 1996), and it is possible that a similar number of

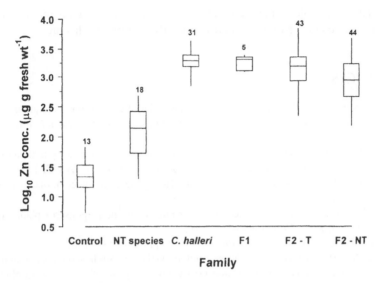

FIGURE 13.5 Box and whisker plots of leaf zinc concentration (Log_{10} µg g fresh wt[-1]) of plants grown in 100 µm Zn for 7 days, and 250 µm for 11 days. Control plants were maintained in 1 µm Zn. NT species: *A. thaliana*, *C. petraea*, and *C. neglecta*. The number tested in each class is indicated above the box. The difference between NT (nontolerant) and T (tolerant) F_2 plants was determined subjectively by visual inspection.

genes would be involved in *C. halleri*. What is clear from these data is that the two characters are largely independent of each other in this F_2. It is possible to find hyperaccumulators which are either tolerant or nontolerant. This result may be unexpected by some. Chaney et al. (1997) asserted that "hypertolerance is the key property which makes hyperaccumulation possible," while Krämer et al. (1997) suggested that nickel hyperaccumulation is simply a manifestation of nickel tolerance in *Thlaspi goesingense*. However, it would support the speculation of Ingrouille and Smirnoff (1986) who suggested that these characters might be independent in *T. caerulescens*. The finding also supports the argument developed above that, in general, tolerance and accumulation may not be related characters.

Since *C. halleri* is closely related to *A. thaliana*, it should be possible to use the power of genetic analysis developed in this model species to determine the number and location of genes for hyperaccumulation and ultimately to clone them. The development of genetically engineered high biomass crops showing this character would then be possible. The potential economic benefit of such plants might make the investment in such a program worthwhile. Speculation on whether it will be necessary for hyperaccumulating plants to be tolerant also (i.e. that both characters will need to be transferred) is interesting. Obviously, if the plants are to be grown on highly contaminated soils (e.g., for biomining), then tolerance will also be necessary. It is not necessarily true that plants would have to be tolerant in order to be able to grow and clean up the less highly contaminated soils that would be the target of phytoextraction. It will depend on the form in which the metal is bound in the plant, and whether the high metal concentrations accumulated in the aerial parts are actually accessible to the membranes, organelles, and proteins that are affected by these toxic metals. More research is urgently required in this area.

REFERENCES

Al-Hiyaly, S.A., T. McNeilly, and A.D. Bradshaw. The effects of zinc contamination from electricity pylons — evolution in a replicated situation. *New Phytol.* 110, 571-580, 1988.

Allen. W.R. and P.M. Sheppard. Copper tolerance in some Californian populations of the monkey flower, *Mimulus guttatus. Proc. R. Soc. London, Ser. B* 177, 177-196, 1971.

Antonovics, J., A.D. Bradshaw, and R.G. Turner. Heavy metal tolerance in plants. *Adv. Ecol. Res.* 7, 1-85, 1971.

Baker, A.J.M. Accumulators and excluders — strategies in the response of plants to heavy metals. *J. Plant Nutr.* 3, 643-654, 1981.

Baker, A.J.M. Metal tolerance. *New Phytol.* 106 (Suppl.), 93-111, 1987.

Baker, A.J.M., R.R. Brooks, A.J. Pease, and F. Malaisse. Studies on copper and cobalt tolerance in three closely related taxa within the genus *Silene* L. (Carophyllaceae) from Zaïre. *Plant Soil* 73, 377-385, 1983.

Baker, A.J.M., R.D. Reeves, and A.S.M. Hajar. Heavy-metal accumulation and tolerance in British populations of the metallophyte *Thlaspi caerulescens* J.& C. Presl. (Brassicaceae). *New Phytol.* 127, 61-68, 1994.

Brooks, R.R. Plants that hyperaccumulate heavy metals, in *Plants and the Chemical Elements,* Farago, M.E., Ed., VCH, Weinheim, 87-105, 1994.

Cary, E.E., W.A. Norvell, D.L. Grunes, R.M. Welch, and W.S. Reid. Iron and manganese accumulation by the brz pea mutant grown in soils. *Agron. J.* 86, 938-941, 1994.

Chaney, R.L., M. Malik, Y.M. Li, S.L. Brown, E.P. Brewer, J.S. Angle, and A.J.M. Baker. Phytoremediation of soil metals. *Curr. Opin. Biotechnol.* 8, 279-284, 1997.

Chapin, F.S. Integrated responses of plants to stress. *BioScience* 41, 29-36, 1991.

Clark, R.B. Plant genotype differences in the uptake, translocation, accumulation, and use of mineral elements required for plant growth. *Plant Soil* 72, 175-196, 1983.

Cumming, J.R. and A.B. Tomsett. Metal tolerance in plants: signal transduction and acclimation mechanisms, in *Biogeochemistry of Trace Metals,* Adriano, D.C., Ed., Lewis Publishers, Boca Raton, FL, 329-364, 1992.

Cunningham, S.D., W.R. Berti, and J.W. Huang. Phytoremediation of contaminated soils. *Trends Biotechnol.* 13, 393-397, 1995.

Delhaize, E. A metal-accumulator mutant of *Arabidopsis thaliana. Plant Physiol.* 111, 849-855, 1996.

Delhaize, E., P.R. Ryan, and P.J. Randall. Aluminium tolerance in wheat (*Tricicum aestivum* L.). II. Aluminium-stimulated excretion of malic acid from root apices. *Plant Physiol.* 103, 695-702, 1993.

Ernst, W.H.O. *Schwermetallvegetation der Erde.* Gustav Fischer Verlag, Stuttgart, 1974.

Ernst, W.H.O., H. Schat, and J.A.C. Verkleij. Evolutionary biology of metal resistance in *Silene vulgaris. Evol. Trends Plants* 4, 45-50, 1990.

Gregory, R.P.G. and A.D. Bradshaw. Heavy metal tolerance in populations of *Agrostis tenuis* Sibth. and other grasses. *New Phytol.* 64, 131-143, 1965.

Harper, F.A., S.E. Smith, and M.R. Macnair. Where is the cost in copper tolerance in *Mimulus guttatus*? Testing the trade-off hypothesis. *Functional Ecol.* 11, 764-774, 1997a.

Harper, F.A., S.E. Smith, and M.R. Macnair. Can an increased copper requirement in copper-tolerant *Mimulus guttatus* explain the cost of tolerance? 1. Vegetative growth. *New Phytol.* 136, 455-467, 1997b.

Hickey, D.A. and T. McNeilly. Competition between metal tolerant and normal plant populations: a field experiment. *Evolution* 29, 458-464, 1975.

Ingrouille, M.J. and N. Smirnoff. *Thlaspi caerulescens* J. & C. Presl. (*T. alpestre* L.) in Britain. *New Phytol.* 102, 219-233, 1986.

Krämer, U., J.D. Cotter-Howells, J.M. Charnock, A.J.M. Baker, and J.A.C. Smith. Free histidine as a metal chelator in plants that accumulate nickel. *Nature* 379, 635-638, 1996.

Krämer, U., R.D. Smith, W.W. Wenzel, I. Raskin, and D.E. Salt. The role of metal transport and tolerance in nickel hyperaccumulation by *Thlaspi goesingense* Halacsy. *Plant Physiol.* 115, 1641-1650, 1997.

Lin, S.L. and L. Wu. Effects of copper concentration on mineral nutrient-uptake and copper accumulation in protein of copper-tolerant and copper-nontolerant *Lotus purshianus* L. *Ecotoxicol. Environ. Safety.* 29, 214-228, 1994.

Lolkema, P.C., M.H. Donker, A.J. Schouten, and W.H.O. Ernst. The possible role of metal-lothioneins in copper tolerance of *Silene cucubalus.* *Planta* 162, 174-179, 1984.

Macnair, M.R. The uptake of copper by plants of *Mimulus guttatus* differing in genotype primarily at a single major copper tolerance locus. *New Phytol.* 88, 723-730, 1981.

Macnair, M.R. The genetic control of copper tolerance in the yellow monkey flower, *Mimulus guttatus.* *Heredity* 50, 283-359, 1983.

Macnair, M.R. The genetics of tolerance in natural populations, in *Heavy Metal Tolerance in Plants: Evolutionary Aspects,* Shaw, J., Ed., CRC Press, Boca Raton, FL, 235-254, 1990.

Macnair, M.R. Tansley Review No. 49: The genetics of metal tolerance in vascular plants. *New Phytol.* 124, 541-559, 1993.

Macnair, M.R., S.E. Smith, and Q.J. Cumbes. Heritability and distribution of variation in degree of copper tolerance in *Mimulus guttatus* at Copperopolis, California. *Heredity* 71, 445-456, 1993.

McNeilly, T. Evolution in closely adjacent populations. III. *Agrostis tenuis* on a small copper mine. *Heredity* 23, 99-108, 1968.

Meharg, A.A. and M.R. Macnair. An altered phosphate uptake system in arsenate tolerant *Holcus lanatus* L. *New Phytol.* 116, 29-35, 1990.

Meharg, A.A. and M.R. Macnair. Genetic correlation between arsenic tolerance and the rate of uptake of arsenate and phosphate in *Holcus lanatus.* *Heredity* 69, 336-341, 1992.

Morishima, H. and H.I. Oka. The impact of copper pollution on barnyard grass populations. *Jpn. J. Genet.* 52, 357-372, 1977.

Ouzounidou, G., L. Symeonidis, D. Babalonas, and S. Karataglis. Comparative responses of a copper-tolerant and a copper-sensitive population of *Minuartia hirsuta* to copper toxicity. *J. Plant Physiol.* 144, 109-115, 1994.

Ow, D.W. Heavy metal tolerance genes: prospective tools for bioremediation. *Res. Conserv. Recyc.* 18, 135-149, 1996.

Pollard, A.J. and A.J.M. Baker. The quantitative genetics of zinc hyperaccumulation in *Thlaspi caerulescens.* *New Phytol.* 132, 113-118, 1996.

Prat, S. Die Erblichkeit der Resistenz gegen Kupfer. *Berich. Deutsch. Bot. Gesellschaft.* 102, 65-67, 1934.

Schat, H. and W.M. Ten Bookum. Genetic control of copper tolerance in *Silene vulgaris.* *Heredity* 68, 219-229, 1992a.

Schat, H. and W.M. Ten Bookum. Metal specificity of metal tolerance syndromes in higher plants, in *The Ecology of Ultramaphic (Serpentine) Soils,* Proctor, J., Baker, A.J.M., and Reeves, R.D., Eds., Intercept, Andover, 337-352, 1992b.

Schat, H. and R. Vooijs. Multiple tolerance and co-tolerance to heavy metals in *Silene vulgaris:* a co-segregation analysis. *New Phytol.* 136, 489-496, 1997.

Schat, H., E. Kuiper, W.M. Ten Bookum, and R. Vooijs. A general model for the genetic control of copper tolerance in *Silene vulgaris:* evidence from crosses between plants from different tolerant populations. *Heredity* 70, 142-147, 1993.

Schat, H., R. Vooijs, and E. Kuiper. Identical major gene loci for heavy-metal tolerances that have independently evolved in different local-populations and subspecies of *Silene vulgaris. Evolution* 50, 1888-1895, 1996.

Smith, S.E. and M.R. Macnair. Hypostatic modifiers cause variation in degree of copper tolerance in *Mimulus guttatus. Heredity* 80, 760-768, 1998.

Strange, J. and M.R. Macnair. Evidence for a role for the cell membrane in copper tolerance of *Mimulus guttatus. New Phytol.* 119, 383-388, 1991.

Tilstone, G.H. "The significance of multiple metal tolerance in *Mimulus guttatus* Fischer ex DC," Ph.D. dissertation, University of Exeter, U.K., 1996.

Tilstone, G.H. and M.R. Macnair. Nickel tolerance and copper-nickel co-tolerance in *Mimulus guttatus* from copper mine and serpentine habitats. *Plant Soil* 191, 173-180, 1997a.

Tilstone, G.H. and M.R. Macnair. The consequence of selection for copper tolerance on the uptake and accumulation of copper in *Mimulus guttatus. Ann. Bot.* 80, 747-751, 1997b.

Tilstone, G.H., M.R. Macnair, and S.E. Smith. Does copper tolerance give cadmium tolerance in *Mimulus guttatus? Heredity* 79, 445-452, 1997.

Vazquez, M.D., C. Poschenrieder, J. Barcelo, A.J.M. Baker, P. Hatton, and G.H. Cope. Compartmentation of zinc in roots and leaves of the zinc hyperaccumulator *Thlaspi caerulescens* J. and C. Presl. *Bot. Acta* 107, 243-250, 1994.

Wilson, J.B. The cost of heavy metal tolerance: an example. *Evolution* 42, 408-413, 1988.

Wu, L. and J. Antonovics. Zinc and copper uptake by *Agrostis stolonifera,* tolerant to both zinc and copper. *New Phytol.* 75, 231-237, 1975.

14 ECOLOGICAL GENETICS AND THE EVOLUTION OF TRACE ELEMENT HYPERACCUMULATION IN PLANTS

A. Joseph Pollard, Keri L. Dandridge, and Edward M. Jhee

CONTENTS

INTRODUCTION

Many authors have described the potential uses of phytoremediation technology, as well as the need to understand the factors that control plant uptake of trace elements in developing this technology (e.g., Baker et al., 1994a; Salt et al., 1995). Plants that sequester trace metals at extremely high concentrations (hyperaccumulators), will probably play some role in phytoremediation, whether it be direct use as phytoremediation crops, indirect sources of genes for bioengineering of phytoremediation crops, or even more indirect physiological models through which we increase our knowledge of basic uptake processes. While much attention is focused on the physiological mechanisms of metal accumulation and tolerance, relatively little is known about the genetic and ecological factors that have led to the evolution of hyperaccumulation in nature. This chapter will attempt to summarize the current state of knowledge regarding these aspects of hyperaccumulation, present some new and previously unpublished findings, and generate hypotheses that may be relevant to future studies.

The central question we will address in this chapter is: Why do plants hyperaccumulate? or more formally: What is the selective advantage of hyperaccumulation and what evolutionary and adaptive processes have led to the development of this trait? The wide geographic and taxonomic ranges over which hyperaccumulation occurs (Baker and Brooks, 1989) imply that it has not arisen randomly in a single lineage, but has evolved independently in several taxa and localities; thus, it is reasonable to investigate the selective force or forces that may have led to this evolution. Possible ecological advantages of hyperaccumulation were reviewed in an excellent paper by Boyd and Martens (1993). We will attempt to expand on their foundation by including recent findings that have arisen since the publication of their review and by including both genetic and ecological perspectives on evolution.

The central paradigm of evolutionary ecology is that adaptation occurs primarily as a result of natural selection. This is a process whereby in a genetically variable population some individuals are more suited to their environment than others, and consequently reproduce more successfully and leave more offspring to future generations. Those offspring are likely to inherit the features that made their parents successful; therefore, over time, the traits conferring fitness in that environment will become more common in the population. Thus, evolution is an interaction between genetic and ecological factors. Currently, the genetics and ecology of hyperaccumulation are both active subjects for research.

GENETIC VARIABILITY IN HYPERACCUMULATION

This book is intended to be read by specialists from a wide range of disciplines. Therefore, it may be appropriate to provide a brief general review of some basic ideas in population and quantitative genetics before proceeding to explore the genetics of hyperaccumulation in natural populations. Readers wishing more detailed background should consult one of the many general texts on this topic (e.g., Falconer, 1989; Briggs and Walters, 1984; Crow, 1986; Silvertown and Lovett Doust, 1993) which act as references for much of the discussion that follows.

SOURCES OF VARIATION IN POPULATIONS

Phenotypic differences among plants are consequences of both genetic variation and the modifications imposed by the particular environments in which individuals are growing. Variation may be either discrete or continuous. Discrete phenotypic classes generally result from genetic control by a small number of loci. Genetic studies of discrete variation are typically conducted using classical Mendelian and population genetics, in which specific allele and genotype frequencies are estimated. Such studies may employ very elaborate schemes analyzing the offspring of controlled crosses (see Chapter 13).

A continuous spectrum of variation usually results from a polygenic system of inheritance, in which many loci affect the same trait. Because of the large number of genes involved, continuous variation is typically studied through the methods of quantitative genetics, which concentrate on phenotypic measurements rather than gene frequencies. This approach is particularly appropriate when environmental influences are large compared to the effect of any one gene. Quantitative geneticists use a variety of cultivation and breeding schemes to attempt to separate and quantify the genetic and environmental determinants of the phenotype.

Total phenotypic variance (V_P) may be partitioned according to the equation:

$$V_P = V_G + V_E = \left(V_A + V_N\right) + V_E$$

in which V_G represents genetic variation and V_E represents environmentally induced variation. In the second part of the equation, the genetic variation has been further subdivided into V_A, additive variation, and V_N, nonadditive variation. Additive genetic variation involves characteristics that are transmitted in a simple manner from parents to offspring. Nonadditive variation represents differences among individuals caused by various genetic interactions such as dominance, epistasis, maternal effects, and genotype-by-environment interactions. Although the underlying causes of nonadditive variation are genetic, it results from complex interactions among genes and thus does not necessarily predict simple parent–offspring resemblance.

It is often useful to estimate the relative contributions of genotype and environment to the phenotype. This is done using a statistic called heritability, symbolized h^2, which varies between zero and one. The ratio V_G/V_P, known as "broad-sense heritability," reflects the fraction of the population's variability that is caused genotypically and, by extension, the probable fraction of an individual's phenotype determined by its genes. The ratio V_A/V_P is termed "narrow-sense heritability." Because of the definition of additive variation, narrow-sense heritability can be said to reflect the degree to which phenotypes are determined by the genes of parents (i.e., the importance of inheritance in controlling the phenotype). In either case, heritability is a characteristic of a particular population in a particular environmental setting. Heritability estimates made under uniform conditions will usually be higher than those measured in a variable environment, because of the decreased contribution of V_E to the denominator, V_P.

EVOLUTION

Evolution is defined as change-over time in the genetic makeup of a population. The population genetic models of the Hardy-Weinberg law describe the factors that can potentially change gene frequencies in a population and thus drive evolution. These include genetic drift in small populations; mutations; migration or gene flow; non-random mating, including inbreeding; and natural selection, or differential fitness as measured through reproductive success. Most of these factors are essentially random; only natural selection directs change in such a way that a population becomes more suited to its environment over time.

Differences in fitness depend on the ecological interactions between particular phenotypes and particular environments. However, only the genetic component of the phenotype can be passed on to the next generation and thus affect the evolution of the population. Natural selection cannot operate on traits which have no genetic variation, and will operate slowly on traits for which phenotypic variation derives predominantly from the environment and only slightly from genes. This is expressed by the equation:

$$R = h^2 \bullet S$$

which indicates that response to selection (R) is equal to the product of heritability times the intensity of selection (S). Low heritability thus can impede the response of the population to even strong selection pressures.

Variability within a species can exist at many spatial scales, including differences between populations and differences among individuals within a population. Variability within a local population is subject to natural selection, allowing the population to adapt to its local conditions. If two populations experience different environmental conditions, selection thus can result in evolutionary divergence between them. However, differences between populations could also result from random, nonselective factors such as genetic drift.

RESEARCH ON GENETICS OF HYPERACCUMULATION

The existence of phenotypic variation in shoot metal concentration within species of hyperaccumulators has been recognized in many studies. The majority of such work (e.g., Reeves and Brooks, 1983a,b; Reeves et al., 1983a,b; Reeves, 1988) has examined the metal content of plants collected in their native field sites and thus includes both genetic and environmental sources of variation. Especially in cases where herbarium specimens have been analyzed, the elemental content of the soil in which the plants were growing is usually unknown.

There have now been several studies in which hyperaccumulating plants from more than one population have been compared for metal content after being grown from seed under uniform and controlled conditions. Perhaps through sheer coincidence, all these studies involve species of *Thlaspi* that have the potential to hyperaccumulate nickel and zinc, and perhaps cadmium. Investigations of *T. goesingense* (Reeves and Baker, 1984), *T. montanum* var. *montanum* (Boyd and Martens, 1998), and *T. caerulescens* (Baker et al., 1994a,b) all found few statistically significant differences between populations in their ability to hyperaccumulate. In a much broader survey of variability in hyperaccumulation, Lloyd-Thomas (1995) compared populations of *T. caerulescens* from sites in Britain, Belgium, and Spain using both soil and hydroponic media. He reported statistically significant differences between populations in their ability to hyperaccumulate zinc, nickel, and cadmium, as well as a number of other metals accumulated to lower concentrations. Other recent studies (Pollard and Baker, 1996; Chaney et al., 1997; Meerts and Van Isacker, 1997) also confirm the existence of interpopulation variability in metal uptake among plants grown from seed in a common environment. Thus, these recent studies imply that ability to hyperaccumulate is not a completely uniform property within a species, but differs from population to population.

Most of the studies described above did not use methods that would allow calculation of genetic statistics such as heritability. However, within-population variability is of great importance to questions of evolution and natural selection. Pollard and Baker (1996) examined zinc hyperaccumulation in *T. caerulescens* from two populations on Zn/Pb mine spoil in central Britain. In order to assess genetic trends, seeds were collected as sib families, i.e., as sets of seeds from a common mother plant. The seeds were germinated and plants were grown hydroponically in nutrient solution containing 10 mg l^{-1} Zn. Statistically significant differences in zinc concentration were found between populations and among the sib families within one population. It was possible to estimate broad-sense heritability based on resemblances among siblings. In the variable population (Black Rocks, Derbyshire, U.K.), variation in zinc accumulation had a heritability of 0.179. The character of shoot dry weight was also analyzed and found to show significant within-population variability, with h^2 = 0.382. These findings imply that significant genetic variation in ability to accumulate metals may exist at the within-population level.

Recent work in our laboratory has extended the analysis of sib families to examine zinc and nickel hyperaccumulation in populations of *T. caerulescens* from a variety of soil types (Table 14.1). Seeds from five populations, collected as sib families, were germinated and grown hydroponically on nutrient solutions supplemented with either 10 mg l^{-1} Zn or 0.5 mg l^{-1} Ni. Leaves were removed from plants for analysis by atomic absorption spectrometry. (Full details of methods will be described in a future journal publication.) We found statistically significant differences between populations in ability to accumulate both metals. Genetic variation

---

TABLE 14.1

Characteristics of Source Populations for *Thlaspi caerulescens* Seeds Used in Heritability Studies

Population	Location	Description	Soil Zn	Leaf Zn	Soil Ni	Leaf Ni
BD	England	Pb/Zn mine spoil	8714	43,090	50	11
HF	Wales	Pb/Zn mine spoil	35,200	47,601	44	1
CH	England	Alluvial deposit (downstream from Pb/Zn mines)	2214	19,384	50	6
PB	Spain	Serpentine outcrop	58	1198	2918	18,357
PE	Spain	Alpine pasture	158	7777	48	0

Note: Soil concentrations are total µg g^{-1} based on aqua regia digests. Leaf concentrations are µg g^{-1} dry weight from field-collected leaves.

among families within populations, as reflected in heritability values significantly greater than zero, was found in three populations for zinc and in one population for nickel (Table 14.2). Of particular interest was the population growing on soil without high metal content, in the Picos de Europa of northern Spain. Plants in the field accumulated zinc concentrations that were below the 10,000 $\mu g \ g^{-1}$ criterion for hyperaccumulation, but were nonetheless remarkable for plants growing on "normal" soil (Table 14.1). In the laboratory, plants from this population displayed strong ability to hyperaccumulate zinc and nickel, but they also harbored highly significant between-family variation in ability to accumulate both metals (Table 14.2). Comparisons between *T. caerulescens* populations from metal-enriched soils in Belgium and populations from unmineralized sites in Luxembourg (Meerts and Van Isacker, 1997) have also demonstrated the existence of variation in hyperaccumulation, both between and within populations.

TABLE 14.2

Variation Between and Within Populations of *Thlaspi caerulescens* in Ability to Accumulate Metals

	Zinc		Nickel	
Population	Pop. Mean ($\mu g \ g^{-1}$ dry wt.)	h²	Pop. Mean ($\mu g \ g^{-1}$ dry wt.)	h²
BD	12.958	NS	1950	NS
HF	20,149	0 36	743	NS
CH	11,456	0.11	686	NS
PB	15,413	NS	830	NS
PE	21,351	0 82	1066	0.67

Note: Plants were grown in nutrient solution with addition of either 10 mg l⁻¹ Zn or 0.5 mg l⁻¹ Ni (as sulfates). Differences among population means were statistically significant for each metal, based on nested ANOVA Estimates of broad-sense heritability (h²) are reported for each metal, in populations where one-way ANOVA revealed significant differences between sib families (NS = not significant). Populations are described in Table 14.1.

GENETIC CONCLUSIONS

It appears, at least in the genus *Thlaspi*, that genetic variation in the ability to hyperaccumulate metals is demonstrated both between populations and within populations. There have been no signs of large, discrete polymorphisms expressed in natural populations; rather, there appears to be continuous variation, as would be expected from polygenic inheritance. Such systems may be truly quantitative if many loci control production of a single gene product that behaves in a dosage-dependent manner. Alternatively, polygenic inheritance can involve genes independently controlling several different aspects of physiology, such as mobilization, uptake, loading, transport, unloading, and storage of metals. Polygenic inheritance may be an obstacle to those attempting to isolate and manipulate a "hyperaccumulation gene." However, the results described above do not rule out the possibility that a major gene for hyperaccumulation exists, but is fixed throughout *T. caerulescens*, and thus displays no variability (unless interspecific hybrids were generated; cf. Chapter 13). In such a situation, the variation expressed in natural populations could result from multiple modifier genes that might accompany the major gene.

In an applied sense, the importance of the heritability values described above stems from the relationship between intensity of selection and response to selection ($R = h^2 \cdot S$). The presence of a reservoir of variation with significant heritability implies that attempts to improve metal accumulation in potential phytoremediation crops through artificial selection may be fruitful. Pollard and Baker (1996) found no evidence for a trade-off between plant size and metal concentration, which might limit selection on total metal yield. Recent results regarding populations of hyper-accumulators from nonmetalliferous sites (Table 14.2, also Meerts and Van Isacker, 1997) suggest that such plants may be particularly valuable resources, because they may possess *both* a strong ability to accumulate metals (as a population average), and high levels of heritable variation that could indicate a potentially rapid response to selection for further increases in uptake.

ECOLOGICAL SIGNIFICANCE OF HYPERACCUMULATION

Reviewing the literature, Boyd and Martens (1993) grouped the published suggestions regarding the adaptive value of hyperaccumulation into five major hypotheses: (1) that hyperaccumulation functions to increase the metal tolerance of the plant, perhaps by aiding in the disposal of excess metals; (2) that hyperaccumulation increases the drought resistance of leaves; (3) that hyperaccumulation benefits plants through allelopathic interactions with other plants (e.g., creating a zone of toxic soil that suppresses competitors); (4) that hyperaccumulation is an inadvertent consequence of high-affinity uptake of other elements that may be scarce in mineralized substrates; and (5) that hyperaccumulation benefits plants through defense against herbivores or pathogens. In the years following their review, several investigations have supported the fifth hypothesis.

HYPERACCUMULATION AS A PLANT DEFENSE

Boyd and Martens (1994) showed that nickel hyperaccumulation in *T. montanum* var. *montanum* can be acutely toxic to larvae of *Pieris rapae* (Lepidoptera). Findings such as these are perhaps better described as antibiosis (an interaction that harms the herbivore), rather than as defense (an interaction that benefits the plant). As discussed by Pollard (1992), the two interactions are not necessarily synonymous.

Martens and Boyd (1994) demonstrated that nickel hyperaccumulated by *Strep-tanthus polygaloides* causes similar acute antibiosis toward three species of insect herbivores. The same study also documented benefits to the plant (functional defense — Pollard, 1992) through deterrence of feeding in choice situations, resulting in greater plant growth and survival. Nickel also reduces bacterial and fungal growth, consequently improving plant growth and flowering in the same species (Boyd et al., 1994). Zinc hyperaccumulation can also have a deterrent role, as shown for insect and slug herbivory in *T. caerulescens* by Pollard and Baker (1997).

The studies described in the preceding paragraph compared plants grown on high-metal vs. low-metal substrates; thus, they measured the response of herbivores to environmentally induced variation (V_E). In order to conclude that herbivore feeding pressures could select for the evolution of hyperaccumulation, it is necessary to show that herbivores discriminate in response to heritable genetic variation (V_G). In other words, to demonstrate that a feature of a plant represents an adaptation against herbivory, rather than an effective but coincidental defense evolved under selection by forces other than herbivory, requires the documentation of feeding deterrence in response to genetic variation that exists in nature (Jones, 1971; Pollard, 1992).

We have recently approached this issue by using plants from our screening of genetic variation in zinc hyperaccumulation, as described earlier. The harvest of leaves for chemical analysis did not involve complete destruction of the plants. Thus, after characterizing the metal-accumulating ability of individuals grown in a common environment, we could subsequently use additional leaves for presentation to herbivores.

Leaves were removed from plants chosen to represent a contrast between high zinc accumulation and low zinc accumulation, in the common environment of culture solution with 10 mg L^{-1} Zn. Two contrasting leaves were placed in a 6-cm plastic petri dish. Across the whole experiment, the mean difference between the high-zinc and low-zinc leaves was 28,135 μg g^{-1}, and in no dish was the difference less than 20,000 μg g^{-1}. The area of each leaf was measured before the experiment using a digital leaf-area meter.

The herbivore used for these experiments was the larva of *P. napi oleracea* (Lepidoptera, Pieridae), the veined white butterfly. This animal was chosen as a bioassay of palatability (Pollard and Baker, 1997), based on availability and willingness to eat *T. caerulescens* grown in low-zinc media. It is not known to feed on *T. caerulescens* in the wild, although Rocky Mountain populations do feed on the closely related *T. montanum* (on nonmetalliferous sites and thus containing low metal concentrations). Eggs (obtained from F. S. Chew at Tufts University) were allowed to hatch, and hatchlings were fed on radish leaves until large enough to be transferred to experimental dishes. One caterpillar was placed in each dish described above; 181 replicate trials were conducted. After 2 h of feeding, the remaining area of the leaves was measured, and leaf area consumed was determined by subtraction.

Results of feeding trials are shown in Figure 14.1. Young larvae (less than 5 mm long) showed a slight preference for low-zinc leaves, but this difference was not statistically significant (paired t = 1.55, df = 62, p = 0.13). However, later-instar larvae showed very strong and significant preferences for the low-zinc leaves (paired t = 7.22, df = 117, p <0.001).

Greater discrimination by later-instar larvae in choices among foodplant species was shown for *P. napi* larvae by Chew (1980). This appears to represent a behavioral reflection of the fact that larvae must eat immediately after hatching, and are able to become mobile foragers only during later instars. The same pattern was reflected

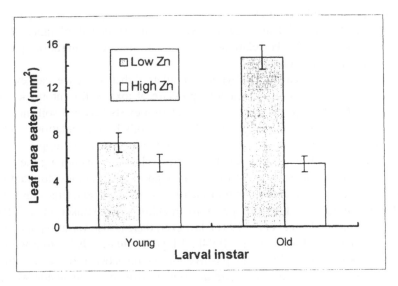

FIGURE 14.1 Leaf area of *Thlaspi caerulescens* consumed by *Pieris napi* larvae. All plants were grown in nutrient solution with 10 mg L⁻¹ Zn. High-Zn and low-Zn plants were chosen based on prior chemical analysis of Zn content. In petri dish feeding trials, caterpillars were presented a choice between a high-Zn and low-Zn leaf. Mean leaf area consumed (±SE) is shown for early-instar larvae (N = 63) and late-instar larvae (N = 118).

here, in terms of intraspecific variation in leaf chemistry. Foodplant choices for young larvae might be made maternally through the oviposition preferences of adult females. However, Martens and Boyd (1994) could not find evidence that *P. rapae* oviposition was influenced by nickel content of *S. polygaloides*.

These results confirm, using genetic variation in a common environment, the conclusion of previous herbivory studies using environmentally induced variation: that hyperaccumulation of metals in plants could have evolved under selection pressure from herbivores. Future studies will need to address the ability of herbivores to discriminate among even more subtle differences among phenotypes, especially if variation in metal accumulation ability is shown to be polygenic. However, it is important to note that phenotypic differences in metal concentration of the magnitude used in these experiments do occur within populations, both in the field and in controlled conditions (Lloyd-Thomas, 1995; Dandridge and Pollard, unpublished).

NONDEFENSIVE HYPOTHESES

Of the five hypotheses on the adaptive role of hyperaccumulation listed above, only the defensive hypothesis has received direct experimental study since the review of Boyd and Martens (1993). The drought-tolerance and allelopathic hypotheses remain relatively unexplored, although Boyd and Martens (1998) have recently argued in favor of the inadvertent uptake hypothesis.

The idea that hyperaccumulation is a mechanism to provide metal tolerance remains pervasive. Krämer et al. (1996) demonstrated at a physiological level that free histidine plays a role in both metal accumulation and metal tolerance in *Alyssum* species. On the other hand, studies of variation in natural populations do not support the existence of a clear linkage between hyperaccumulation and tolerance. We will discuss this by examining correlations between soil metal concentration, plant tolerance, and hyperaccumulation.

It is well established for nonhyperaccumulators that tolerance can evolve in populations under the localized selective pressure of toxic soil (Antonovics et al., 1971). Thus, for these plants, a positive correlation exists between soil metal concentration and tolerance in that nonmetalliferous soils bear populations with low metal tolerance (on average), while metal-contaminated sites support tolerant ecotypes. It is less clear whether this trend of correlation exists among the tolerant populations (i.e., whether the most toxic soils tend to bear the most tolerant populations), but limited data seem to support such a trend (Gregory and Bradshaw, 1965).

Hyperaccumulating taxa are generally metal-tolerant (Baker et al., 1994b; Krämer et al., 1996; Homer et al., 1991); however, there is also variation in tolerance among populations (Baker et al., 1994b; Lloyd-Thomas, 1995). Ingrouille and Smirnoff (1986) reported significant positive correlation between zinc concentrations in the soil and zinc tolerance in *T. caerulescens* in Britain, a conclusion which has been recently substantiated for the same species in continental Europe (Meerts and Van Isacker, 1997). A common feature of these studies was that they included plants from populations on nonmetalliferous soils.

If hyperaccumulation represents a mechanism of metal tolerance, then we would expect relative ability to hyperaccumulate (based on studies in a uniform environment) to be positively correlated with both soil toxicity and plant tolerance. The few data sets in which this analysis is possible (Lloyd-Thomas, 1995; Meerts and Van Isacker, 1997; Dandridge and Pollard, unpublished) consistently fail to support this prediction for the case of zinc in *T. caerulescens*. No positive correlations were detected, either between soil concentration and hyperaccumulation ability or between degree of tolerance and hyperaccumulation ability. All included populations from nonmetalliferous (thus, nontoxic) soils; these plants not only had the ability to hyperaccumulate, but did so more strongly than populations from zinc-mine spoil and other contaminated areas.

Curiously, the data of Lloyd-Thomas (1995) do show a significant correlation between nickel tolerance and the ability to accumulate nickel in *T. caerulescens* across a range of populations mostly collected from zinc-mine spoil. The importance of this finding for a species that only rarely occurs on high-nickel substrates like serpentine remains to be investigated. Working with North American *T. montanum* var. *montanum*, which occurs both on serpentine and on normal soils, Boyd and Martens (1998) found no significant differences in the ability of serpentine and nonserpentine populations to take up nickel from uniform soil media (tolerance was not measured directly in their study).

ECOLOGICAL AND EVOLUTIONARY CONCLUSIONS

There is mounting support for the hypothesis that hyperaccumulation may have direct benefits for the plant, especially protection against herbivores and pathogens. The finding that plants from low-zinc soils possess strong powers of zinc accumulation is consonant with this suggestion in that generating effective defenses in such habitats would require movement of metal ions against a strong concentration gradient.

In general, data from studies of interpopulation variation do not support the hypothesis that hyperaccumulation is an adaptation that confers metal tolerance. It is worth noting in this regard that the great majority of known hyperaccumulators are serpentine endemics taking up nickel (Baker and Brooks, 1989); however, the ecological importance of nickel toxicity on serpentine soils has never been clearly established. Plants that are "serpentine-intolerant" may be restricted by many different physiochemical factors other than metal toxicity (Brooks, 1987; Baker, 1987). Apart from hyperaccumulators, few plants with demonstrable nickel tolerance are known from serpentine.

The observation that hyperaccumulators are generally metal tolerant implies that there is some connection between the two traits. If it is not true that hyperaccumulation confers tolerance, then two possible evolutionary pathways can be hypothesized. One is that plants evolve tolerance in response to soil toxicity, permitting them to accumulate low levels of metals in their tissues (Baker and Walker, 1990); subsequently, selection by herbivores or some other ecological advantage drives the evolution of increasing metal concentration, resulting in hyperaccumulation. The alternative hypothesis is that a selective advantage such as defense against herbivory is primary, and that tolerance evolves secondarily in order to detoxify accumulated metals. While this last suggestion may seem radical, it is supported circumstantially by several observations, including the association of hyperaccumulation with serpentine soils where metal toxicity is not particularly acute and the pattern of constitutive metal tolerance in hyperaccumulators contrasted with local evolution of tolerance as found in most other species (Antonovics et al., 1971; Baker 1987). The presence of high metal transport ability but low metal tolerance in *T. arvense* (Krämer et al., 1997) also agrees with this interpretation.

SUMMARY AND APPLIED CONCLUSIONS

Studies of intraspecific variation in natural plant populations have proved to be a useful tool for understanding the genetics, ecology, and evolution of hyperaccumulation. This field is still young, so broad generalizations must be regarded as tentative until more findings can be compared. At present, it appears that the ability to accumulate metals is a constitutive property of hyperaccumulating species (i.e., it is expressed in all members of the species regardless of native soil), but it may also demonstrate continuous, heritable genetic variation between and within populations. Hyperaccumulation may evolve for several reasons, but it is increasingly clear that defense against herbivores and pathogens can be an important selective force. On the other hand, the evolutionary relationship between metal tolerance and metal hyperaccumulation remains unclear, and requires further study to determine how these two properties interact.

Investigations such as those reviewed here can have a number of important implications for phytoremediation. In any biotechnological application, it is important to understand the underlying genetic diversity of the biological resource base,

which can be utilized in manipulations such as artificial selection, controlled hybridization, and genetic engineering (see Genetic Conclusions above). Genetic polymorphisms can also be useful tools for understanding the regulation of plant physiology.
Knowledge of the ecological role of hyperaccumulation in natural populations will
be relevant in attempts to plant monocultures of metal-accumulating crops for
phytoremediation. Apart from these benefits to phytoremediation research, study of
the unusual phenomenon of hyperaccumulation may provide a general model system
for study of many aspects of plant adaptation.

ACKNOWLEDGMENTS

We wish to thank F.S. Chew for supplying caterpillar eggs, A.J.M. Baker and R.D.
Reeves for help and advice with seed collecting, and D.H. Lloyd-Thomas, R.S.
Boyd, and N. Van Isacker for access to unpublished results during the preparation
of the manuscript.

REFERENCES

Antonovics, J., A.D. Bradshaw, and R.G. Turner. Heavy metal tolerance in plants. *Adv. Ecol.
Res.* 7, 1-85, 1971.
Baker, A.J.M. Metal tolerance. *New Phytol.* 106 (Suppl.), 93-111, 1987.
Baker, A.J.M. and R.R. Brooks. Terrestrial higher plants which hyperaccumulate metallic
elements — a review of their distribution, ecology, and phytochemistry. *Biorecovery*
1, 81-126, 1989.
Baker, A.J.M., S.P. McGrath, C.M.D. Sidoli, and R.D. Reeves. The possibility of *in situ* heavy
metal decontamination of polluted soils using crops of metal-accumulating plants.
Res. Conserv. Recyc. 11, 41-49,1994a.
Baker, A.J.M., R.D. Reeves, and A.S.M. Hajar. Heavy metal accumulation and tolerance in
British populations of the metallophyte *Thlaspi caerulescens* J. & C. Presl (Brassi-
caceae). *New Phytol.* 127, 61-68, 1994b.
Baker, A.J.M. and P.L. Walker. Ecophysiology of metal uptake by tolerant plants, in *Heavy
Metal Tolerance in Plants: Evolutionary Aspects,* Shaw, A.J., Ed. CRC Press, Boca
Raton, FL, 155-177, 1990.
Boyd, R.S. and S.N. Martens. The raison d'être for metal hyperaccumulation by plants, in
The Vegetation of Ultramafic (Serpentine) Soils, Baker, A.J.M., J. Proctor, and R. D.
Reeves, Eds. Intercept, Andover, U.K., 279-289, 1993.
Boyd, R.S. and S.N. Martens. Nickel hyperaccumulated by *Thlaspi montanum* var. *montanum*
is acutely toxic to an insect herbivore. *Oikos* 70, 21-25, 1994.
Boyd, R.S. and S.N. Martens. Nickel hyperaccumulation by *Thlaspi montanum* var. *montanum*
(Brassicaceae): a constitutive trait. *Am. J. Bot.* 85, 259-265, 1998.
Boyd, R.S., J.J. Shaw, and S.N. Martens. Nickel hyperaccumulation defends *Streptanthus
polygaloides* (Brassicaceae) against pathogens. *Am. J. Bot.* 81, 294-300, 1994.
Briggs, D. and S.M. Walters. *Plant Variation and Evolution.* Cambridge University Press,
Cambridge, U.K., 1984.
Brooks, R.R. *Serpentine and Its Vegetation: A Multidisciplinary Approach.* Dioscorides Press,
Portland, OR, 1987.

Chaney, R.L., M. Malik, Y.M. Li, S.L. Brown, E.P. Brewer, J.S. Angle, and A.J.M. Baker. Phytoremediation of soil metals. *Curr. Opin. Biotechnol.* 8, 279-284, 1997.

Chew, F.S. Foodplant preferences of *Pieris* caterpillars (Lepidoptera). *Oecologia* 46, 347-353, 1980.

Crow, J.F. *Basic Concepts in Population, Quantitative, and Evolutionary Genetics.* W.H. Freeman, New York, 1986.

Dandridge, K.L. and A.J. Pollard, unpublished data, 1996.

Falconer, D.S. *Introduction to Quantitative Genetics.* Longman, Harlow, U.K., 1989.

Gregory, R.P.G. and A.D. Bradshaw. Heavy metal tolerance in populations of *Agrostis tenuis* Sibth. and other grasses. *New Phytol.* 64, 131-143, 1965.

Homer, F.A., R.S. Morrison, R.R. Brooks, J. Clemmens, and R.D. Reeves. Comparative studies of nickel, cobalt and copper uptake by some nickel hyperaccumulators of the genus *Alyssum. Plant Soil* 138, 195-205, 1991.

Ingrouille, M.J. and N. Smirnoff. *Thlaspi caerulescens* J. & C. Presl (*T. alpestre* L.) in Britain. *New Phytol.* 102, 219-233, 1986.

Jones, D.A. Chemical defense mechanisms and genetic polymorphism. *Science* 173, 945, 1971.

Krämer, U., J.D. Cotter-Howells, J.M. Charnock, A.J.M. Baker, and J.A.C. Smith. Free histidine as a metal chelator in plants that accumulate nickel. *Nature* 379, 635-638, 1996.

Krämer, U., R.D. Smith, W.W. Wenzel, I. Raskin, and D. Salt. The role of metal transport and tolerance in nickel hyperaccumulation by *Thlaspi goesingense* Hálácsy. *Plant Physiol. (Rockville).* 115, 1641-1650, 1997.

Lloyd-Thomas, D.H. "Heavy Metal Hyperaccumulation by *Thlaspi caerulescens* J. & C. Presl," thesis presented to the University of Sheffield, U.K., in partial fulfillment of the requirements for the Ph.D. degree, 1995.

Martens, S.N. and R.S. Boyd. The ecological significance of nickel hyper-accumulation: a plant chemical defense. *Oecologia* 98, 379-384, 1994.

Meerts, P. and N. Van Isacker. Heavy metal tolerance and accumulation in the metallicolous and non-metallicolous populations of *Thlaspi caerulescens* from continental Europe. *Plant Ecol.* 133, 221-231, 1997.

Pollard, A.J. The importance of deterrence: responses of grazing animals to plant variation, in *Plant Resistance to Herbivores and Pathogens,* Fritz, R.S. and E.L. Simms, Eds. University of Chicago Press, Chicago, 216-239, 1992.

Pollard, A.J. and A.J.M. Baker. Quantitative genetics of zinc hyperaccumulation in *Thlaspi caerulescens. New Phytol.* 132, 113-118, 1996.

Pollard, A.J. and A.J.M. Baker. Deterrence of herbivory by zinc hyperaccumulation in *Thlaspi caerulescens* (Brassicaceae). *New Phytol.* 135, 655-658, 1997.

Reeves, R.D. Nickel and zinc accumulation by species of *Thlaspi* L., *Cochlearia* L., and other genera of the Brassicaceae. *Taxon* 37, 309-318, 1988.

Reeves, R.D. and A.J.M. Baker. Studies on metal uptake by plants from serpentine and non-serpentine populations of *Thlaspi goesingense* Hálácsy (Cruciferae). *New Phytol.* 98, 191-204, 1984.

Reeves, R.D. and R.R. Brooks. European species of *Thlaspi* L. (Cruciferae) as indicators of nickel and zinc. *J. Geochem. Explor.* 18, 275-283, 1983a.

Reeves, R.D. and R.R. Brooks. Hyperaccumulation of lead and zinc by two metallophytes from mining areas of central Europe. *Environ. Pollut. (Ser. A).* 31, 277-285, 1983b.

Reeves, R.D., R.R. Brooks, and T.R. Dudley. Uptake of nickel by species of *Alyssum, Bornmuellera,* and other genera of old world tribus alysseae. *Taxon* 32, 184-192, 1983a.

Reeves, R.D., R.M. Macfarlane, and R.R. Brooks. Accumulation of nickel and zinc by western North American genera containing serpentine-tolerant species. *Am. J. Bot.* 70, 1297-1303, 1983b.

Salt, D.E., M. Blaylock, P.B.A. Nanda Kumar, V. Dushenkov, B.D. Ensley, I. Chet, and I. Raskin. Phytoremediation: a novel strategy for removal of toxic metals from the environment using plants. *Biotechnology* 13, 468-474, 1995.

Silvertown, J. and J. Lovett Doust. *Introduction to Plant Population Biology.* Blackwell Science, Oxford, 1993.

15 The Role of Bacteria in the Phytoremediation of Heavy Metals

D. van der Lelie, P. Corbisier, L. Diels, A. Gilis,
C. Lodewyckx, M. Mergeay, S. Taghavi,
N. Spelmans, and J. Vangronsveld

CONTENTS

INTRODUCTION

The metabolic capacity of plant-associated bacteria may be used to develop new phytoremediation strategies. In the rhizosphere, many pesticides as well as trichloro-ethylene, polycyclic aromatic compounds, and petroleum hydrocarbons are degraded at accelerated rates (Hsu and Bartha, 1979; Nichols et al., 1997). Scientists are using plants for phytoextraction and phytoimmobilization of heavy metals from polluted soils and wastewater. Although plant-associated bacteria have dynamic and varied metabolic capacities, current strategies do not capitalize on the physiology of these microbes in phytoremediation processes. Plants stimulate the growth of microorganisms due to secretion of organic molecules by their roots. This results in higher population densities of bacteria in the rhizosphere (Anderson and Coats, 1995; Rovira et al., 1979). In addition, endophytic bacteria colonize the interior of root and stem tissues. In this chapter, the potential impact of soil bacteria on heavy metal extraction by plants is discussed. We have given special attention to heterologous expression of heavy metal resistance genes by plant associated baceria. In addition, the development of reporter gene systems to determine the level of available heavy metals is explored, this in relation to phytotoxicity.

EFFECTS OF SIDEROPHORE PRODUCTION BY RHIZOSPHERE BACTERIA ON PLANT GROWTH

Rhizosphere bacteria, including those which carry plasmid-borne resistances to heavy metals, may be expected to play an important role in the bioavailability of metals to the plant. They can produce metal-chelating agents, like siderophores, that possess a high affinity for Fe^{3+}. In oxygenated environments such as the soil, iron is basically unavailable for uptake and limiting to microbial growth, even though it is the fourth most abundant element in the Earth's crust (Lindsay, 1979). Therefore, aerobic and facultative anaerobic microorganisms have evolved various systems to overcome the low solubility of external iron (reviewed by Neilands, 1981; Crichton and Charloteaux-Wauters, 1987; Crosa, 1989; Guerinot, 1994).

Siderophores produced by plant growth-promoting rhizobacteria (PGPR) and bacterial biological control agents are associated with improved plant growth, either through a direct effect on the plant, through control of noxious organisms in the soil, or via some other route (Kloepper et al., 1980). The PGPR include members of the genera *Arthrobacter*, *Alcaligenes* (Ralstonia), *Serratia*, *Pseudomonas*, *Rhizobium*, *Agrobacterium*, and *Bacillus* (O'Sullivan and O'Gara, 1992). The most promising group for application in the biocontrol of plant diseases is that of the fluorescent pseudomonads, especially *P. fluorescens* and *P. putida* (O'Sullivan and O'Gara, 1992). Several mechanisms are known to account for the growth-promoting and disease-suppressing impact of fluorescent siderophores on the plant. One mechanism by which antagonists control disease is by competition for iron, mediated by fluorescent siderophores, which have an extremely high affinity for ferric iron. Production of siderophores by antagonists further limits the already limited supply of iron in the environment (Neilands, 1981; Neilands and Leong, 1986). Growth of pathogens is therefore inhibited in the presence of siderophore-producing strains due to an inadequate supply of iron.

Some plants utilize microbial ferric iron-siderophore complexes. Microbial hydroxamate type siderophores were used by plants, such as sunflower and sorghum, for the uptake of iron (Cline et al., 1982, 1983, 1984). Furthermore, siderophores from fluorescent pseudomonads have also been implicated in iron uptake by tomato (Duss et al., 1986), carnations, barley (Duijff et al., 1991), and in the reversion of lime-induced chlorosis by peanut (Jurkevitch et al., 1988). In contrast, the fluorescent siderophore from *Pseudomonas* sp. B10 inhibited iron uptake by peas and maize plants (Becker et al., 1985).

INTERACTIONS OF SIDEROPHORES WITH HEAVY METALS

Siderophore production can be stimulated by the presence of heavy metals. Since most siderophores also show a lower but significant affinity for bivalent heavy metal ions, they affect the bioavailability of the heavy metals. For instance, in *Azotobacter vinelandii* siderophore production is increased in the presence of zinc (Huyer and Page, 1988). Schizoken production in *Bacillus megaterium* is increased by exposure to copper, chromate, cadmium, zinc, and aluminium (Beyers et al., 1967; Hu and Boyer, 1996). Zinc, cadmium, nickel, and aluminium were found to increase siderophore production in *P. aeruginosa* strains (Gilis, 1993; Hassan, 1996). The same effect was found for zinc and aluminium in *P. fluorescens* ATCC17400 (Gilis, 1993). In the plant growth beneficial strain *P. aeruginosa* 7NSK2, zinc was found to stimulate pyoverdine production even in the presence of iron (Höfte et al., 1993). Cadmium was shown to have a similar effect on *P. aeruginosa* PAO and *P. fluorescens* strain 6.2 (Mergeay et al., 1978). Different explanations have been offered for the stimulating effect of metals other than iron on siderophore production. First, the metal may be directly involved in the siderophore biosynthesis pathway or its regulation. For example, in *P. aeruginosa* 7NSK2, a site-specific recombinase (similar to XerC of *E. coli*) encoded by a gene called *sss* (stress-induced siderophore synthesis) was found to be involved in the regulation of pyoverdine synthesis by zinc (Höfte et al., 1994). Alternatively, the free siderophore concentration in the medium may be reduced by complex formation with metal ions other than Fe(III). Because iron limitation stimulates siderophore production, more siderophores would be produced. Zinc- and cadmium-induced pyoverdine production in a family of heavy metal-resistant *P. aeruginosa* strains (Sss+ phenotype; Hassan, 1996). These strains showed very high resistance levels for zinc and cadmium, with MIC values for these metals of 8.0 and 2.0 mM, respectively. Genes for resistance to zinc and cadmium are encoded by the *czr* operon and are induced by these metals. Thus, induction of siderophore biosynthesis genes may act in combination with metal resistance genes to overcome heavy metal toxicity.

The siderophore alcaligin E produced by the soil bacterium *Alcaligenes eutrophus* CH34 effects the bioavailability of cadmium (Gilis et al., 1996; 1997). The addition of exogenous alcaligin E overcomes the inhibition of growth of a siderophore minus mutant (Sid⁻) by cadmium. The presence of 1mM cadmium was sufficient to inhibit growth of the Sid⁻ strain AE1595, when no exogenous source of alcaligin E was added. In contrast, when 0.8 or 8 μm alcaligin E was added the bacteria grew at higher concentrations of cadmium, 4 or 8 mM, respectively (Figure 15.1). The same beneficial effect of alcaligin E is observed with cadmium under iron-limiting and -replete conditions, indicating that alcaligin E directly interacted with cadmium, thereby decreasing the availability of cadmium and consequently its toxic effects.

Although cadmium-uptake studies showed that alcaligin E does not effect the cellular concentration of cadmium, other evidence suggests that alcaligin E alters the bioavailability of cadmium. A reporter gene system designated to determine the level of cadmium available biologically demonstrated that cadmium availability

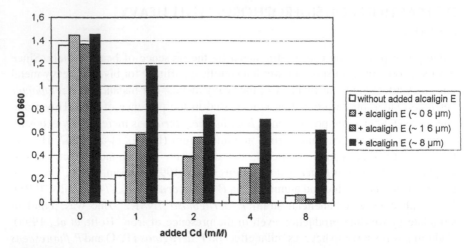

FIGURE 15.1 Influence of alcaligin E on the growth of the alcaligin E-deficient strain AE1595 in the presence of cadmium in precipitating conditions, after 44 h of growth in Schatz lactate medium. Due to the high phosphate content in this medium, most cadmium will be present in the form of $CdHPO_4$. (From Gilis, A., P. Corbisier, W. Baeyens, S. Taghavi, M. Mergeay, and D. van der Lelie, *J. Ind. Microbiol. Biotechnol.* 20, 61-68, 1998. With permission.)

decreases in the presence of alcaligin E (Figure 15.2). Alcaligin E also shifted the toxicity of cadmium to higher concentrations, while altering the morphology of precipitated cadmium crystals observed by SEM. These data suggest that alcaligin E may provide protection against heavy metal toxicity as well as function as an iron transport vehicle for *A. eutrophus.*

It is not clear what role siderophore production by rhizosphere bacteria have on plant growth in the presence of heavy metals. Siderophores might directly interact with the heavy metals, making them less bioavailable for the plants or may help plants to overcome metal-induced iron limitation. In addition, production of siderophores might protect plant roots from pathogens. Siderophore production might contribute to a competitive advantage resulting in dominance of pseudomonads in the rhizosphere. Importantly, the heavy metal resistance mechanisms of these bacteria combined with post-efflux metal binding might decrease the bioavailable metal fraction and, consequently, the metal toxicity for the plants.

METAL-RESISTANT SOIL BACTERIA

In addition to other benefits that rhizosphere bacteria confer to their hosts, plant-associated bacteria may protect plants from heavy metal toxicity. A prerequisite in protecting plants from heavy metals is resistance of the bacteria to these metals.

Metal-tolerant bacteria are isolated from many metal-rich biotopes, of either manmade or natural origins. The bacteria isolated from metal-rich biotopes of anthropogenic origin (mining and industrial sites) mainly belong to the genus *Ralstonia* (Yabuuchi et al., 1995), of which the former species *Burkholderia pickettii,*

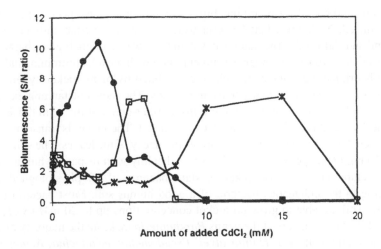

FIGURE 15.2 Bioluminescence profile of the *ale*-1595 (Sid⁻) cadmium, zinc, and lead biosensor (AE2350) in the presence of increasing concentrations of cadmium in precipitating conditions, with and without addition of alcaligin E-containing supernatant. (●) without alcaligin E, (□) with approximately 1.6 μm alcaligin E-containing supernatant, and (∗) with approximately 8 μm alcaligin E-containing supernatant. The light production, expressed in RLU (relative light units) was measured after 4 h and related to the OD_{660} of the culture. (From Gilis, A., P. Corbisier, W. Baeyens, S. Taghavi, M. Mergeay, and D. van der Lelie, *J. Ind. Microbiol. Biotechnol.* 20, 61-68, 1998. With permission.)

B. solanacearum (formerly *Pseudomonas solanacearum*), and *A. eutrophus* are members.

Metal-tolerant *A. eutrophus* strains are a strongly related group that are well adapted to environments polluted by heavy metals and/or organic xenobiotics. Similar bacteria are isolated from desertified soils (Maatheide) and low-grade ore deposits in Belgium and Zaïre (Diels and Mergeay, 1990). All these strains carry one or two megaplasmids which contain genes for multiple resistances to heavy metals. In addition, they exhibit similar resistance and substrate utilization patterns. The type-strain of this family is *A. eutrophus* strain CH34 (Mergeay et al., 1985). A similar strain, *A. eutrophus* strain KT02, was isolated from the wastewater treatment plant of Göttingen, and carries three plasmids (Schmidt et al., 1991). *A. eutrophus* strain 31A was isolated from the metal-working industry in Holzminden. It is an organotrophic bacterium highly resistant to nickel and carries two plasmids (Schmidt and Schlegel, 1994). *A. denitrificans* strain 4a-2 was isolated from the wastewater treatment plant of Dransfeld (Kaur et al., 1990). It is highly resistant to nickel and carries the heavy metal resistance determinants on the chromosome (Stoppel et al., 1995; Stoppel and Schlegel, 1995).

The bacteria isolated from natural metal-rich biotopes belonged to various genera. Bacteria resistant to nickel were isolated from soils of two kinds of ecosystems, both naturally laden with nickel. Soils from both ecosystems developed from nickeliferous rocks. One ecosystem is characterized by serpentine or ultramafic soils which originate from weathering of serpentine rocks frequently enriched with nickel, chromium, and cobalt. They are inhabited with endemic plant species (herbs) resistant to nickel. Nickel-resistant bacteria were isolated from such soils collected in Scotland and California. The other ecosystem in New Caledonia is also characterized by rocks rich in iron and manganese and enriched with nickel, chromium, and cobalt. The soils originating from such rocks are inhabited by many nickel-hyperaccumulation plants, among them many trees and shrubs. The most outstanding example is the tree *Sebertia acuminata*. It contains a blue-green milk with 25% nickel (weight/dry weight) and leaves with 1 to 2% nickel. The nickel is transiently stored in the vacuoles and released into the soil during decay of the leaves. The soil contains about 10 mg nickel per gram soil. The soil is healthy with respect to humus content and structure and contains nickel-resistant bacteria of all physiological groups belonging to the indigenous bacteria of a good humus soil. Most bacteria isolated from this soil are able to grow on nickel concentrations up to 20 mM (Stoppel and Schlegel, 1995). Members of the genus *Burkholderia* make up the majority of nickel-resistant strains collected. *Hafnia alvei*, *Pseudomonas mendocina*, *Acinetobacter*, *Comamonas acidovorans*, and *Agrobacterium tumefaciens* are other nickel-resistant strains collected. All the strains collected among New Caledonian soil samples are physiologically different from the European strains. Gram-positive bacteria were isolated from a soil sample (serpentine) in California. They were identified as *Arthrobacter ramosus* strain 60-6 and *Arthrobacter aurescens* strain 59-6 (Stoppel and Schlegel, 1995).

BACTERIAL HEAVY METAL RESISTANCE: THE *czc* (CADMIUM, ZINC, COBALT) OPERON

Among the heterotrophic bacteria, members of the β-Proteobacteria have the highest levels of resistance to heavy metals. *A. eutrophus* is a member of this group. The type strain *A. eutrophus* CH34 was originally isolated from a decantation tank of a zinc factory; the genetics, physiology, and biochemistry of this organism are the best studied.

Strain CH34 harbors two endogenous megaplasmids encoding multiple heavy metal resistance genes. Plasmid pMOL28 is 180 kb and codes for resistance to cobalt, nickel, chromate, mercury, and thallium. Resistance genes are organized into operons with the *chr* and *mer* operons, coding for resistances to chromate and mercury, respectively. The *mer* operon coding for mercury resistance is derived from Tn*4378*. Resistance genes for both cobalt and nickel are present in the *cnr* operon (Mergeay et al., 1985; Taghavi et al., 1997). In addition to these operons, the *tll*A locus is involved in thallium resistance (Collard et al., 1994). The second plasmid from strain CH34, pMOL30, is 240 kb and is responsible for resistance to some of the same heavy metals for which plasmid pMOL28 has resistance genes. On this plasmid, resistance genes are also organized into operons. The *mer* (this time from Tn*4380*), *cop*, and *pbr* operons encode resistance to mercury, copper, and lead. The *czc* operon encodes for cadmium, zinc, and cobalt resistance. Again, a single locus, *tll*B, is involved in thallium resistance (Mergeay et al., 1985; Collard et al., 1994).

These resistance genes have been used as probes to detect of a large number of related strains with resistance to heavy metals in mining areas or industrial sites in Congo and Belgium (Diels and Mergeay, 1990).

The *czc* operon of *A. eutrophus* CH34 is the most completely studied heavy metal resistance operon of *A. eutrophus*. The CzcABC structural resistance proteins form an efflux pump that functions as a chemiosmotic cation/proton antiporter (Figure 15.3) (Nies, 1995; Nies and Silver, 1995). The proteins involved have become the prototype for a new family of three-component chemiosmotic exporters, including members that efflux toxic cations or organic compounds (Diels et al., 1995a; Dong and Mergeay, 1994).

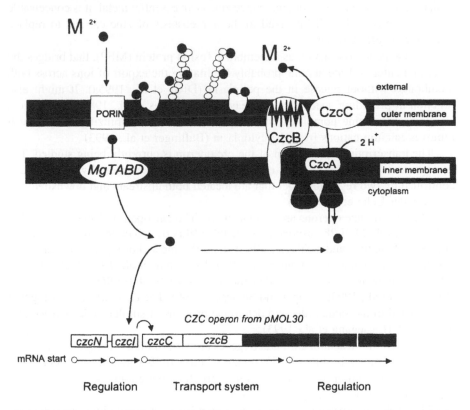

FIGURE 15.3 Schematic presentation of the Czc efflux system and a working model for heavy metal uptake, processing by efflux, and post-efflux metal fixation on polysaccharides and proteins. (From Taghavi, S., M. Mergeay, D. Nies, and D. van der Lelie, *Res. Microbiol.* 148, 536-551, 1997. With permission.)

CzcA, the central component of the system, functions as an inner membrane transport protein. CzcA is a chemiosmotic cation/proton antiporter belonging to the RND (resistance/nodulation/division) family (Nies and Silver, 1995; Nies, 1995). Such exporters, which are not driven by ATP and are thus different from traffic

ATPases, transport various metabolites, antibiotics, or drugs to the extracellular space (Saier et al., 1994).

CzcC is thought to function as an outer membrane protein. The N-terminal part of CzcC is typical of a signal for peptide secretion (Diels et al., 1995a). Additionally, there are sequence homologies and topological similarities between CzcC and other accessory proteins of ABC exporter systems. High homology is seen with a family of outer membrane factors (OMF; Diels et al., 1995a; Dong and Mergeay, 1994). CzcC is required in the process to complete the efflux of cadmium to the extracellular medium. Transport data (Nies and Silver, 1989) suggest that the export of the very toxic cadmium to the extracellular medium requires the participation of CzcC, which is dispensable for the efflux of zinc. Since zinc is an essential metal, it is conceivable that a chromosomal OMF involved in the homeostasis of zinc can act to replace CzcC for the efflux of zinc.

CzcB appears to function as a membrane fusion protein (MFP), that bridges the inner and outer cell membrane, probably facilitating the export of ions across both membranes without release in the periplasm (Diels et al., 1995a). It might also function in providing specificity for heavy metals (Nies et al., 1989). In addition, CzcB displays some homology with calphotin, a metal "mobilizing" protein that removes calcium cations from the cytoplasm (Ballinger et al., 1993).

The transcriptional regulation of the *czc* operon is currently being studied. The *czc* operon is inducible by zinc, cadmium, and to a lesser extent by cobalt and is controlled by two regulatory loci that are located both upstream and downstream of *czc* (van der Lelie et al., 1997).

The *cnr* and *ncc* operons are similar to *czc*. The *cnr* operon of the *A. eutrophus* CH34 plasmid pMOL28 ensures the inducible efflux of cobalt and nickel (Varma et al., 1990; Sensfuss and Schlegel, 1988). It is the most thoroughly studied nickel resistance determinant (Siddiqui et al., 1989; Liesegang et al., 1993; Collard et al., 1993). Sequencing analysis revealed that *cnr* consists of 6 ORFs, *cnr*YXHCBA (Liesegang et al., 1993; Stoppel and Schlegel, 1995). The *cnr*CBA structural genes are arranged in the same order and determine proteins of similar molecular weights as *czc*CBA (Liesegang et al., 1993).

Although the structural *cnr* and *czc* genes are very similar, the regulation of both operons is completely different. Upstream of the *cnr*CBA genes, three loci, *cnr*Y, *cnr*X, and *cnr*H, grouped in an operon-like structure, were found to be required for regulation of *cnr* (Liesegang et al., 1993).

The *cnr*H gene product seems to be a member of a novel sigma factor group, belonging to the extracytoplasmic (ECF) subfamily of sigma 70 (Liesegang et al., 1993). Some regulators involved in siderophore biosynthesis are also belonging to this subfamily. The CnrY protein has been suggested to function as an (auto)repressor: inactivation of *cnr*Y led to highly increased constitutive cobalt and nickel resistance and resulted in zinc resistance (ZinB phenotype; Collard et al., 1993; Liesegang et al., 1993). This observation led to the conclusion that the *cnr* system may also be involved in zinc efflux in the wildtype strain but at levels too low to confer a detectable resistance phenotype. This hypothesis was confirmed by zinc efflux studies after induction of the *cnr* operon by nickel. These data show that the stuctural genes of the *cnr* and *czc* systems are fundamentally the same. The role of the *cnr*X locus is not clear. However, recent data suggest that the CnrX protein is secreted into the periplasm where it might function as a metal sensor (C. Tibazarwa, personal communication). The *ncc* operon of *A. eutrophus* 31A is very similar to the *cnr* operon (Schmidt and Schlegel, 1994). In addition to the six genes, *ncc*YX-HCBA, whose homologs were also present in the *cnr* operon, a seventh gene *ncc*N was identified. The expression of *ncc*N is required for full nickel resistance: when *ncc*N is deleted, the nickel resistance drops from 30 to 5 m*M*, the same resistance level as found with the *cnr* operon. Downstream of *ncc*, a second nickel resistance determinant, *nre*, was identified.

Similar three component efflux systems were also identified in *E. coli* where they encode silver resistance (*sil* operon; S. Silver) and copper resistance (*cur* operon; T. V. O'Halloran). Also, a three-component efflux system similar to *czc* that conferred resistance to Cd and Zn was identified and cloned form *P. aeruginosa* CMG103 (Hassan, 1996). The three structural genes, *czr*CBA, were very similar to *czc*CBA. However, *czr* regulation seems to be under control of a classical two-component regulatory system.

Despite the similarity in structural genes, the heterologous expression patterns of *ncc-nre*, *czr*, and *czc* are quite different. This might be a consequence of the differences between the regulatory loci associated with the structural resistance genes. Many of these genes have been expressed in heterologous organisms, however. The *ncc-nre* determinant could be expressed in *E. coli*, *Sphingobacterium heparium*, *Rhodobacter sphaeroides*, *Thiobacillus versutus* (Q. Dong, personal communication), and the plant associated bacteria *Pseudomonas putida*, *Pseudomonas stutzeri*, *Burkholderia cepacia*, and *Herbaspirillum* sp. (S. Taghavi and C. Lodewyckx, personal communication). The *ncc-nre* determinant was also introduced on a broad host-range plasmid into an activated sludge system. Natural transfer of this plasmid into the activated sludge system resulted in a rapid adaptation of the system to treat nickel-containing wastewater (Q. Dong, personal communication). Analysis of the nickel-resistant strains that were isolated after gene transfer indicated that the *ncc-nre* determinant could be expressed in a range of Gram⁻ and Gram⁺ species. The *czr* operon of *P. aeruginosa* could be expressed in *A. eutrophus* (Hassan, 1996), *Herbaspirillum* sp., and in some *P. fluorescens* strains (S. Taghavi and C. Lodewyckx, personal communication). The *czc* operon has never been expressed outside *A. eutrophus*.

The broad range of organisms which express the *ncc-nre* determinant opens the possibility to increase the nickel (and cadmium and cobalt) resistance of plant-associated bacteria, while the *czr* determinant could be used to increase resistance to cadmium and zinc of plant-associated *Pseudomonas* sp. The presence of heavy-metal-resistant, plant-associated bacteria might have beneficial effects on plant growth in the presence of heavy metals. The speciation of the heavy metals and, consequently, their bioavailability for plants might alter due to the activity of bacteria expressing heavy metal resistance.

POSSIBLE EFFECTS OF HEAVY METAL-RESISTANT SOIL BACTERIA ON HEAVY METAL SPECIATION

A. eutrophus effects the concentration of heavy metal ions present in culture and may alter the species present in the environment. The presence of *A. eutrophus*-like bacteria in soils contaminated with heavy metals might have long-term effects on the speciation of the heavy metals and consequently their bioavailability and toxicity for plants.

In cultures of *A. eutrophus* CH34, grown in the presence of high concentrations of cadmium (2 mM) or zinc (5 mM), metal concentrations decreased drastically (up to 99%) in the late log phase. This effect was always accompanied by a progressive pH increase (up to 9), and precipitation and sequestration of metals (Diels, 1990). The alkalization is thought to be a consequence of the proton influx during the *czc*-mediated proton antiporter efflux of cations (Diels et al., 1995a). The mechanisms of the bioprecipitation and biological sequestration are still poorly understood. X-ray diffraction spectroscopy of the precipitated material showed that carbonates precipitate with cadmium ions to form $Cd(HCO_3)_2$ and $CdCO_3$ (Diels et al., 1995a). The precipitation of cadmium seems to occur at defined nucleation foci around the cell surface as shown by transmission electron spectroscopy (Diels et al., 1995a). It is thought that extracellular polysaccharides (which are able to bind high amounts of metals and whose concentration has been shown to increase as a consequence of cadmium resistance) and outer membrane proteins are important post-efflux functions, which avoid re-entry of metal ions in the cell (Diels et al., 1995a).

APPLICATIONS OF *A. eutrophus* CH34 HEAVY METAL RESISTANCES FOR THE REMOVAL OF HEAVY METALS

The observation of bioprecipitation as a physiological consequence of plasmid-mediated efflux of cations led to the possible application of metal sequestration/bioprecipitation as a tool to remove heavy metals from polluted effluents. A special reactor was developed involving immobilization of cells, a crystal collection system for collecting the heavy metal precipitates, and a nutrient system to keep bacteria alive. This resulted in the design of a novel reactor named BICMER (Bacteria Immobilized Composite MEmbrane Reactor; Diels et al., 1993a; Diels et al., 1995b). Once operational, this reactor provides a final residual concentration of zinc or cadmium (output) lower than 1 ppm. A simple modification of the growth medium allowed the efficient sequestration of cobalt, nickel, and copper ions, only after induction by cadmium or zinc (Diels et al., 1993a,b, 1995b).

Treatment of soils contaminated by heavy metals is another potential application for metal-resistant *A. eutrophus* CH34 and related bacteria. In a BMSR (Bacteria Metal Sludge Reactor), polluted sandy soils and bacteria were mixed together, and the fate of the heavy metals was studied. The bacterial treatment modified the colloidal behavior of sandy soil suspensions in such a way that soil particles were sedimenting much faster than soils without bacterial treatment. This allowed the easy separation of the bacterial suspension from the soil particle fraction (Diels, 1997). In this way, 30 to 70% of the metals from these polluted soils could be extracted after 48 h. However, soil bioremediation is a more difficult challenge than effluent bioremediation because of the prohibitive costs and problems with upscaling (Collard et al., 1994).

HEAVY METAL BIOSENSORS BASED ON BACTERIAL HEAVY METAL RESISTANCE GENES

Genes for resistance to heavy metals are not constitutively expressed by bacteria but are expressed in response to changes in the environment. In particular, heavy metals act as environmental signals to activate genes coding resistance to those metals. Thus, resistance is a response to the available concentration of heavy metals in the environment. Although the concentration of heavy metals can be determined from an environmental sample, the amount of the metals, which are biologically available, cannot be determined by simple chemical analysis. Reporter gene systems have been used to determine the bioavailability of iron in the environment (Loper et al., 1991). Genes for resistance to heavy metals may be useful in constructing biosensors for determining the bioavailability of heavy metals because expression of these genes is directly related to the concentration of heavy metals. The challenge in constructing such a biosensor is to chose a reporter gene system whose product can easily be detected and assayed in laboratory studies.

The molecular study of heavy metal-resistance genes of *A. eutrophus* CH34 and insight in their regulation mechanisms opened the way to an application as biosensors. In particular, the *lux*CDABE genes from *Vibrio fisheri* were placed as reporter genes under the control of specific bacterial genes involved in the regulation of heavy metal resistances. Two metal-*lux* fusions were constructed in CH34. In the resulting strains, named AE1239 and AE1433 (Corbisier et al., 1996; van der Lelie et al., 1997), the *lux*-genes were fused with a *cop*-sensitive promoter and a *czc*-promoter, respectively. Strain AE1239 emits light in the presence of copper ions and AE1433 in the presence of zinc, cadmium, and lead ions. Both strains were used as microbial bioluminescent sensors for the evaluation of fly ashes, vitrified wastes, soils, and sludge contaminated by heavy metals (Corbisier et al., 1996). The biosensors were able to successfully assess the bioavailability of heavy metals in a simple and rapid assay. Other metal-*lux* fusions have been constructed, each responding to a particular metal, e.g., to chromate, thallium, arsenite, cobalt, lead, and nickel, so that a specific biosensor can be used to trace a particular metal of major concern (Corbisier et al., 1993; Corbisier et al., 1994).

USE OF BIOMET SENSORS TO PREDICT HEAVY METAL PHYTOTOXICITY

The detection of the leachable heavy metal fraction of heavy metal-contaminated incinerator fly ashes, soils, and sludge traditionally relies on highly sensitive and specific methods in analytical biochemistry. However, such methods cannot distinguish between those heavy metal constituents contained in inert or complexed forms and those that are present in leachable forms. Inert and complexed metal forms can

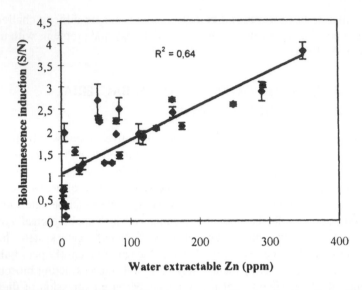

FIGURE 15.4 Relationship between the water-extractable zinc fraction from different soil samples and the bioluminescence (expressed as signal to noise ratio) of the Biomet sensor strain AE1433, whose light production is specifically inducible by zinc.

exist relatively safely in the environment, while leachable forms present an environmental hazard. This is of particular concern with respect to toxic heavy metals. Biosensors can complement analytical chemistry methods by detecting biologically available metals in environmental samples. The presence of metals in the environment, such as zinc, can now be measured by analytical chemistry methods, phytotoxicity experiments, and biosensors to give a clearer picture of the relevance of the metals in the environment.

There was a positive correlation between the concentration of water-extractable zinc from different soil samples and the expression of bioluminescence by strain AE1433 (Figure 15.4). This correlation is not surprising since bioluminescence in this strain is under the regulation of a zinc-inducible promotor. Induction of glutamate dehydrogenase activity in beans can also be compared to the presence of zinc as determined by chemical extraction or by bioluminescence (Figures 15.5 and 15.6). From these data we conclude that a positive relationship exists between the bioavailability of heavy metals in soils and the induction of glutamate dehydrogenase activity in beans grown on the same soils. This is not completely unexpected, since both biological systems are mainly induced by free metal ions.

The relationship between the bioavailability of heavy metals in soils as determined using bacterial biosensors and the induction of glutamate dehydrogenase activity in beans opens the possibility of using these biosensors to predict phytotoxicity. Instead of carrying out time-consuming plant-growth experiments to determine phytotoxicity, a simple biosensor assay could be used to assess potential phytotoxicity. This could be of great importance to predict the feasibility of phytoremediation strategies for treating sites contaminated with heavy metals.

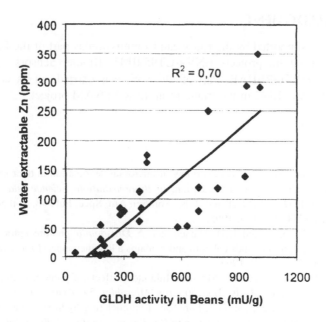

FIGURE 15.5 Relationship between the induction of glutamate dehydrogenase (GLDH) activity in beans (expressed in mUnits/g dry weight) and the water-extractable fraction of zinc (in ppm) as was determined by sequential extraction.

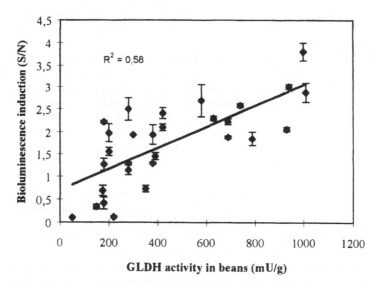

FIGURE 15.6 Relationship between the induction of glutamate dehydrogenase activity (GLDH) in beans (expressed as mUnits/g dry-weight) and the fraction of zinc, as was determined using the Biomet.

ACKNOWLEDGMENT

This work was supported by the European Commission as part of the Environment and Climate program projects ENV4-CT95-0141, Biomet Sensors, and ENV4-CT95-0083, PHYTOREHAB. Part of the work was carried out as collaboration between VITO and LUC in the frame of an EFRO/OVAM project.

REFERENCES

Anderson, T.A. and J.R. Coats, An overview of microbial degradation in the rhizosphere and its implications for bioremediation, in *Bioremediation: Science and Applications,* Skipper, H.D. and R.F. Turco, Eds., Soil Sci. Am. Spec. Pub. 43, Soil Sci. Soc. Am, Madison WI, 135-143, 1995.

Ballinger, DG., N. Xue, and K.D. Harschman, A *Drosophila* photoreceptor cell specific protein, calphotin, binds calcium and contains a leucine zipper. *Proc. Natl. Acad. Sci. USA* 90, 1536-1540, 1993.

Becker, J.O., R.W. Hedges, and E. Messens, Inhibitory effects of pseudobactin on the uptake of iron by higher plants. *Appl. Environ. Microbiol.* 54, 1090-1093, 1985.

Beyers, B.R., M.V. Powell, and C.E. Lankfort, Iron-chelating hydroxamic acid (schizoken) active in initiation of cell division in *Bacillus megaterium. J. Bacteriol.* 93, 286-294, 1967.

Cline, G.R., P.E. Powell, P.J. Szaniszlo, and C.P.P. Reid, Comparison of the abilities of hydroxamate, synthetic and other organic acids to chelate iron and other ions in nutrient solution. *Soil. Sci. Soc. Am. J.* 46, 1158-1164, 1982.

Cline, G.R., P.E. Powell, P.J. Szaniszlo, and C.P.P. Reid, Comparison of the abilities of hydroxamate and other organic acids to chelate iron and other ions in soil. *Soil. Sci.* 136, 145-157, 1983.

Cline, G.R., C.P.P. Reid, P.E. Powell, and P.J. Szaniszlo, Effects of hydroxamate siderophore on iron adsorption by sunflower and sorghum. *Plant Physiol.* 76, 36-39, 1984.

Collard, J.-M., A. Provoost, S. Taghavi, and M. Mergeay, A new type of *Alcaligenes eutrophus* CH34 zinc resistance generated by mutations affecting regulation of the *cnr* cobalt-nickel resistance system. *J. Bacteriol.* 175, 779-784, 1993.

Collard, J.-M., P. Corbisier, L. Diels, Q. Dong, C. Jeanthon, M. Mergeay, S. Taghavi, D. van der Lelie, A. Wilmotte, and S. Wuertz, Plasmids for heavy metal resistance in *Alcaligenes eutrophus* CH34: mechanisms and applications. *FEMS Microbiol. Rev.* 14, 405-414, 1994.

Corbisier, P., G. Ji, G. Nuyts, M. Mergeay, and S. Silver, *luxAB* gene fusions with the arsenic and cadmium resistance operon of *Staphylococcus aureus* plasmid pI258. *FEMS Microbiol. Lett.* 110, 231-238, 1993.

Corbisier, P., E. Thiry, and L. Diels, Bacterial biosensors for the toxicity assessment of solid wastes. *Environ. Toxicol. Water Qual.* 11, 171-177, 1996.

Corbisier, P., E. Thiry, A. Masolijn, and L. Diels, Construction and development of metal ion biosensors, in *Bioluminescence and Chemoluminescence: Fundamentals and Applied Aspects,* Campbell, A.K., Cricka, L.J., and Stanley, P.E., Eds., John Wiley & Sons, Chichester, 1994.

Crichton, R.R. and M. Charloteaux-Wauters, Iron transport and storage. *Eur. J. Biochem.* 164, 485-506, 1987.

Crosa, J.H., Genetics and molecular biology of siderophore-mediated iron-transport in bacteria. *Microbiol. Rev.* 53, 517-530, 1989.

Diels, L., Accumulation and precipitation of Cd and Zn ions by *Alcaligenes eutrophus* strains, in *Biohydrometallurgy 89, Proc. Int. Symp. Jackson Hole, Wyoming,1990,* Salby, J., McCready, R.G.I., and Wichlacz, P.Z., Eds., 369-377, 1990.

Diels, L., Heavy metal bioremediation of soil, in *Methods in Biotechnology, Vol. 2: Bioremediation Protocols,* Sheehan, D., Ed., Humana Press, Totowa, NJ, 283-295, 1997.

Diels, L., Q. Dong, D. van der Lelie, W. Baeyens, and M. Mergeay, The *czc* operon of *Alcaligenes eutrophus* CH34: from resistance mechanism to the removal of heavy metals. *J. Ind. Microbiol.* 14, 142-153, 1995a.

Diels, L., S. Van Roy, K. Somers, I. Willems, W. Doyen, M. Mergeay, D. Springael, and R. Leysen, The use of bacteria immobilized in tubular membrane reactors for heavy metal recovery and degradation of chlorinated aromatics. *J. Membr. Sci.* 100, 249-258, 1995b.

Diels, L. and M. Mergeay, DNA probe mediated detection of resistant bacteria from soils highly polluted by heavy metals, *Appl. Environ. Microbiol.* 57, 3301-3309, 1990.

Diels, L., A. Sadouk, and M. Mergeay, Large plasmids governing multiple resistance to heavy metals: a genetic approach. *Toxicol. Environ. Chem.* 23, 79-89, 1989.

Diels, L., S. Van Roy, M. Mergeay, W. Doyen, S. Taghavi, and R. Leysen, Immobilisation of bacteria in composite membranes and development of tubular membrane reactors for heavy metal recuperation, in *Effective Membrane Processes: New Perspectives,* Paterson, R., Ed., Mechanical Engineering Publications Limited, London, 275-293, 1993a.

Diels, L., S. Van Roy, S. Taghavi, W. Doyen, R. Leysen, and M. Mergeay, The use of *Alcaligenes eutrophus* immobilised in a tubular membrane reactor for heavy metal recuperation, in *Biohydrometallurgy 93, Proc. Int. Symp. Jackson Hole, Wyoming, 1993,* 22-25, 1993b.

Dong, Q. and M. Mergeay, Czc/Cnr efflux: 3 component export pathway with 12 transmembrane helix exporter. *Mol. Microbiol.* 14, 185-187, 1994.

Duijff, B.J., P.A.H.M. Bakker, and B. Schippers, Influence of pseudobactin-358 on the iron nutrition of plants, in *Abstr. 6th Int. Fe Symp.,* 31, 1991.

Duss, F., A. Moazfar, J.J. Oertli, and W. Jaeggi, Effect of bacteria on the iron uptake by axenically-cultured roots of Fe-efficient and Fe-inefficient tomatoes (*Lycopersicon esculentum* Mill.). *J. Plant Nutr.* 9, 587-598, 1986.

Gilis, A., Interactie tussen verschillende potentieel toxische metalen (Zn, Cd, Ni en Al) en siderofoor-afhankelijke ijzer-opname in verschillende fluorescerende *Pseudomonas* stammen, Licentiaatsthesis, departement Algemene Biologie, Vrije Universiteit Brussel, 1993.

Gilis, A., P. Corbisier, W. Baeyens, S. Taghavi, M. Mergeay, and D. van der Lelie, Effect of the siderophore alcaligin E on the bioavailability of Cd to *Alcaligenes eutrophus* CH34. *J. Ind. Microbiol. Biotechnol.* 20, 61-68, 1998.

Gilis, A., M.A. Khan, P. Cornelis, J.M. Meyer, M. Mergeay, and D. van der Lelie, Siderophore-mediated iron uptake in *Alcaligenes eutrophus* CH34 and identification of *aleB* encoding the ferric iron-alcaligin E receptor. *J. Bacteriol.* 178, 5499-5507, 1996.

Guerinot, M.L., Microbial iron transport. *Annu. Rev. Microbiol.* 48, 743-772, 1994.

Hassan, M.-E.-T., Genetic mechanism of heavy metal resistance of *Pseudomonas aeruginosa* CMG103, PhD thesis, University of Karachi, Pakistan, 1996.

Höfte, M., S. Buysens, N. Koedam, and P. Cornelis, Zinc affects siderophore-mediated high affinity iron uptake systems in the rhizosphere *Pseudomonas aeruginosa* 7NSK2. *BioMetals* 6, 85-91, 1993.

Höfte, M., Q. Dong, S. Kourambas, V. Krishnapillai, D. Sherratt, and M. Mergeay, The *sss* gene product, which affects pyoverdine production in *Pseudomonas aeruginosa* 7NSK2, is a site-specific recombinase. *Mol. Microbiol.* 14, 1011-1020, 1994.

Hsu, T.S. and R. Bartha, *Appl. Environ. Microbiol.* 37, 36-41, 1979.

Hu, X. and G.L. Boyer, Siderophore-mediated aluminium uptake by *Bacillus megaterium* ATCC 19213. *Appl. Environ. Microbiol.* 62, 4044-4048, 1996.

Huyer, M. and W. Page, Zn^{2+} increases siderophore production in *Azotobacter vinelandii*. *Appl. Environ. Microbiol.* 54, 2625-2631, 1988.

Jurkevitch, E., Y. Hadar, and Y. Chen, Involvement of bacterial siderophores in the remedy of lime-induced chlorosis in peanut. *Soil Sci. Soc. Am. J.* 52, 1032-1037, 1988.

Kaur, P., K. Ross, R.A. Siddiqui, and H.G. Schlegel, Nickel resistance of *Alcaligenes denitrificans* strain 4a-2 is chromosomally coded. *Arch. Microbiol.* 154, 133-138, 1990.

Kloepper, J.W., J. Leong, M. Teintze, and M.N. Schroth, Enhanced plant growth by siderophores produced by plant growth promoting rhizobacteria. *Nature* 286, 885-886, 1980.

Liesegang, H., K. Lemke, R. Siddiqui, and H.G. Schlegel, Characterisation of the inducible nickel and cobalt resistance determinant cnr from pMOL28 of *Alcaligenes eutrophus* CH34. *J. Bacteriol.* 175, 767-778, 1993.

Lindsay, W.L., *Chemical Equilibria in Soils,* John Wiley & Sons, Inc., New York, 1979.

Loper, J.E. and J.S. Buyer, Siderophores in microbial interactions on plant surfaces. *Mol. Plant-Microbe Interact.* 4, 5-13, 1991.

Neilands, J.B., Microbial iron compounds. *Annu. Rev. Biochem.* 50, 715-731, 1981.

Neilands, J.B. and S.A. Leong, Siderophores in relation to plant growth and disease. *Annu. Rev. Plant Physiol.* 37, 187-208, 1986.

Nichols, T.D., D.C. Wolf, H.B. Rogers, C.A. Beyrouty, and C.M. Reynolds, Rhizosphere microbial populations in contaminated soils. *Water, Air Soil Pollut.* 95, 165-178, 1997.

Mergeay, M., J. Gerits, and C. Houba, Facteur transmissible de la résistance au cobalt chez un *Pseudomonas* du type *hydrogenomonas*. *C. R. Soc. Biol.* 172, 575-579, 1978.

Mergeay, M., D. Nies, H.G. Schlegel, J. Gerits, P. Charles, and F. Van Gijsegem, *Alcaligenes eutrophus* CH34 is a facultative chemolithotroph with plasmid bound resistance to heavy metals. *J. Bacteriol.* 162, 328-334, 1985.

Nies, D.H. The cobalt, zinc and cadmium efflux system CzcABC from *Alcaligenes eutrophus* functions as a cation-proton antiporter in *Escherichia coli. J. Bacteriol.* 177, 2707-2712, 1995.

Nies, D.H., A. Nies, L. Chu, and S. Silver, Expression and nucleotide sequence of a plasmid-determined divalent cation efflux system from *Alcaligenes eutrophus. Proc. Natl. Acad. Sci. USA* 86, 7351-7355, 1989.

Nies, D.H. and S. Silver, Plasmid-determined inducible efflux is responsible for resistance to cadmium, zinc and cobalt in *Alcaligenes eutrophus. J. Bacteriol.* 171, 896-900, 1989.

Nies, D.H., and S. Silver, Ion efflux systems involved in bacterial resistances. *J. Ind. Microbiol.* 14, 186-199, 1995.

O'Sullivan, D.J. and F. O'Gara, Traits of fluorescent *Pseudomonas* spp. involved in suppression of plant root pathogens. *Microbiol. Rev.* 56, 662-676, 1992.

Rovira, A.D., R.C Foster, and J.K. Martin, Note on terminology: origin, nature and nomenclature of organic materials in the rhizosphere, in *Soil-borne Plant Pathogens,* Schippers, B. and Gams, W., Eds., Academic Press, New York, 1-4, 1979.

Saier, M.H. Jr., R. Tam, A. Reizer, and J. Reizer, Two novel families of bacterial membrane proteins concerned with nodulation, cell division and transport. *Mol. Microbiol.* 11, 841-847, 1994.

Schmidt, T., R.D. Stoppel, and H.G. Schlegel, High-level nickel resistance in *Alcaligenes xylosoxidans* 31A and *Alcaligenes eutrophus* KTO2. *Appl. Environ. Microbiol.* 57, 3301-3309, 1991.

Schmidt, T. and H.G. Schlegel, Combined nickel-cobalt-cadmium resistance encoded by the *ncc* locus of *Alcaligenes xylosoxidans* 31A. *J. Bacteriol.* 176, 7045-7054, 1994.

Sensfuss, C. and H.G. Schlegel, Plasmid pMOL28-encoded resistance to nickel is due to specific efflux. *FEMS Microbiol. Lett.* 55, 295-298, 1988.

Siddiqui, R.A., K. Benthin, and H.G. Schlegel, Cloning of pMOL28 encoded nickel resistance genes and expression of the genes in *Alcaligenes eutrophus* and *Pseudomonas* spp. *J. Bacteriol.* 171, 5071-5078, 1989.

Stoppel, R.D. and H.G. Schlegel, Nickel resistant bacteria from anthropogenically nickel-polluted and naturally nickel-percolated ecosystems. *Appl. Environ. Microbiol.* 61, 2276-2285, 1995.

Stoppel, R.D., M. Meyer, and H.G. Schlegel, The nickel resistance determinant cloned from the enterobacterium *Klebsiella oxytoca*: conjugal transfer, expression, regulation and DNA homologies to various nickel resistant bacteria. *Biometals* 8, 70-79, 1995.

Taghavi, S., M. Mergeay, and D. van der Lelie, Genetic and physical maps of the *Alcaligenes eutrophus* CH34 megaplasmid pMOL28 and its derivative pMOL50 obtained after temperature-induced mutagenesis and mortality. *Plasmid* 37, 22-34, 1997.

van der Lelie, D., T. Schwuchow, U. Schwidetzky, S. Wuertz, W. Baeyens, M. Mergeay, and D.H. Nies, Two-component regulatory system involved in transcriptional control of heavy metal homoeostasis in *Alcaligenes eutrophus*. *Mol. Microbiol.* 23, 493-503, 1997.

Varma, A.K., C. Sensfuss, and H.G. Schlegel, Inhibitor effects of the accumulation and efflux of nickel ions in plasmid pMOL28-harbouring strains of *Alcaligenes eutrophus*. *Arch. Microbiol.* 154, 42-49, 1990.

Yabuuchi, E., Y. Kosako, I. Yano, H. Hotta, and Y. Nishiuchi, Transfer of two *Burkholderia* and an *Alcaligenes* species to *Ralstonia* gen. nov.: proposal of *Ralstonia pickettii* (Ralston, Palleroni and Doudoroff 1973) comb. nov., *Ralstonia solanacearum* (Smith 1896) comb. nov. and *Ralstonia eutropha* (Davis 1969) comb. nov. *Microbiol. Immunol.* 39, 897-904, 1995.

16 Microphyte-Mediated Selenium Biogeochemistry and its Role in *In Situ* Selenium Bioremediation

Teresa W.-M. Fan and Richard M. Higashi

CONTENTS

INTRODUCTION

Only during the last 2 decades has the complexity of the selenium (Se) biogeochemical cycle begun to be realized. Based on the limited information available, Se biogeochemistry appears to be largely analogous to that of sulfur. For example, similar to sulfur, the oxyanions of Se (selenite and selenate) are often the dominant forms in oxic aquatic environments and groundwater (Sugimura et al., 1977; Cutter and Bruland, 1984; White and Dubrovsky, 1994). Volatilization of both Se and S via biomethylation represents a major process by which these two elements enter the atmosphere (Duce et al., 1975; Craig, 1986; Cooke and Bruland, 1987). Furthermore, like sulfur, Se is extensively metabolized by organisms into amino acids and antioxidant compounds (Lewis, 1976). However, many of the processes involved in the Se cycle are yet to be elucidated.

1-56670-450-2/00/$0.00+$.50
© 2000 by CRC Press LLC

This complex biogeochemical cycling of Se, together with an unusually narrow tolerance between nutritional requirement and toxicity, plus a highly heterogeneous distribution have all contributed to the difficult problem of Se pollution. This is exemplified by the environmental problems associated with California's agricultural drainage disposal systems since the Kesterson Reservoir (California) incident in the early 1980s. The wildlife deformities observed there and in the present agricultural drainage evaporation basins have been attributed to Se bioaccumulation and biotransformation through the food chain (Skorupa and Ohlendorf, 1991; Maier and Knight, 1994). However, the actual Se biotransformations through the aquatic and terrestrial foodweb are still largely unknown. The Se-related ecotoxic effect is rapidly becoming a serious concern for agricultural operations in seleniferous soils of the western 17 states and for industrial discharges throughout the U.S. (Skorupa and Ohlendorf, 1991; Reash et al., 1996; Fairbrother et al., 1996). Long-term economic solutions to the problem need to be explored to make these agricultural and industrial activities sustainable.

Many Se removal schemes that are physically and/or chemically based have been tested but they were shown to be cost-prohibitive or ineffective for waterborne Se concentrations of 10 µg/l or lower (Mudder, 1997). Biologically based remediation schemes have also been proposed that involve precipitation and/or volatilization of selenium by soil bacteria and vascular plants (Gerhardt et al., 1991; Thompson-Eagle and Frankenberger, 1991; Terry et al., 1992). The efficacy of these bioremoval schemes has yet to be evaluated.

In this chapter, we describe an approach that utilizes the biotransformation activities of aquatic microphytes intrinsic to the Se-laden agricultural evaporation basins for "natural" remediation. Similar to the vascular plant-based scheme, this process is solar-driven and does not require extensive maintenance; thus the economic advantage is self-evident. In addition, evaporation basins are currently employed for disposing large volumes of agricultural drainage waters, some of which have been in operation for over 2 decades. Moreover, we illustrate the importance of acquiring a molecular-level understanding of the Se fate in the environment to achieve not just Se removal but, more importantly, reduction in ecotoxic risk, which is the real goal of remediation.

SITE DESCRIPTION AND RATIONALE

The evaporation basin system of the Tulare Lake Drainage District (TLDD) located near Tulare, CA has been in continuous use for disposing large volumes of agricultural drainage waters (e.g., 12,000 acre-ft/yr) for more than 2 decades. These basins are arranged as a sequence of shallow (e.g., 0.7 to 1.5 m) cells, where drainage waters are channeled sequentially from the first to the terminal cell to optimize water evaporation. Since the drainage water itself is moderately saline (e.g., 7 ‰), salinity (primarily Na_2SO_4 and NaCl) in successive cells increases progressively due to evaporation (Figure 16.1A), reaching levels as high as 300 ‰ in the summer months. Contaminant salts including Se oxyanions can be predicted to increase based on evaporite chemistry (Tanji, 1989), as was observed in systems such as the Peck Pond basin, located in Fresno County, CA (Figure 16.1B). However, waterborne Se

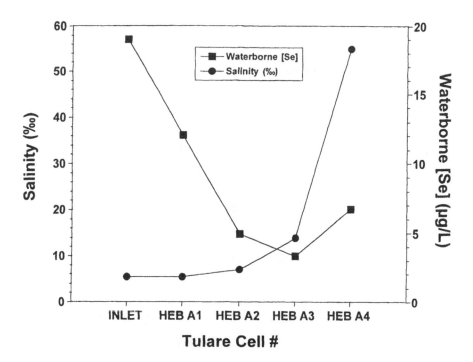

FIGURES 16.1A and B Relationship between waterborne Se concentration and salinity at two multicell evaporation basins currently used for agricultural drainage disposal. Waterborne Se concentration ([Se]) is often observed to increase progressively with increasing salinity through a squence of evaporation cells as in the Peck basin system, CA (Figure 16.1B). (Modified from Tanji, K.K. *Toxic Substances in Agricultural Water Supply and Drainage*, 1989, 109-121.) This behavior is consistent with that predicted from evaporite chemistry as shown. However, an opposite trend was observed for the TLDD basins, CA (Figure 16.1A), where waterborne [Se] decreased with increasing salinity from inlet to the terminal cell (HEB A4). This trend is persistent year-round (Fan et al., 1998b). Cell waters were collected monthly in acid-washed polyethylene bottles by TLDD staff at the designated location of each cell. After removing particulate matters by centrifugation, waters were acidified to <pH 2 with HCl and analyzed for total Se via microdigestion and fluorescence measurement.

concentrations at the TLDD basins often exhibit a *decreasing* trend, despite the large buildup of salinity across the sequence of cells as illustrated in Figure 16.1A. Although selenate is often the dominant form in the subsurface source waters, a significant fraction (up to 60%) of the waterborne Se has been reported to be in the selenite form during residence at these basins (Tanji and Gao, personal communication; Fan and Higashi, unpublished data; Maier, 1997).

Since the disappearance of waterborne Se in the presence of 10^6-fold higher sulfate concentration cannot be readily accounted for by abiotic physical and chemical mechanisms, we focused our attention on the biological mechanism(s). A close examination of the TLDD basin waters revealed that they are abundant in microphytes (with chlorophyll *a* content up to 1 µg/ml water) including coccoid and filamentous cyanobacteria, diatoms, and green phytoplankton, while vascular plants

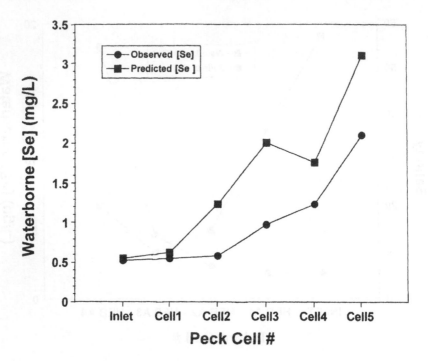

FIGURE 16.1B (Continued)

and aquatic macrophytes are absent. Except for the most saline cell (i.e., terminal cell), microphytes persist in these waters year-round. However, there has been little information regarding the Se biogeochemistry associated with these microphytes, or even microphytes in general. Our recent investigation of a green coccoid phytoplankton (*Chlorella* sp.) isolated from a similar system (Pryse Pond, Tulare County, CA) demonstrated that aquatic microphytes are active in transforming Se oxyanions into organic Se metabolites including volatile alkylselenides (Fan et al., 1997a). Further studies on microphyte-mediated biogeochemistry should help reveal the Se dissipation mechanism(s) against the strong salinity gradient at TLDD basins.

More importantly, by tracing the environmental fate of the biotransformed products, particularly through the foodweb, the ecotoxic consequence of Se contamination may be better understood. This knowledge should also lead to a general understanding of the role(s) of aquatic microphytes in the biogeochemical cycling of Se or other elements in surface waters. These organisms have been postulated to be a major driver of Se biogeochemistry (Cutter and Bruland, 1984; Cooke and Bruland, 1987), as they are in the case of the sulfur cycling (e.g., Andreae, 1986; Bates et al., 1987). Recent findings in the marine environment corroborate this hypothesis (Amouroux and Donard, 1996).

Because it is unlikely that the natural phenomenon occurring at TLDD basins is fortuitously optimal in Se removal, understanding the removal mechanism(s) may help develop a more efficient scheme for reducing the load of Se or other contaminants in these and other basins. At the same time, by integrating the ecotoxic considerations into the removal scheme, we hope to achieve the true goal of *in situ* bioremediation — wildlife protection.

APPROACH

We have taken a parallel laboratory and field approach to investigate the Se biotransformation activities in microphytes occurring at the TLDD basins. Here, we will describe mainly the laboratory studies which involved fractionating the commonly occurring microphyte species from the basin waters, establishing monocultures of these species, measuring their Se volatilization kinetics, and characterizing the biotransformation products of Se.

ISOLATION AND CULTURING OF MICROPHYTES

Microphyte species were isolated from the basin waters according to the procedure described by Fan et al. (1997a). Briefly, 10 to 100 µl of water was streaked onto a 1% agarose plate prepared in f/2 seawater medium and incubated at 20 to 22°C under a light/dark cycle of 16/8 h. Distinct and isolated colonies were then inoculated aseptically into the f/2 seawater medium and incubated similarly as above to establish stock cultures. The plating procedure was repeated using the stock culture to minimize bacterial contamination. The stock culture was maintained by periodic inoculation into fresh medium. Under careful microscopic examination, only a single species of microphyte was visible in each isolated culture with no apparent bacterial contamination. One phytoplankton species (*Chlorella* sp.) was isolated from the Pryse evaporation basin water and two cyanophyte species were isolated from the TLDD basin water. One TLDD basin species was a filamentous cyanophyte possibly of the LPP (*Lyngbya*, *Phormidium*, and *Plectonema*) group (Rippka et al., 1979; T. Hanson, personal communication) while the other was tentatively assigned to be a *Synechocystis* sp. We are conducting further species identification of the microphytes based on the 16S ribosomal RNA method.

MEASUREMENT OF SE VOLATILIZATION KINETICS

The rate of Se volatilization from growing microphyte cultures supplemented with selenite or selenate was measured as described previously (Fan et al., 1997a). Briefly, 0.8 liter of f/2 seawater medium was supplemented with 10 µg/l to 100 mg/l Se and inoculated with microphyte stocks. The culture was then constantly bubbled with 0.22 µm filtered air at 30°C under continuous fluorescent light with a light intensity of approximately 360 cd. The air aeration provided the CO_2 needed for microphyte growth, and purged the volatile Se compounds (e.g., alkylselenides) out of the medium into an alkaline peroxide trap containing 50 mM NaOH plus 30% H_2O_2, 4/1 (v/v) for 18 to 22 h. The volatile Se compounds were also trapped in their original forms into a 1/4 in. Teflon tube kept at liquid nitrogen temperature. Total Se in the alkaline peroxide trap was measured by GC-ECD or fluorescence while the liquid nitrogen-trapped Se forms were analyzed by GC-MS (see below). The rate of Se volatilization was calculated from the total Se and the duration of the trapping. Suspended microphyte cell density was monitored by taking visible spectra from 350 to 800 nm, from which the optical density at 680 nm was obtained.

SE ANALYSIS

Medium water, the trap, and biomass samples were digested using a microdigestion method described previously (Fan et al., 1997a), which substantially simplified the procedure and reduced sample requirement compared with reported macrodigestion methods (e.g., McCarthy et al., 1981). Total Se in the digest was determined by using the GC-ECD method (Fan et al., 1997a) or by fluorescence. The fluorescence method was modified from the Analytical Methods Committee (1979; Fan et al., 1998b). Briefly, microdigestion with either nitric acid or alkaline peroxide converted various Se forms into selenate which were then reduced to selenite by 6 N HCl at 105°C, followed by derivatization with 4-nitrophenylene-o-diamine or 2,3-diaminonaphthalene to form the corresponding piazselenol derivatives, which were quantified by GC-ECD or fluorescence, respectively. The detection limit for the GC-ECD method was 1 μg/l while that for the fluorescence method was in the sub-μg/l range, both based on a 250 μl of water sample. The fluorescence method is more convenient to perform with a broader linear range of concentrations, while the GC-ECD method should be less subject to matrix interference. A typical correlation coefficient obtained for the fluorescence-based standard curve of 12 Se concentrations ranging from 0-250 μg/l was better than 0.99. Known addition method was used to check interference from sample matrix, which was negligible in all cases.

The biotransformed products including volatile alkylselenides, selenonium compounds, and selenoamino acids were analyzed using a combination of GC-MS and NMR techniques (Fan et al., 1997a and 1998b). Alkylselenides were trapped at liquid nitrogen temperature and analyzed directly on an open-tubular DB-1 column in a Varian 3400 gas chromatograph interfaced to a Finnegan ITD 806 mass spectrometer. Selenonium compounds and selenoamino acids were extracted from microphyte biomass using 5% perchloric acid (PCA) or 10% trichloroacetic acid (TCA). Selenoamino acids in the extract were then silylated with MTBSTFA (N-methyl-N-[$tert$-butyldimethylsilyl]trifluoroacetamide) before GC-MS analysis, or removed of paramagnetic ions by passing through Chelex-100 resin (BioRad) column before NMR analysis. The combined MTBSTFA derivatization and GC-MS analysis enabled a simultaneous determination of selenomethionine, selenocysteine, and methylselenocysteine at trace levels plus a wide range of other metabolites with structure confirmation (Fan et al., 1998a). Selenonium compounds in the biomass or extracts were analyzed indirectly by GC-MS for dimethylselenide (DMSe) which was liberated by 5 M NaOH treatment at room temperature or at 105 to 110°C (Fan et al., 1997a). The precursor to DMDSe (e.g., methylselenocysteine) was similarly determined by alkaline treatement at 105 to 110°C, followed by GC-MS analysis for DMDSe (Fan et al., 1998a). Selenonium compounds such as dimethylselenonium propionate (DMSeP) and methylselenomethionine (CH$_3$-Se-Met) may be the biological precursors of DMSe, while selenoamino acids such as Se-Met are considered to be the toxicity surrogate to wildlife (e.g., Heinz et al., 1996).

Protein-bound Se-Met was analyzed using the following procedure. The protein fraction was first obtained by extracting the biomass with a Tris-SDS buffer, followed by dialyzing the extract against a 3.5-kDa molecular weight cutoff membrane (MWCO) to remove low molecular mass components (Fan et al., 1997b). The resulting protein-rich fraction was digested in 6 N HCl, followed by MTBSTFA derivatization and GC-MS analysis (Fan et al., 1998b). The HCl digestion allowed a full recovery of Se-Met but was unsuited for recovering Se-Cys and selenocystine (Fan et al., 1998a). The GC-MS method provided trace-level analysis for selenoamino acids with rigorous structure confirmation, while circumventing the use of radiotracer [75]Se (e.g., Wrench, 1978). The latter coupled with liquid or thin-layer chromatography is also suited for trace analysis but impractical for the analysis of samples collected from the field. Without ^{75}Se radiotracers, conventional chromatographic methods do not allow sensitive detection and fall short of structure confirmation (Bottino et al., 1984). The ability to analyze various food chain components for protein-bound Se-Met is needed since this Se form may be the vehicle through which Se is biotransferred and bioconcentrated via the food chain, thereby causing ecotoxic effects at the higher trophic levels.

FINDINGS AND IMPLICATIONS

Se Volatilization by Isolated Microphytes

All three microphyte species could volatilize Se from selenite-supplemented f/2 seawater media, as shown in Figure 16.2. The *Chlorella* culture also volatilized Se from selenate-supplemented medium, albeit at a lower rate (Fan et al., 1997a). The Se volatilization activity of the two cyanophytes in selenate media has not been tested. The peak rate of volatilization from selenite for the filamentous cyanophyte was at least fourfold higher than that for the *Chlorella* sp. at 1 mg/l Se supplement. Assuming that the volatilization rate is proportional to the Se concentration of the medium, this rate may be comparable to that for the *Synechocystis* sp.

In addition, the rate of Se volatilization by the three microphytes was dependent on the suspended cell density (optical density at 680 nm), although the growth-dependent time course of the volatilization rate differed among the three species. The rate of Se volatilization by the *Chlorella* sp. tracked the suspended cell density and chlorophyll content, while that by the filamentous cyanophyte exhibited a sharp peak coincident with the suspended cell and chlorophyll density, followed by another more sustained peak during cell senescence (Figure 16.2 and data not shown). The Se volatilization kinetics by the *Synechocystis* sp. also showed two peaks, with one during the exponential growth and the other during early senescing period. In all cases, the cells aggregated and precipitated to the bottom while the optical density of the culture dropped to near zero during senescence (Figure 16.2). There was no increase at any point during senescence in the visible spectrum from 350 to 800 nm, indicating the absence of heterotrophic bacterial growth in suspension.

The close association of Se volatilization kinetics with the exponential growth of microphytes in all three cultures indicates that this activity was intrinsic to the

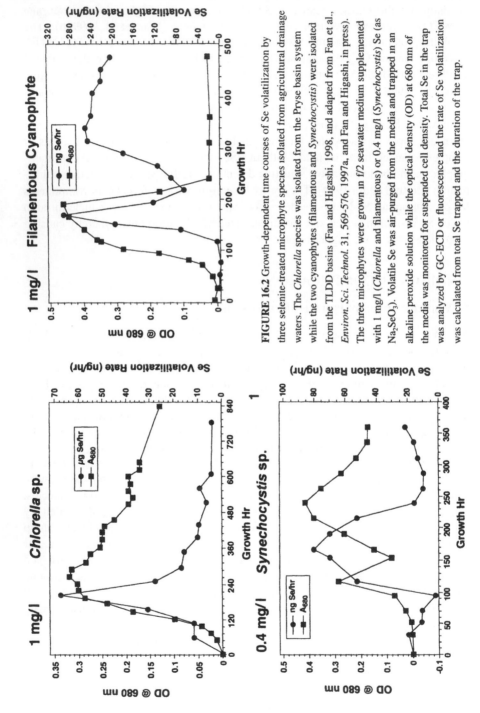

FIGURE 16.2 Growth-dependent time courses of Se volatilization by three selenite-treated microphyte species isolated from agricultural drainage waters. The *Chlorella* species was isolated from the Pryse basin system while the two cyanophytes (filamentous and *Synechocystis*) were isolated from the TLDD basins (Fan and Higashi, 1998, and adapted from Fan et al., *Environ. Sci. Technol.* 31, 569-576, 1997a, and Fan and Higashi, in press). The three microphytes were grown in f/2 seawater medium supplemented with 1 mg/l (*Chlorella* and filamentous) or 0.4 mg/l (*Synechocystis*) Se (as Na_2SeO_3). Volatile Se was air-purged from the media and trapped in an alkaline peroxide solution while the optical density (OD) at 680 nm of the media was monitored for suspended cell density. Total Se in the trap was analyzed by GC-ECD or fluorescence and the rate of Se volatilization was calculated from total Se trapped and the duration of the trap.

microphytes. Because the microphyte cultures investigated have not been proven to be axenic, the possibility of some contribution of heterotrophic bacterial activity to Se volatilization during the senescing period cannot be eliminated. However, this contribution is unlikely to be major based on the lack of suspended growth of heterotrophic bacteria and the predominance of senescent microphyte cells in the aggregate biomass. The more sustained peak of Se volatilization by the two cyano-phyte cultures during the senescing period could result from a switch of Se metabolism in the cyanophytes and/or interaction with closely associated heterotrophic bacteria which could not be differentiated from microscopic examination.

The relationship between Se depletion from the medium and rates of Se volatilization as a function of selenite concentrations by the filamentous cyanophyte culture was also investigated (Fan et al., 1998b). From treatments of 20 µg/l to 1 mg/l Se, the rate of Se volatilization followed a qualitatively similar time course as that in Figure 16.2 and was roughly proportional to the treatment concentrations (Fan et al., 1998b). Meanwhile, Se depletion from the medium exhibited an opposite time course (e.g., Figure 16.3). Even at treatment of 10 µg/l Se, Se was volatilized from the medium by the filamentous cyanophyte culture as the medium Se concentration decreased with time to 2 µg/l (Figure 16.3). The volatilization process accounted for a major fraction of the Se depletion from the medium (up to 77%), although Se incorporation into the biomass was also significant, as described below. These results indicate that both processes are capable of contributing to the decrease in waterborne Se concentrations of the TLDD basins (cf. Figure 16.1).

Moreover, the chemical form(s) of the volatilized Se trapped at liquid nitrogen temperature were examined using GC-MS. Figure 16.4 illustrates the total ion

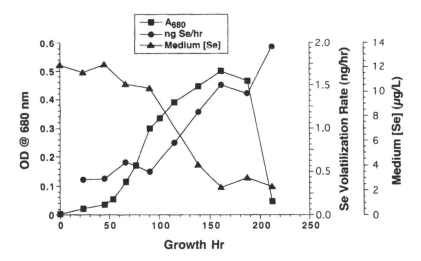

FIGURE 16.3 Growth-dependent time courses of Se volatilization and depletion from medium by selenite-treated filamentous cyanophytes. The filamentous cyanophyte was grown in f/2 seawater medium supplemented with 10 µg/l Se (as Na_2SeO_3). (Reprinted with permission from Fan et al., 1998a.) Total Se volatilized and suspended cell density were measured as in Figure 16.2. Medium [Se] was analyzed by the fluorescence method (Fan et al., 1998b).

chromatograms of volatile compounds released by a *Chlorella* culture grown at 100 mg/l Se (Figure 16.4A) and a filamentous cyanophyte culture grown at 10 mg/l Se (Figure 16.4B). For both microphytes, three volatile Se compounds were identified: DMSe, DMDSe (dimethyldiselenide), and DMSeS (dimethylselenenyl sulfide; Fan et al., 1998b). Dimethylsulfide (DMS) and dimethyldisulfide (DMDS) were also liberated from the *Chlorella* (Fan et al., 1997a) and filamentous cyanophyte cultures (Fan et al., 1998b), respectively. However, at lower Se treatment concentrations (e.g., ≤1 mg/l), the only detectable product from both species was DMSe (data not shown). It is likely that the production of DMSe and DMDSe shares a similar mechanism as that of DMS and DMDS, respectively. As for the release of DMSeS, the mechanism may involve a cross-reaction between the production of DMDS and DMDSe (see below).

Se Allocation in Microphyte Biomass

To examine Se allocation into the microphyte biomass, we developed a fractionation scheme that involved extraction of the biomass with 10% TCA (Fan et al., 1997a) and with Tris-SDS buffer plus dialysis. The TCA-soluble fraction generally includes water-soluble, small molecular mass metabolites including small peptides, while the residue fraction is expected to be comprised of proteins, lipids, other macromolecules such as cell wall constituents, and Se^0. The combination of Tris-SDS buffer extraction and dialysis against a 3.5-kDa MWCO membrane should yield a protein-enriched fraction. The Se distribution in these four fractions and the whole biomass of *Chlorella* and filamentous cyanophyte is summarized in Figure 16.5. Clearly, a major fraction (60%) of the Se in *Chlorella* biomass resided in the TCA residue when cells were grown in the 100 mg/l Se (as selenite) medium, while the TCA-soluble fraction accounted for 15% (Figure 16.5A). It should be noted that the TCA residue may contain Se^0, since a red amorphous material codeposited with the biomass and was not extracted by TCA (Fan et al., 1997a). The Tris-SDS extract constituted only 6%, while the protein-rich fraction (3.5-kDa retentate) contributed even less (1%) to total Se in biomass (Figure 16.5A).

In contrast, the Tris-SDS extract contained the majority of the Se (89%) in the filamentous cyanophyte biomass when grown at 10 mg/l Se (Figure 16.5B; Fan et al., 1998b). Selenium present in the protein-rich fraction (3.5-kDa retentate) also contributed to a significant fraction (17%) of the biomass Se. Moreover, very little Se was extracted with TCA (0.17%), while 52% of the biomass Se remained in the TCA residue. The TCA extract and residue combination did not account fully for the biomass Se. It is possible that the missing Se was a result of degradation of labile Se compounds during the TCA extraction.

Nonvolatile Forms of Se in Microphyte Biomass

The fractionation of Se into TCA and protein extracts allowed further analysis of the chemical form(s) of Se present in these extracts. Our present focus is on selenonium metabolites and selenoamino acids, since the former may be indicative of

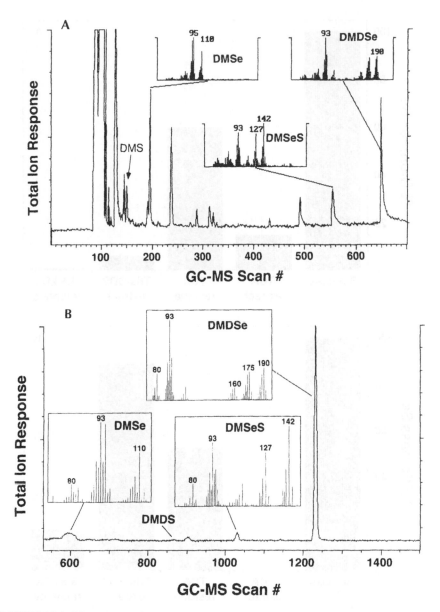

FIGURE 16.4 GC-MS analysis of liquid N_2 temperature trap of selenite-treated microphyte cultures. The liquid N_2 temperature traps were obtained from the 100 mg/l Se-treated *Chlorella* (Figure 16.4A, from Fan et al., *Environ. Sci. Technol.*, 31, 569-576, 1997a. With permission.) and 10 mg/l Se-treated filamentous cyanophyte cultures (Figure 16.4B). GC-MS analysis was performed as described by Fan et al. (1997a). DMSe, DMDSe, and DMS was identified based on GC retention times, molecular ions, and mass fragmentation patterns (insets) as compared with authentic standards. The assignment of DMSeS was deduced from the molecular ion, mass fragmentation pattern, and known GC elution order.

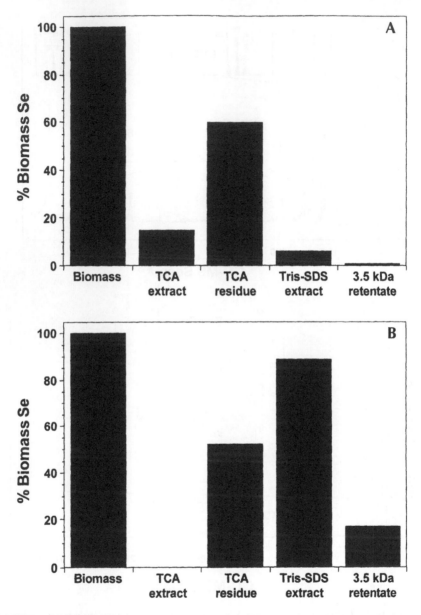

FIGURE 16.5 Distribution of total Se in microphyte biomass and subfractions. The biomass obtained from the *Chlorella* (grown at 100 mg/l Se, Figure 16.5A) and filamentous cultures (grown at 10 mg/l Se, Figure 16.5B) were extracted with TCA and Tris-SDS. The Tris-SDS extract was further fractionated by dialysis against a 3.5-kDa MWCO membrane. Se in microphyte biomass and subfractions were determined by GC-ECD or fluorescence via micodigestion and expressed as a percentage (%) of the biomass Se. The total Se content in the *Chlorella* and filamentous cyanophyte biomass was 566 and 2397 µg/g dry weight, respectively.

the Se volatilization potential while the latter may be a biomarker of ecotoxic risk to wildlife. However, we should add that volatilization and ecotoxic effects are not necessarily mutually exclusive; there is the possibility that parts of the biochemical pathways are shared for both processes.

For the characterization of selenonium compounds, an alkaline hydroelimination procedure, analogous to that for the sulfonium compounds (White, 1982), was developed for both biomass and TCA fractions (Fan et al, 1997a; Fan et al., 1998a). The procedure uses GC-MS to measure volatile Se and S compounds released from treating biomass and TCA extracts with NaOH in a sealed vial at both room temperature (e.g., Figure 16.6) and 105 to 110°C. The pattern of volatile Se and S compounds released by the treatment can be used as a convenient diagnostic test for different methylated metabolites. For example, we found that DMSe and DMS were liberated from standards of dimethylselenonium propionate (DMSeP) and dimethylsulfonium propionate (DMSP) at room temperature, respectively, while DMSe and DMDSe were released, respectively, from standards of methylseleno methionine (CH_3-Se-Met) or trimethylselenonium ion (TMSe[+]) and methylseleno cysteine (CH_3-Se-Cys) at 105 to 110°C (Fan et al., 1998a). It should be noted that the alkaline hydroelimination test gave clues to but did not provide the identity of the alkylselenide precursor. Additional structure characterization (e.g., by NMR or GC-MS) would be required to determine identity.

In the case of *Chlorella*, DMSe was released from Se-treated biomass and TCA extracts without heating (Figure 16.6A), indicating the presence of a DMSeP-like selenonium compound and ruling out the presence of methylselenomethionine (CH_3-Se-Met). The absence of CH_3-Se-Met was also reported for a marine *Chlorella* sp. (Bottino et al., 1984). Totaling all selenonium forms accounted for all of the Se in the TCA extract, in turn the TCA extract contained all of the selenonium metabolite(s) in the entire biomass (Fan et al., 1997a). The latter result would be expected of DMSeP, since it was very water-soluble. To further characterize the selenonium metabolite(s), the TCA extract was analyzed by ^{77}Se NMR, which revealed the presence of one dominant Se compound (Figure 16.7). However, the [77]Se resonance of this compound did not correspond to any of the known standards (DMSeP, CH_3-Se-Met, CH_3-Se-Cys, and TMSe[+]; Fan et al., 1998a). We are currently utilizing two-dimensional NMR techniques (e.g., Fan, 1996) to determine the detailed structure of this metabolite.

As for the filamentous cyanophyte, DMSe was detected from the 1 mg/l Se-treated biomass (Figure 16.6B) while DMSe, DMDSe, and DMSeS were liberated from the 10 mg/l Se-treated biomass only upon heating with the alkaline treatment (Fan et al., 1998b). This pattern of release suggests that DMSe may arise from CH_3-Se-Met and/or TMSe[+], while DMDSe may be derived from CH_3-Se-Cys and DMSeS may result from a cross-reaction between the DMDSe and DMDS precursors (i.e., CH_3-Se-Cys and methylcysteine, respectively; Chasteen, 1993). Subsequent ^{1}H NMR analysis of the TCA extract prepared from the 1 mg/l Se-treated biomass indicates that CH_3-Se-Met, instead of TMSe[+], may be the DMSe precursor in the filamentous cyanophyte.

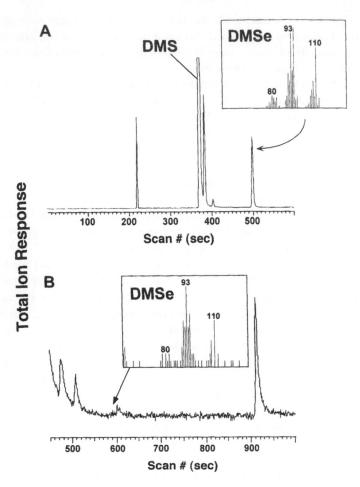

FIGURE 16.6 GC-MS analysis of volatile alkaline hydroelimination products of selenite-treated microphyte biomass. Lyophilized *Chlorella* (grown at 100 mg/l Se, A) and filamentous cyanophyte biomass (grown at 1 mg/l Se, B) was subject to alkaline treatment at room temperature and at 110°C, respectively. The volatile Se compounds liberated were analyzed by GC-MS as described by Fan et al. (1997a). The identity of DMSe and DMS was confirmed by GC retention time, molecular ion, and mass fragmentation pattern (inset), and their content in the *Chlorella* biomass was approximately 5 ng DMSe/mg and 185 ng DMS/mg, respectively. (From Fan et al., *Environ. Sci. Technol.* 31, 569-576, 1997a.)

From the above NMR analysis, it is clear that selenoamino acids were not present at any appreciable concentration in the TCA extract of either *Chlorella* or filamentous cyanophytes. We thus turned to GC-MS for trace-level analysis (Fan et al., 1997a). The TCA extracts were silylated with MTBSTFA to make selenomethionine (Se-Met), selenocysteine, and CH_3-Se-Cys analyzable by GC-MS. The MTBSTFA method was not suited for selenocystine analysis while other selenoamino acids were not screened for lack of authentic standards.

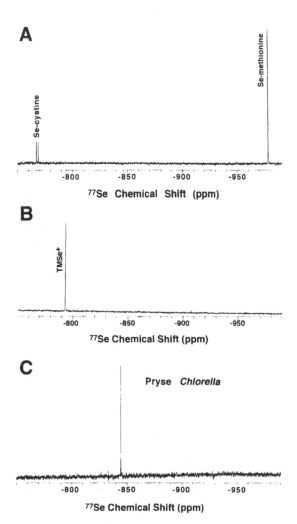

FIGURE 16.7 1-D ⁷⁷Se NMR spectra of *Chlorella* extract and Se-standards. The PCA extract of a 100 mg/l Se-treated *Chlorella* biomass was processed through a Chelex 100 column before ⁷⁷Se NMR analysis on a Bruker AM-400 NMR spectrometer operating at 76.3 MHz. The ⁷⁷Se NMR spectrum of the *Chlorella* extract is displayed along with that of the Se-Met, Se-cystine, and trimethylselenonium ion (TMSe⁺) standards in D_2O. (Reprinted with permission from Fan and Higashi, 1998). The dominant resonance with a chemical shift of -845 ppm in the *Chlorella* spectrum did not correspond to that of selenate, selenite, Se-Met, Se-cystine, TMSe⁺, CH₃-Se-Met, or DMSeP (Fan et al., 1998a). However, this chemical shift is consistent with that of a selenonium compound. (From Duddeck, H. *Progr. NMR Spectrosc.* 27: 1-323, 1994. With permission.)

As shown in Figure 16.8, Se-Met was not observable in the total or selected ion chromatogram of *Chlorella* extract with the electron impact mode detection. However, using the 40-fold more sensitive isobutane chemical ionization mode of the iontrap MS, a peak with m/z 427 appearing at the GC retention time of Se-Met

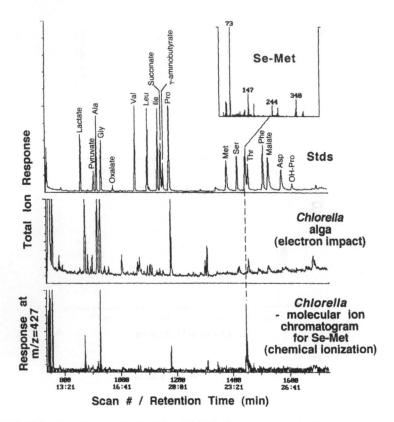

FIGURE 16.8 GC-MS analysis of silylated *Chlorella* extract and standards. The same *Chlorella* extract as in Figure 16.7 was silylated with MTBSTFA and analyzed by GC-MS as described by Fan et al. (1997a). The silylated derivatives of a mixed standard were also analyzed by GC-MS for comparison (top panel). Se-Met in the *Chlorella* extract was confirmed by GC retention time (noted by dash line for Se-Met) and by the detection of the molecular ion of silylated Se-Met (m/z = 427) using chemical ionization method (bottom panel). Se-Met was present only at trace (sub µg/l) level in the *Chlorella* extract such that no corresponding peak in the electron-impact ion chromatogram (mid panel) was detected. (From Fan et al., 1997a, *Environ. Sci. Technol.*, 31, 569–576. With permission.)

was detected, providing evidence for the presence of free Se-Met in *Chlorella* biomass at the sub-µg/l level. Neither selenocysteine nor CH$_3$-Se-Cys was observed through this analysis. However, a number of other amino acids, organic acids, and phosphate metabolites were simultaneously measured with this simple one-step extraction. More importantly, the approach is based on GC-ion trap MS analysis, widely accepted as one the few reliable methods of structure confirmation and quantification when faced with trace levels of organic compounds. An earlier study on a marine *Chlorella* sp. also reported the presence of Se-Met, along with selenocystine, selenocysteic acid, and CH$_3$-Se-Cys in the protein-free extract (Bottino et al., 1984). Because these Se metabolites were largely assigned based on chromatographic coelution with standard compounds, their identity remained tentative until further structure confirmation. The issue of mistaken identity for selenoamino acids based on chromatographic comigration has been raised previously (Huber and Criddle, 1967). In any rate, species and treatment differences may account for the different observations. No Se-Met, selenocysteine, and CH_3-Se-Cys was detected in the TCA extract of the filamentous cyanophytes (data not shown). Thus, the free selenoamino acids were at best a very minor component of the microphyte biomass.

In contrast to the findings above, when the protein extracts (3.5-kDa retentates) of both *Chlorella* and filamentous cyanophytes were analyzed for Se-Met (e.g., Figure 16.9), up to to 70% of the proteinaceous Se was in the Se-Met form (Fan et al., 1998a; Fan et al., 1998b). No protein-derived Se-Met was reported in the marine *Chlorella* sp. by Bottino et al. (1984). We also noted a major difference in the extent

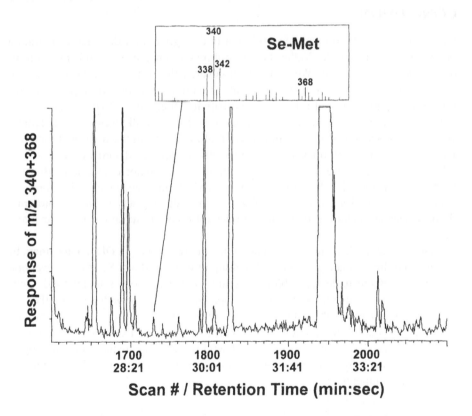

Scan # / Retention Time (min:sec)

FIGURE 16.9 Proteinaceous Se-Met analysis of filamentous cyanophytes. The 5 mg/l Se-treated filamentous cyanophyte biomass was extracted with the Tris-SDS buffer, followed by dialysis against a 3.5-kDa MWCO membrane. The 3.5-kDa retentate was then digested in 6 N HCl, silylated with MTBSTFA, and analyzed by GC-MS (Fan et al., 1998b). The GC-MS tracing shown represents the chromatogram of the sum of two selected ions (m/z 340 + 368). Se-Met was identified based on GC retention time and mass fragmentation ion (inset). It should be noted that, using GC-MS and 1 mg of proteins, it was practical to analyze for proteinaceous Se-Met present at μg/g protein level.

of Se allocation into proteins between the filamentous cyanophyte and *Chlorella*. About 17 to 80% of the biomass Se resided in the proteinaceous fraction of the filamentous cyanophytes when grown at 20 μg/l to 10 mg/l Se concentrations (Fan et al., 1998b), while the *Chlorella* protein fraction only contributed to 0.7% of the biomass Se when grown at 100 mg/l Se (Fan et al., 1998a). Because in bird-feeding experiments, dietary Se-Met has been shown to mimic Se toxicity (Heinz et al., 1996), this difference in proteinaceous Se-Met between microphytes could mean that the microphyte community structure may be a significant factor in ecotoxic consequences. Armed with this information, a system could someday be devised to promote microphyte volatilization of Se while avoiding microphyte compositions that pose ecotoxic risk.

CONCLUSION

Selenium contamination and accumulation in the agricultural drainage evaporation basins of the San Joaquin Valley has been a major concern for California's multi-billion dollar agriculture. We recently discovered that waterborne Se in some of these basins did not accumulate and that Se biotransformation by aquatic microphytes may play a critical role in Se dissipation. Three microphytes isolated from these basins were active in volatilizing Se in the forms of alkylselenides. Selenium volatilization accounted for a major fraction of Se loss from the treatment medium. However, a significant fraction of Se was also incorporated into the biomass, of which proteinaceous Se-Met sometimes represented a major component. Because Se-Met is considered to be an important ecotoxic form, incorporation into protein-aceous Se-Met will need to be examined when evaluating any *in situ* Se bioreme-diation scheme, including those that are heterotrophic bacteria and vascular plant-based.

Nevertheless, this "natural attenuation" of Se at the TLDD basins may be enhanced or better managed to avoid buildup of ecotoxic-level Se at other evapora-tion basins, if the underlying biogeochemical mechanism(s) is understood. Such knowledge should also facilitate the development of *in situ* remediation of other Se-laden environments such as fly-ash ponds resulting from power plant operations. Because the goal of remediation — whether it involves Se or other contaminants — is to minimize risk to human health and wildlife, a mechanistic understanding of the role of microphytes is necessary for an efficacious solution to Se problems in aquatic environments.

ACKNOWLEDGMENT

We thank Dr. Andrew Lane for his NMR expertise and valuable discussion. We are grateful to Douglas Davis and his staff at TLDD for the water samples and general assistance in the field. This work was supported in part by the UC Salinity/Drainage program and U.S. EPA (R819658) UC-Davis Center for Ecological Health Research. The authors also acknowledge the MRC Biomedical NMR center at Mill Hill, U.K. for providing valuable NMR instrumentation.

REFERENCES

Amouroux, D. and Donard, O.F.X. Maritime emission of selenium to the atmosphere in eastern Mediterranean seas. *Geophys. Res. Lett.* 23, 1777-1780, 1996.

Andreae, M.O. The ocean as a source of atmospheric sulfur compounds, in Buat-Ménard, P., Ed., *The Role of Air-Sea Exchange in Geochemical Cycling,* D. Reidel Publishing Company, Dordrecht, The Netherlands, 1986, 331-362.

Bates, T.S., Charlson, R.J., and Gammon, R.H. Evidence for the climate role of marine biogenic sulphur. *Nature* 329, 319-321, 1987.

Bottino, N.R., Banks, C.H., Irgolic, K.J., Micks, P., Wheeler, A.E., and Zingaro, R.A. Selenium containing amino acids and proteins in marine algae. *Phytochemistry* 23, 2445-2452, 1984.

Chasteen, T.G. Confusion between dimethyl selenenyl sulfide and dimethyl selenone released by bacteria. *Appl. Organometall. Chem.* 7, 335-342, 1993.

Cooke, T.D. and Bruland, K.W. Aquatic chemistry of selenium: evidence of biomethylation. *Environ. Sci. Technol.* 21, 1214-1219, 1987.

Craig, P.J. Occurrence and pathways of organometallic compounds in the environment — general considerations, in Craig, P.J., Ed., *Organometallic Compounds in the Environment,* Longman, U.K., 1986, 1-64.

Cutter, G.A. and Bruland, K.W. The marine biogeochemistry of selenium: a re-evaluation. *Limnol. Oceanogr.* 29, 1179-1192, 1984.

Duce, R.A., Hoffman, G.L., and Zoller, W.H. *Science* 187, 59-61, 1975.

Duddeck, H. Selenium-77 nuclear magnetic resonance spectroscopy. *Progr. NMR Spectrosc.* 27, 1-323, 1994.

Fairbrother, A., Bennett, R.S., Kapustka, L.A., Dorward-King, E.J., and Adams, W.J. Risk from mining activities to birds in southshore wetlands of Great Salt Lake, Utah, in *Abstracts of the 17th Annual Meeting of the Society of Environmental Toxicology and Chemistry,* Society of Environmental Toxicology and Chemistry, Pensacola, FL, 1996, 97.

Fan, T. W.-M. and Higashi, R.M. Biochemical fate of selenium in microphytes: natural bioremediation by volatilization and sedimentation in aquatic environments, in Frankenberger, W.T., and Engberg, R.A., Eds., *Environmental Chemistry of Selenium,* Marcel Dekker, Inc., New York, 545-563, 1998.

Fan, T. W.-M., Lane, A.N., Martens, D., and Higashi, R.M. Synthesis and structure characterization of selenium metabolites. *Analyst,* 123, 875-884, 1998a.

Fan, T. W.-M., Higashi, R.M., and Lane, A.N. Biotransformation of selenium oxyanions by halophytic filamentous cyanophytes. *Environ. Sci. Technol.* 32, 3185-3193, 1998b.

Fan, T.W.-M., Lane, A.N., and Higash, R.M. Selenium biotransformations by a euryhaline microalga isolated from a saline evaporation pond. *Environ. Sci. Technol.* 31, 569-576, 1997a.

Fan, T. W.-M., Higashi, R.M., Frenkiel, T.A., and Lane, A.N. Anaerobic nitrate and ammonium metabolism in flood-tolerant rice coleoptiles. *J. Exp. Bot.* 48, 1655-1666, 1997b.

Fan, T.W.-M. Metablite profiling by one and two-dimensional NMR analysis of complex mixtures. *Prog. NMR Spectrosc.* 28, 161-219, 1996.

Gerhardt, M.B., Green, F.B., Newman, R.D., Lundquist, T.J., Tresan, R.B., and Oswald, W.J. Removal of selenium using a novel algal-bacterial process. *Res. J. Water Pollut. Control Fed.* 63, 799-805, 1991.

Heinz, G.H., Hoffman, D.J., and LeCaptain, L.J. Toxicity of seleno-L-methionine, seleno-DL-methionine, high selenium wheat, and selenized yeast to mallard ducklings. *Arch. Environ. Contam. Toxicol.* 30, 93-99, 1996.

Huber, R.E. and Criddle, R.S. Comparison of the chemical properties of selenocysteine and selenocystine with their sulfur analogs. *Arch. Biochem. Biophys.,* 122, 164, 1967.

Lewis, B.-A.G. Selenium in biological systems, and pathways for its volatilization in higher plants, in Nriagu, J.O., Ed., *Environmental Biogeochemistry. Carbon, Nitrogen, Phosphorus, Sulfur and Selenium Cycles,* Ann Arbor Science Publishers, Ann Arbor, MI, Vol. 1, 1976, 389-409.

Maier, K.J. and Knight, A.W. Ecotoxicology of selenium in freshwater systems. *Rev. Environ. Contam. Toxicol.* 134, 31, 1994.

Maier, K.J. Bioaccumulation and conversion of selenium to organo-forms. Presentation at the Symposium on *Understanding Selenium in the Aquatic Environment,* Salt Lake City, UT, March 1997.

McCarthy, T.P., Brodie, B., Milner, J.A., and Bevill, R.F. Improved method for selenium determination in biological samples by gas chromatography. *J. Chromat.* 225, 9-16, 1981.

Mudder, T.I. Selenium treatment technologies. Presentation at the Symposium on *Understanding Selenium in the Aquatic Environmen,* Salt Lake City, UT, March 1997.

Reash, R., Lohner, T., Wood, K., and Leveille, R. Selenium in fish inhabiting a fly ash receiving stream: implications for National Water Quality Criteria, in *Abstracts of the 17th Annual Meeting of the Society of Environmental Toxicology and Chemistry.* Society of Environmental Toxicology and Chemistry, Pensacola, FL, 1996, 12.

Rippka, R., Deruelles, J., Waterbury, J.B., Herdman, M., and Stanier, R.Y. Genetic assignments, strain histories and properties of pure cultures of cyanobacteria. *J. Gen. Microbiol.* 111, 1-61, 1979.

Skorupa, J.P. and Ohlendorf, H.M. Contaminants in drainage water and avian risk thresholds, in Dinar, A., and Zilberman, D., Eds., *The Economy and Management of Water and Drainage in Agriculture.* Kluwer Academic Publishers, Norwell, MA, 1991, 345.

Sugimura, Y., Suzuki, Y., and Miyake, Y. The content of selenium and its chemical form in seawater. *J. Oceanogr. Soc. Jpn.* 32, 235-241, 1977.

Tanji, K.K. Chemistry of toxic elements (As, B, Mo, Se) accumulating in agricultural evaporation ponds, in Summers, J.B. Ed., *Toxic Substances in Agricultural Water Supply and Drainage.* An International Environmental Perspective. Abstracts of The Second Pan-American Regional Conference of the International Commission on Irrigation and Drainage, Ottawa, Canada, U.S. Committee on Irrigation and Drainage, Denver, CO, 1989, 109-121.

Terry, N., Carlson, C., Raab, T.K., and Zayed, A.M. Rates of selenium volatilization among crop species. *J. Environ. Qual.* 21, 341-344, 1992.

Thompson-Eagle, E.T., and Frankenberger, Jr., W.T. Selenium biomethylation in an alkaline, saline environment. *Wat. Res.* 25, 231-240, 1991.

White, R.H. Analysis of dimethyl sulfonium compounds in marine algae. *J. Mar. Res.* 40, 529-536, 1982.

White, A.F. and Dubrovsky, N.M. Chemical oxidation-reduction controls on selenium mobility in groundwater systems, in Frankenberger, Jr., W.T. and Benson, S., Eds., *Selenium in the Environment,* Marcel Dekker, New York, 1994, 185-221.

Wrench, J.J. Selenium metabolism in the marine phytoplankters *Tetraselmis tetrathele* and *Dunaliella minuta. Marine Biol.* 49, 231-236, 1978.

17 *In Situ* Gentle Remediation Measures for Heavy Metal-Polluted Soils

S.K. Gupta, T. Herren, K. Wenger, R. Krebs, and T. Hari

CONTENTS

INTRODUCTION

Over the course of recent decades, industrial and agricultural activities have led to a considerable increase in heavy metal levels in different environmental compartments, especially in soil. A large number of sites throughout the world are classified as polluted. Although in most of these sites the risk for man, plants, and animals is at present not very acute, soil quality and groundwater are severely affected. On certain polluted sites, there is the hazard of entry of pollutants into the food chain. Besides reducing emissions, the development of a concept of risk management for these polluted sites is an important task for soil protection. The hazard-alleviating measures can be classified into three categories (Figure 17.1): (1) gentle *in situ* remediation measures, (2) harsh soil use restrictive measures, and (3) harsh soil destructive measures. The main goal of the last two harsh alleviating measures is to avert hazards either to man, plant, or animals. The main goal of gentle *in situ* remediation is to restore the multifunctionality of soil (soil fertility), which allows a safe use of the soil (Krebs et al., 1998). The category of gentle remediation measures consists of two main groups, stabilization (immobilization) and decontamination.

Hazard Alleviating Measures for Heavy Metal Polluted Soils

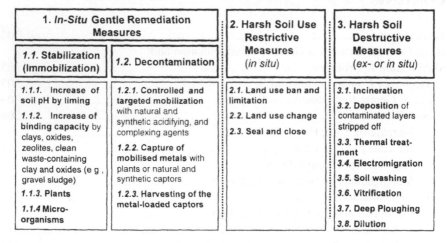

FIGURE 17.1 Possible measures to reduce the hazard of a soil polluted with heavy metals.

Gentle remediation techniques are applied *in situ* and, therefore, no excavation or transport of soil is necessary. The physical soil structure is maintained and may even be improved during the remediation process. Besides their ecological advantages, gentle remediation techniques may be economically advantageous. Costs for gentle remediation may be orders of magnitude less than costs associated with harsh physicochemical technologies. One problem associated with gentle remediation techniques is that a longer curing time is generally required as compared to conventional harsh techniques. Another restriction may be the use of living organisms (e.g., plants or microorganisms) for gentle processes; the soil texture, pH, salinity, pollutant concentrations, and the presence of other toxins must be within the limits of tolerance of such organisms (Cunningham et al., 1995a; Bollag and Bollag, 1995).

The main objectives of this chapter are to discuss and evaluate *in situ* gentle remediation measures, to discuss the complementary relationship between stabilization and decontamination, and to evaluate the possibilities of decontamination with techniques other than the use of plants (phytoremediation).

BASIC CONCEPT FOR *IN SITU* GENTLE DECONTAMINATION AND STABILIZATION APPROACHES

DEGREE OF CONTAMINATION AND SEVERITY OF RISK

The decision to remediate a site or not is made on the basis of the degree of the risk posed by a further spread of heavy metals. For polluted sites which pose a severe risk of further spread of the contaminants, only harsh methods are suitable, but for larger areas and soils with diffuse sources of pollution below a certain degree of contamination, gentle remediation techniques are ecologically and economically reasonable alternatives.

Three levels are considered important to assess the effects of any potentially toxic metal species in soil (Tadesse et al., 1994): (1) background levels, (2) tolerable levels, and (3) harmful pollutant levels. Soils with background levels contain no or small amounts of anthropogenic trace elements. Background levels of heavy metals in soils are highly variable and therefore detailed knowledge is fundamental for regulations based on these levels (Frink, 1996). In the range of tolerable levels, soils contain increased amounts of anthropogenic heavy metals. In this case, the multi-functionality or the fertility of soil might be affected and toxicity symptoms on vegetation or crop plants will become visible. Such soils are a potential risk to plants, animals, or men and pathway-specific measures have to be taken. For soils with harmful pollutant levels, an immediate remediation is required, because such soils are a hazard for any use.

A concept was proposed by Gupta et al. (1996) for risk assessment and risk management of heavy metal-polluted soils based on threshold values representing the limits between the ranges defined by Tadesse et al. (1994; Figure 17.2). In soils exceeding the "guide values" (tolerable levels), the long-term functionality of the soil is no longer assured. In this case, the location of the heavy metal source and a reduction of the emissions are the appropriate measures. Soils with heavy metal concentrations above the "cleanup values" (limit between tolerable levels and harmful polluting levels) are a hazard, and fast and stringent measures have to be taken. A third value was inserted within the range of tolerable levels, the "trigger values." In soils exceeding the trigger values, either the land use must be changed or remediation measures have to be taken according to the results of subsequent site-specific investigations. This concept has been implemented in the Swiss Ordinance regarding pollutant impacts on the soil (VBBO, 1998).

Gentle remediation methods may be used in soils with heavy metal concentrations between the trigger value and the cleanup value. To choose the appropriate method, determination of the type and extent of soil contamination is necessary. Measurements of total heavy metal levels of a soil include both metal species available to the biota and metal fixed in minerals that is normally not available to plants or animals (Phillips and Chapple, 1995; Sims et al., 1997). Only the mobile fraction of cations is available for plant uptake and poses a risk of being leached to the groundwater. The mobile fraction may be defined as the fraction that may enter a living receptor when in contact with it. In context of risk assessment, this fraction

Soil Protection Concept & Gentle Remediation

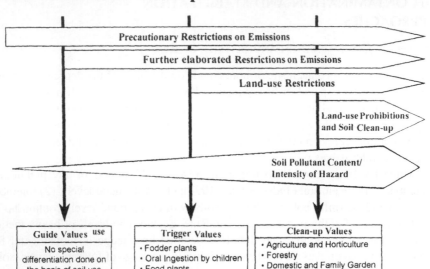

FIGURE 17.2 Concept of soil protection and values for remediation measures.

may induce a toxic effect or an impairment of quality on plants. There are different approaches to estimate the extent of this fraction. In our laboratory, we use the $NaNO_3$-extractable fraction as approximation of mobile heavy metals. If only the immobile and, therefore, not phytoavailable fraction of heavy metals in soil is high, this may not have toxic effects on plants. In such soils, the contamination is stable, but there might be a need to reduce the total content, because soil properties may change due to natural processes or due to environmental effects such as acid rain, increased decomposition of soil organic matter, or global climate change. Therefore, it is not reasonable to focus decisions concerning remediation measures only on the total content of heavy metals, but also on the different metal fractions in soil (Figure 17.3).

TYPES OF GENTLE REMEDIATION TECHNIQUES

Usually, only the mobile fractions of heavy metals in soil can be directly influenced by gentle remediation methods. The equilibrium between soluble and insoluble fractions may either be shifted toward more insoluble or toward more soluble heavy metals. A decrease in the soluble fraction will stabilize the pollutants in the soil, whereas an increase in the soluble fraction will not only increase the danger of a further spread of the pollutants but will also make them more available for decontamination. Therefore, the soluble fraction plays the central role in the decision on the appropriate decontamination technique. In order to obtain an ecologically safe decontamination, the maintenance of an optimal ratio between soluble and insoluble heavy metals is necessary.

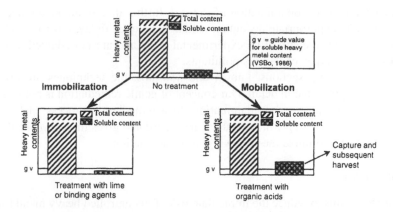

FIGURE 17.3 Illustration of the central role of the soluble fraction of heavy metals in soil.

There are principally two categories of remediation techniques of a contaminated soil: stabilization and decontamination (Figure 17.4). The choice of the principal category is mainly made on different site factors such as soil type, the nature and distribution of pollution as well as the severity of the hazard, current land use, soil pH, and cleanup goals (Gabriel, 1991). With knowledge of these major points, the decision can be made as to whether a stabilization or a decontamination procedure is preferable. Knowledge of the current land use will reveal whether or not changes are needed. If the pollutants should be stabilized, the pH of the soil makes it clear whether liming or another stabilization technique should be applied. When the final

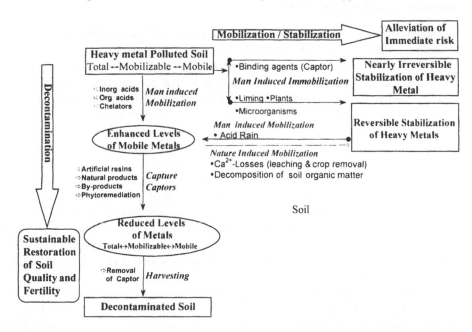

FIGURE 17.4 Concept of immobilization and decontamination of heavy metals in soil.

goal is a complete decontamination of the soil, further investigations are necessary to determine the appropriate decontamination technique. Today, most *in situ* remediation techniques are still at an experimental stage and are not adapted to a large spectrum of soil types or various pollutants.

In the following sections, known and new possible techniques are critically evaluated and presented in detail. Our concept of gentle remediation is not restricted to either stabilization or decontamination. In a remediation process, stabilization may only be the first step which reduces the hazard and gives time to make detailed investigations to optimize the following decontamination.

STABILIZATION

This strategy aims to reduce the immediate risk of uncontrolled heavy metal transfer to the groundwater or to the biosphere (Conner, 1994; Vangronsveld et al., 1995). To attain this aim, the heavy metal fraction available to plants in the soil has to be reduced, which means that heavy metals are immobilized and the equilibrium between soluble and insoluble fraction is intentionally shifted toward more insoluble forms either by increasing soil pH or by increasing the binding capacity of the soil. Nevertheless, the heavy metals remain in the soil. Therefore, the result of stabilization is not a decontaminated but a stabilized soil where metals are transferred into an inactive form. In the next section, recent stabilization (immobilization) techniques that have been tested either under field conditions or under greenhouse conditions are reviewed.

Increase of Soil pH by Liming

Immobilization can be achieved by increasing the soil pH, as described for zinc and cadmium (Alloway and Jackson, 1991). Liming is used in agriculture to increase the pH of acidic soils. Most experiments investigating the effect of liming on the availability of heavy metals were made in soils that had received high doses of sewage sludge. Little is known about the formation of complexes with soluble organic substances and the effects of liming on the complexation.

Krebs et al. (1998) investigated heavy metal uptake of peas in limed and unlimed plots treated with mineral fertilizer (control), sewage sludge, or pig manure. The above-ground parts of field peas grown on limed soils contained lower heavy metal concentrations than plants grown on fertilized, unlimed soils (Figure 17.5). The highest reductions in zinc uptake, due to the addition of lime, was found in plants grown on control plots. The zinc concentration decreased from 73 to 50 mg/kg dry matter (DM) in seeds and from 59 to 19 mg/kg DM in crop residues. Cadmium uptake was reduced even further by liming than zinc uptake. The maximal reduction was again found on control plots, where cadmium concentrations of 213 µg/kg DM were measured in crop residues from unlimed plots and only 67 µg/kg DM from limed plots. Liming also led to a considerable reduction of copper uptake by seeds and crop residues in all treatments, which was unexpected in view of the unchanged $NaNO_3$-extractable (mobile) copper concentrations (data not shown). An explanation may be that the enhanced mobility was due to an increased formation of organic

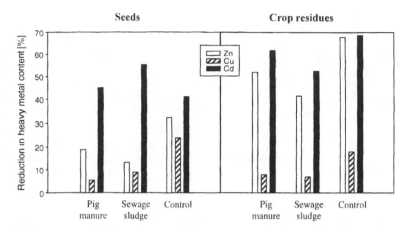

FIGURE 17.5 Reduction of the zinc, copper, and cadmium contents of seeds and crop residue after lime application.

complexes of large molecular size that are less available to plant uptake than free copper ions.

Increase Binding Capacity

Another way to immobilize heavy metals is to increase the metal-binding capacity of the soil by the addition of clay minerals, iron oxides, or waste products such as gravel sludge. Such additives reduce the mobility of heavy metals due to their large specific surface and high cation exchange capacities. However, heavy metals can readily be exchanged by other cations such as calcium and magnesium (van Bladel et al., 1993). It is therefore important that the addition of binding agents does not (or only slightly) affect the availability of nutritional cations for plants.

Several binding agents such as zeolite, beringite, hydrous manganese, or ferrous oxides have been studied for the use as immobilizing agents under pot and field conditions (Czupyrna et al., 1989; Didier et al., 1993; Greinert, 1995; Vangronsveld et al., 1995). Lothenbach et al. (1998) compared the effectiveness of different binding agents in zinc immobilization in batch experiments (Figure 17.6). Al-montmorillo-nite significantly reduced dissolved concentrations of zinc in the pH 5 to 8 range. In most studies, expensive and purified clay minerals were used. For the application on agricultural soils, binding agents must be available in large quantities at a sufficiently low cost.

Gravel sludge is a waste product of the gravel industry and, at least in Switzerland, is available in large quantities at a low price. Normally, this product contains about 45% clay minerals, and the concentrations of heavy metals are much lower than limit values of sewage sludge according to the Swiss Ordinance on Substances (StoV, 1986). Relative to the heavy metal concentration already present in soils, metal input due to the experimental application of the gravel sludge is insignificant.

A comparison between gravel sludge and Na-montmorillonite as binding additives in pot experiments was made by Lothenbach et al. (1998). Both additives

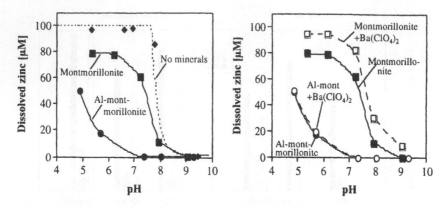

FIGURE 17.6 Dissolved zinc concentrations in the presence of montmorillonite or Al-mont-morillonite before (left) and after the addition of $Ba(ClO_4)_2$ (right) as a function of pH.

reduced the soluble zinc fraction in soils by about a factor of eight (Figure 17.7). In contrast to Na-montmorillonite, gravel sludge only slightly affected soil pH (Figure 17.8). The addition of Na-montmorillonite had a negative effect on the yield of red clover, whereas gravel sludge did not reduce the yield. The application of gravel sludge was also studied in the field (Krebs et al., 1999). At all three experimental sites investigated, the application of gravel sludge led to an increase in soil pH, which can be attributed to the high $CaCO_3$ content (30%) of the gravel sludge. In all treatments, the effects on $NaNO_3$-extractable Cu concentrations were less than on zinc. The concentrations in soil and the total plant uptake of zinc, copper, and cadmium by ryegrass due to gravel sludge application were most strongly reduced at site 1 (Figure 17.9). At the other sites, the effect on the concentrations and the total metal uptake by ryegrass was less evident. Thus, gravel sludge treatments led to a decrease in heavy metal uptake by plants, a varying level depending on the plant and site characteristics.

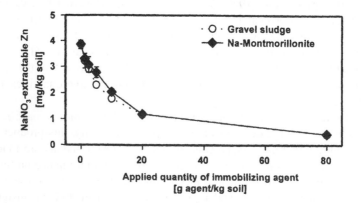

FIGURE 17.7 Decrease of $NaNO_3$-extractable zinc after the addition of binding agents.

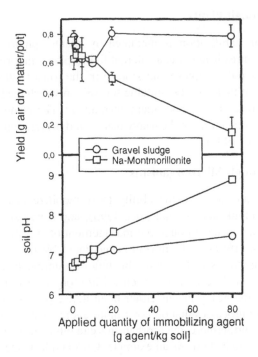

FIGURE 17.8 Effect of binding agents on biomass yield of red clover (*Trifolium pratense*; top) and on soil pH (bottom).

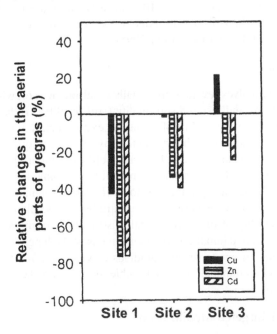

FIGURE 17.9 Relative changes in copper, zinc, and cadmium contents of aerial parts of ryegrass (*Lolium perenne*) due to the addition of gravel sludge.

Stabilization Through Plants

A special form of stabilization is mediated by plants (phytostabilization). This immobilization requires heavy metal-tolerant living plants that reduce the mobility of heavy metals in soil by uptake and storage in the roots (Salt et al., 1995). The stabilization effect of plants is also suitable for soils in which the heavy metals were previously immobilized by methods mentioned above. The plants growing on these soils will increase the stability of the heavy metals and prevent wind or water erosion (Vangronsveld et al., 1995).

Stabilization Through Microorganisms

Different microorganisms have the ability to immobilize heavy metals in soils (Summers, 1992; Frankenberger and Losi, 1995), and it was suggested to use this ability to immobilize metals through the management of specific microbial populations (Morel et al., 1997). Immobilization mediated by microorganisms has the advantage of being more selective in the binding of a unique heavy metal than that associated with synthetic chemical sorbents. Microorganisms are able to stabilize soils by concentrating heavy metals either in an active process, called bioaccumulation, or by uptake processes that do not require energy, called biosorption (Bolton and Gorby, 1995). Furthermore, microorganisms can influence heavy metal solubility by direct or indirect reduction. As an example, Cr(VI) (chromate, CrO_4^{2-}) is mobile and toxic, whereas the reduced form, Cr(III), is relatively nontoxic and nonmobile in the environment (Bolton and Gorby, 1995). Similar reactions which precipitate metals were reported for arsenic (As(III) to As(0)), uranium (U(VI) to U(IV)), or selenium (Se(VI) to Se(0)). Sulfate-reducing bacteria are capable of precipitating metals as metal sulfides (Farmer et al., 1995).

DECONTAMINATION

Decontamination involves several steps, finally resulting in a soil with reduced heavy metal concentrations and restored soil quality (soil fertility). Today, soil decontamination techniques use the ability of plants to extract heavy metals from soil. To remove sufficient amounts of heavy metals by this technique, both high tissue concentrations in the plant and high biomass yields are important. Plants known as hyperaccumulators have high concentrations, but their biomass is usually very small. Plants with a high biomass production normally take up small amounts of heavy metals if only moderate concentrations are available. These plant characteristics, as well as the availability of the heavy metal in soil, strongly influence the length of time required for decontamination. In all known cases, the length of decontamination time is in the range of one to several hundreds of years. An *in situ* decontamination of heavy metal-contaminated soils is feasible if plant uptake of heavy metals is strongly enhanced.

Controlled and Targeted Mobilization

The basic need of the decontamination process is the mobilization of heavy metals to render them more accessible to the captor, which may be plants or natural or artificial exchangers. The mobilization process should be controlled in order to avoid loss of heavy metals by leaching to the groundwater but at the same time provide a maximum of soluble heavy metals available for removal. When used with living organisms, the optimal concentration of soluble heavy metals allows maximal uptake by the captor, but does not induce toxicity symptoms. Depending on the decontamination method, it might be necessary to repeat the mobilization treatment several times at a low dosage throughout the decontamination process (e.g., during the vegetative period in phytoremediation). The interval between each treatment is determined by the rate of the degradation process of the substance used for mobilization (Figure 17.10), but certainly, the highest possible soluble heavy metal concentration in combination with an accumulating plant will drastically shorten the decontamination time.

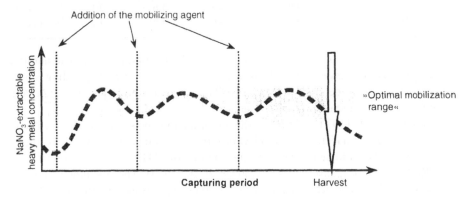

FIGURE 17.10   Maintenance of an optimal mobilization range for heavy metals in soil with no growth restriction for remediation organisms.

An experiment was performed by Wenger et al. (1997) to investigate the uptake capacity of several plant species in a soil with different levels of soluble zinc. In this pot experiment, zinc was not mobilized, but added as $ZnNO_3$. The total extraction of zinc after 25 days was highest by the tobacco plant (Table 17.1).

Mobilization by microorganisms

Microorganisms produce a wide range of different chelating agents which make metals more accessible for plant uptake and therefore may also be useful for remediation applications (Bolton and Gorby, 1995). Other microbial mobilization processes for the mobilization of metals may involve the oxidizing reactions as mentioned previously for stabilization. Some microorganisms gain their energy from the oxidation of reduced inorganic compounds such as sulfur or iron. The oxidation of sulfides produces sulfuric acid that solubilizes the ions of different metals (Bolton and Gorby, 1995) a process known as bioleaching (Tuovinen, 1990).

Organic and inorganic acids or chelators

Heavy metals can be mobilized by decreasing soil pH (Gupta, 1992; Herms and Brümmer, 1980). By this method, the uptake of heavy metals by plants can be

TABLE 17.1

Extraction Capacity of Tobacco, Birch, Knotgrass, and Mustard at Different Levels of Soluble Zinc in a Pot Experiment

Zinc Added to Soil as $Zn(NO_3)_2.6H_2O$ (μg/g soil)	NaNO$_3$- Extractable Zinc Concentration (μg/g soil)	Zinc Extraction after 25 Days (mg/kg soil)			
		Tobacco	Birch	Knotgrass	Mustard
0	6.2	1.9	1.3	1 0	0.7
100	14.2	2.7	1.5	1.3	0.7
170	17.5	3.5	1.8	1.6	0.6
330	31.0	4.7	1.6	2.1	0.4
500	45.3	4.0	1.2	1.7	0.3

increased by a factor of 2 to 3 (Hasselbach and Boguslawski, 1991). The addition of inorganic acids, however, may lead to accumulation of the corresponding anions such as nitrate, chloride, and sulfate in soil. Therefore, it is preferable to use organic agents that are degradable by microorganisms.

Wallace et al. (1974) described the mobilization by synthetic chelates, e.g., ethylenedinitrilotetraacetic acid (EDTA) or nitrilotriacetic acid (NTA). Heavy metal uptake by plants could be increased by the addition of EDTA or NTA to soil (Balmer and Kulli, 1994; Jorgensen, 1993; Wallace et al., 1974). These synthetic ligands, especially EDTA, are barely degradable by microorganisms. For this reason, it is preferable to use soil-borne agents. Depending on their degradation rate, it might be necessary to add such agents several times to maintain an optimal concentration of soluble heavy metals during the decontamination process.

In laboratory studies, the effectiveness of different agents on zinc mobilization was tested (Wenger et al., 1998). The synthetic chelator NTA mobilized about 15 times more zinc than other agents tested (Figure 17.11). Among the natural organic acids, citric acid had the greatest solubilization effect on zinc, about three times more than the other organic acids or nitric acid. The effects of soil treatment with NTA as well as with citric acid on soil pH were very similar, but the metal complex equilibrium constants of NTA are higher than those of the tested natural organic acids (Martell and Smith, 1989). Therefore, it is assumed that the greater mobilizing effect of NTA is based mainly on its greater complexing capacity. The degradation of NTA is much slower than the degradation of the natural organic acids used to mobilize zinc (Figure 17.12). By treatment with natural organic acids at 25 mmol/kg soil, the mobilizing effect of the agents decreased after 17 days and the zinc concentration dropped below the levels measured at the lower acid concentration. Possibly, this effect occurred because the sudden large supply of soil-borne organic agents led to an increased growth of microorganisms that live by degrading these organic substances.

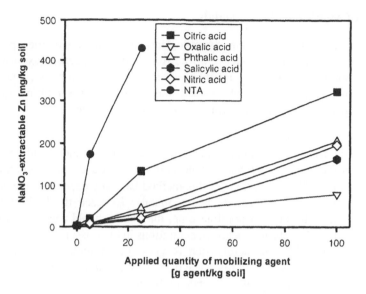

FIGURE 17.11 Mobilizing effect on NaNO$_3$-extractable zinc by several concentrations of synthetic and natural ligands and nitric acid.

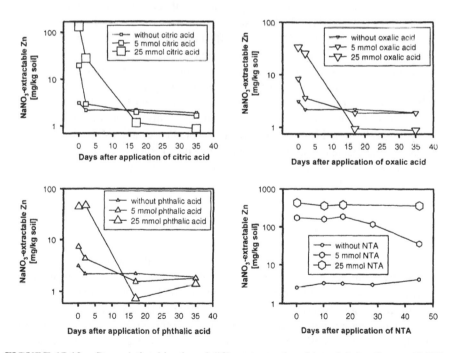

FIGURE 17.12 Degradation kinetics of different organic acids and their effect on NaNO$_3$-extractable zinc concentrations by various doses of mobilizing agents (0, 5, and 25 mmol agent/kg soil).

Capture of Mobilized Heavy Metals

Capture of heavy metals involves specific binding to surfaces or the uptake into living organisms that may be later removed from soil after a certain exposure time. Similar to immobilization, the specificity of this process is very important; large quantities of captured nutritional ions will induce deficiency symptoms for plants or microorganisms, and their growth will be drastically reduced.

Capture by plants (phytoremediation)

Special metal-accumulating plants are used to take up, transport, and concentrate metals from the soil into the harvestable parts of roots and above-ground shoots (Salt et al., 1995). The restrictions of this method basically depend on the ability of plants to take up and concentrate heavy metals. To reach the highest possible extraction within the shortest possible time, the plants should achieve both a high biomass yield and high heavy metal concentrations in their tissue. The biomass yield depends on several different factors such as types of plant roots, yield potential, ion absorption ability of plants, and tolerance of high concentrations. The concentration in plants depends on mobile metal concentrations in the soil, root types, ion interactions, and the hydrogen ion concentration.

Hyperaccumulator plants have an unusually high uptake of heavy metals from polluted sites. However, the biomass production of these hyperaccumulator plants is usually very low and the calculated decontamination times vary from one hundred to several hundreds of years. Therefore, efforts were made to increase the biomass of such plants by crossing them with related plants that have a larger biomass (Cunningham and Ow, 1996). In recent years, clones of high biomass crop plants that are able to accumulate increased levels of heavy metals were selected (Ow, 1993). Transgenic plants containing animal or plant genes encoding for different types of metallothioneins were tested for their ability as phytoextraction plants (Elmayan and Tepfar, 1994; Hattori et al., 1994; Pan et al., 1994). To understand more about the mechanisms involved in heavy metal tolerance, attempts were also made to research molecular mechanisms and genes leading to hyperaccumulation in tolerant species (Brown et al., 1995a,b).

Capture by microorganisms

Similar to the immobilization of heavy metals, the uptake or biosorption by microorganisms may be used to capture heavy metals in soil (Bolton and Gorby, 1995). Until now, such systems have mainly been used in the remediation of water, but attempts are being made to use this system also in soil. For use as captors, microorganisms should not be dispersed in soil, but immobilized in a matrix with a certain mechanical strength (Brierley, 1990; Volesky, 1990). The advantage of using microorganisms to capture metals are the low costs of production and the possibility to adapt the microorganisms either by cultural conditions or genetic manipulations to the environmental conditions at the remediation site (Bolton and Gorby, 1995).

A special type of removal is the volatilization of metals mediated by microorganisms. By this method metals are not concentrated and cannot be captured. The contamination is only transferred from soil to the air. As an example, the soluble

Hg(II) may be reduced directly by microorganisms to the volatile Hg(0) (Robinson and Touvinen, 1984). Another reaction catalyzed by microorganisms is the methylation that also usually leads to a volatilization of metals (Gadd, 1993). The volatilization by methylation was shown for selenium, which was transformed to dimethyl selenide and dimethyl diselenide (McCarty et al., 1995), and arsenic (Bachofen et al., 1995). However, the methylation of metals can greatly enhance their toxicity to humans (Bolton and Gorby, 1995).

Other captors

A first chemical approach is to introduce artificial resins into the soil. These resins may function as ion exchangers or as specific chelators. There is little experience with such applications, and the ideal form of such resins to allow easy removal has yet to be found.

Besides artificial resins, natural products or byproducts may also be able to bind specific heavy metals in the soil. Such products might have the form of a bulb, which could be incorporated into soil and harvested with standard methods used in agriculture. Attempts are being made at our institute to prepare such bulbs made of artificial or natural exchangers. Exchangers needed to fulfill criteria, such as stability during application and harvesting, specificity for heavy metals, have no side effects on soil functionality and soil organisms. Experiments are under way with several artificial and natural exchangers to investigate their heavy metal-binding capacity in the presence of metals and known salt concentrations.

Harvesting of Metal-Loaded Captors

Harvesting strongly depends on the type of captor. Principally, the disturbance of soil structure during harvesting should be as low as possible. The use of standard agricultural harvesting methods would render the whole decontamination process less expensive, especially for larger areas.

Harvesting of phytoremediation plants will fulfill the criterion of standard agricultural methods if most of the captured heavy metal is located in above-ground plant parts. The removal of capturing bulbs or other structures consisting of artificial or natural chelating agents from soil is not yet sufficiently developed. In the development of such new methods, harvesting should be taken into account when these structures are incorporated into the soil. Harvesting methods are available for plants with edible parts below the ground (e.g., potatoes).

GENTLE REMEDIATION: CHANCE OR UTOPIA

As defined previously, gentle techniques involve the *in situ* remediation of heavy metal-polluted soils by a combination of various biophysical and chemical treatments. The concept of gentle remediation does not mean a complete decontamination of a soil, but its aim is to reduce the risk of contamination by balancing between ecological and economical needs. Gentle remediation does not consist only of a single treatment, but of a series of different individual steps that may be optimally adapted to the actual situation at a polluted site.

To date, few worldwide studies have been conducted on gentle remediation techniques under large-scale field conditions. The stabilization of heavy metal-polluted soils by liming or by the addition of different binding substances has been established in several field trials (Vangronsveld et al., 1995; Krebs et al., 1999). The capture and the resulting removal of heavy metals from soil is still at an experimental phase. Several experiments were made with plants for phytoextraction, but only a few on a large scale under field conditions. Furthermore, plants that are able to accumulate high concentrations of heavy metals while possessing a high biomass yield are still lacking. Besides phytoextraction, other captors mentioned in this article are in the conceptual state.

In the future, several steps of the decontamination strategy (see Figure 17.4) will be improved and new procedures will be proposed. A scenario for possible improvement of existing techniques is shown in Table 17.2. First, the time required for a fivefold decrease of the heavy metal concentrations in soil with help of existing crop plants is calculated. The time for zinc is between 200 and 250 years, and for cadmium between 36 and 40 years. The total uptake of heavy metals by crop plants may be increased by enhancing the yield either by breeding or by improved crop production techniques. With an increase in yield of 25%, the remediation time may be reduced to 160 to 200 years for zinc and 24 to 32 years for cadmium. To extend the total uptake further, it is also important to increase the heavy metal concentration in plant tissue. This might either be achieved by approaches to enhance the uptake capacity of the plants or by increasing the availability of heavy metals in the soil by mobilization. By the latter method, a reduction to 32 to 46 years is possible in the case of zinc. For cadmium, the remediation time may even be reduced to 3 to 8 years.

TABLE 17.2

Extraction Capabilities of Maize and Tobacco in Soils Contaminated by Zinc or Cadmium

		High Yield Varieties[a]		Optimized Yield[b] (+25%)		Optimized Yield Heavy Metals Mobilized[c]	
		Zinc	Cadmium	Zinc	Cadmium	Zinc	Cadmium
Tobacco	Yield [t/ha]	16	16	20	20	20	20
	Content [kg/t DM]	0.3	0.011	0.3	0.011	1.3	0.117
	Total Uptake [kg/ha y]	4.8	0.18	6 0	0.22	26.0	2.34
	Reduction Time [y]	**250**	**36**	**200**	**29**	**46**	**3**
Maize	Yield [t/ha]	20	20	25	25	25	25
	Content [kg/t DM]	0.3	0.008	0.3	0.008	1.5	0.033
	Total Uptake [kg/ha y]	6.0	0.16	7.5	0.20	38	0.83
	Reduction Time [y]	**200**	**40**	**160**	**32**	**32**	**8**

Note: The heavy metal concentrations in the soil are assumed fivefold the guide value, that should be reduced to the guide value (750 to 150 ppm for Zn and 4 to 0.8 ppm for Cd). For the calculations, soil bulk density was assumed to be 1 g cm^{-3}. Soil weight per hectare with a depth of 20 cm will therefore be

$$1.0 \text{ g cm}^{-3} \cdot \frac{20 \text{ cm} \cdot (100 \text{ cm m}^{-1})^2 \cdot 10,000 \text{ m}^2 \text{ ha}^{-1}}{1000 \text{ g kg}^{-1}} = 2 \text{ million kg ha}^{-1}.$$

[a] Varieties that have a high yield capacity under good growth conditions.
[b] Assuming that the yield can be increased 25% by breeding and optimal growth conditions
[c] The soluble heavy metal fraction is maintained at the highest possible level.

REFERENCES

Alloway, J.B. and A.P. Jackson. The behavior of heavy metals in sewage sludge-amended soils. *Sci. Total Environ.* 100, 151-176, 1991.

Bachofen, R., L. Birch, U. Buchs, P. Ferloni, I. Flynn, G. Jud, H. Tahedi, and T.G. Chasten, Volatilization of arsenic compounds by microorganisms, in *Bioremediation of Inorganics*, Hinchee, R.E., J.L. Means, and D.R. Burris, Eds., Battelle Press, OH, 1995.

Balmer, M. and B. Kulli, Der Einfluss von NTA auf die Zink- und Kupferaufnahme durch Lattich und Raigras. Diploma work, Institute of terrestrial ecology, ETH Zürich, 1994.

Bollag, J.-M. and W.B. Bollag, Soil contamination and the feasibility of biological remediation, in *Bioremediation: Science and Applications*, Skipper, H.D. and R.F. Turco, Eds., Soil Science Society of America, Madison, WI, 1995.

Bolton, H. and Y.A. Gorby, An overview of the bioremediation of inorganic contaminants, in *Bioremediation of Inorganics*, Hinchee, R.E., J.L. Means, and D.R. Burris, Eds., Battelle Press, OH, 1995.

Brierley, C.L., Bioremediation of metal-contaminated surface and groundwaters. *Geomicrobiol. J.* 8, 201-223, 1990.

Brown, S.L., R.L. Chaney, J.S. Angle, and A.J.M. Baker. Zinc and cadmium uptake by hyperaccumulator *Thlaspi caerulescens* grown in nutrient solution. *Soil Sci. Soc. Am. J.* 59, 125-133, 1995a.

Brown, S.L., R.L. Chaney, J.S. Angle, and A.J.M. Baker. Zinc and cadmium uptake by hyperaccumulator *Thlaspi caerulescens* and metal tolerant *Silene vulgaris* grown on sludge-amended soils. *Environ. Sci. Technol.* 29, 1581-1585, 1995b.

Conner, J.R., Chemical stabilization of contaminated soils, in *Hazardous Waste Site Soil Remediation*, Wilson, D.J. and A.N. Clarke, Eds., Marcel Dekker, New York, 1994.

Cunningham, S.D., W.R. Berti, and J.W. Huang. Phytoremediation of contaminated soils. *Trends Biotechnol.* 13, 393-397, 1995a.

Cunningham, S.D. and C.R. Lee, Phytoremediation: plant-based remediation of contaminated soils and sediments, in *Bioremediation: Science and Applications*, Skipper, H.D. and R.F. Turco, Eds., Soil Science Society of America, Madison, WI, 1995b.

Cunningham, S.D. and D.W. Ow. Promises and prospects of phytoremediation. *Plant Physiol.* 110, 715-719, 1996.

Czupyrna, G., R.D. Levy, A.J. MacLean, and H. Gold, *In situ Immobilization of Heavy-Metal Contaminated Soil*. Noyes Data Corporation, NJ, 1989.

Didier, V., M. Mench, A. Gomez, A. Manceau, D. Tinet, and Ch. Juste, Réhabilitation de sols pollués par le cadmium, évaluation de l'efficacité d'amendements minéraux pour diminuer la biodisponibilité du cadmium. *Contes Rendues Acad. Sci.* 316, Série III, 83-88, 1993.

Elmayan, T. and M. Tepfar. Synthesis of a bifunctional metallothionein/beta-glucuronidase fusion protein in transgenic tobacco plants as a means of reducing leaf cadmium levels. *Plant J.* 6, 433-440, 1994.

Farmer, G.H., D.M. Updegraff, P.M. Radehaus, and E.R. Bates, Metal removal and sulfate reduction in low-sulfate mine drainage, in *Bioremediation of Inorganics*, Hinchee, R.E., J.L. Means, and D.R. Burris, Eds., Battelle Press, OH, 1995.

Frankenberger, W.T. and M.E. Losi, Applications of bioremediation in the cleanup of heavy metals and metalloids, in *Bioremediation: Science and Applications*, Skipper, H.D. and R.F. Turco, Eds., Soil Science Society of America, Madison, WI, 1995.

Frink, C.R., A perspective on metals in soils. *J. Soil Contam.* 5, 329-359, 1996.

Gadd, G.M., Microbial formation and transformation of organometallic and organometalloid compounds. *FEMS Microbiol. Rev.* 11, 297-316, 1993.

Gabriel, P.F., Innovative technologies for contaminated site remediation: focus on bioremediation. *J. Air Waste Manage. Assoc.* 41, 1657-1660, 1991.

Greinert, A., Clay as substances limiting phytotoxic influence of Pb, Zn, and Cd in sandy soils, in *Contaminated Soil '95*, van den Brink, W.J., R. Bosman, and F. Arendt, Eds., Kluwer Academic Publishers, The Netherlands, 1995.

Gupta, S.K., Mobilizable metal in anthropogenic contaminated soils and its ecological significance, in *Impact of Heavy Metals on the Environment*, Vernet, J.P., Ed., Elsevier, Amsterdam, 1992.

Gupta, S.K., M.K. Vollmer, and R. Krebs. The importance of mobile, mobilisable and pseudo total heavy metal fractions in soil for three-level risk assessment and risk management. *Sci. Total Environ.* 178, 11-20, 1996.

Hamby, D.M.. Site remediation techniques supporting environmental restoration activities — a review. *Sci. Total Environ.* 191(3), 203-224, 1996.

Hasselbach, G. and E. von Boguslawski, Bodenspezifische Einflüsse auf die Schwermetalllaufnahme der Pflanzen, in *Auswirkungen von Siedlungsabfällen auf Böden, Bodenorganismen und Pflanzen*, Sauerbeck, D. and S. Lübben, Eds., Berichte aus der ökologischen Forschung, Band 6, 1991.

Hattori, J., H. Labbe, and B.L. Miki. Construction and expression of a metallothionein-beta-glucuronidase gene fusion. *Genome* 37, 508-512, 1994.

Herms, U. and G. Brümmer, Einfluss der Bodenreaktion auf Löslichkeit un tolerierbare Gesamtgehalte an Nickel, Kupfer, Zink, Cadmium und Blei in Böden und kompostierten Siedlungsabfällen. *Landwirtsch. Forsch.* 33, 408-423, 1980.

Jorgensen, S.E., Removal of heavy metals from compost and soil by ecotechnological methods. *Ecol. Eng.* 2, 89-100, 1993.

Krebs, R., S.K. Gupta, G. Furrer, and R. Schulin, Solubility and plant uptake of metals with and without liming sludge applied soils. *J. Environ. Qual.* 27, 18-23, 1998.

Krebs, R., S.K. Gupta, G. Furrer, and R. Schulin, Gravel sludge as binding additive in soils polluted with zinc, copper, and cadmium — a field study. *Water Air Soil Pollut.* 1999. In press.

Lothenbach, B., R. Krebs, G. Furrer, S.K. Gupta, and R. Schulin, Heavy metal immobilization in soil by addition of montmorillonite, Al-montmorillonite, and gravel sludge: batch and pot experiments. *Eur. J. Soil Sci.*, 49, 141-148, 1998.

Martell, A.E. and R.M.Smith, *Critical Stability Constants*, 2nd ed., Plenum Press, New York, 1989.

McCarty, S.L., T.G. Chasteen, V. Stalder, and R. Bachofen, Bacterial bioremediation of selenium oxyanions using a dynamic flow bioreactor and headspace analysis, in *Bioremediation of Inorganics*, Hinchee, R.E., J.L. Means, and D.R. Burris, Eds., Battelle Press, OH, 1995.

Morel, J.L., G. Bitton, C. Schwartz, and M. Schiavon, Bioremediation of soils and waters contaminated with micropollutants: which role for plants?, in *Ecotoxicology: Responses, Biomarkers and Risk Assessment, an OECD Workshop*, Zelikoff, J.T., Ed., SOS Publications, Fair Haven, NJ, 1997.

Ow, D.W., Phytochelatin-mediated cadmium tolerance in Schizosaccharomyces pombe. *In Vitro Cell Dev. Biol.* 29P. 213-219, 1993.

Pan, A., F. Tie, Z. Duau, M. Yang, Z. Wang, L. Li, Z. Chen, and B. Ru, Alpha-domain of human metallothionein I-A can bind to metals in transgenic tobacco plants. *Mol. Gen. Genet.* 242, 666-674, 1994.

Phillips, I. and L. Chapple, Assessment of a heavy metals-contaminated site using sequential extraction, TCLP, and risk assessment techniques. *J. Soil Contam.* 4, 311-325, 1995.

Robinson, J.B. and O.H. Touvinen. Mechanisms of microbial resistance and detoxification of mercury and organomercury compounds: Physiological, biochemical, and genetic analyses. *Microbiol. Rev.* 48, 95-124, 1984.

Salt, D.E., M. Blaylock, N.P.B.A. Kumar, V. Dushenkov, B.D. Ensley, I. Chet, and I. Raskin. Phytoremediation: a novel strategy for the removal of toxic metals from the environment using plants. *Biotechnology* 13, 468-474, 1995.

Sims, J.T., S.D. Cunningham, and M.E. Sumner. Assessing soil quality for environmental purposes: roles and challenges for soil scientists. *J. Environ. Qual.* 26, 20-25, 1997.

StoV, *Verordnung über umweltgefährdende Stoffe*. Verordnung des Schweiz. Bundesrates, Eidg. Drucksachen und Materialzentrale, Bern, Switzerland, SR 814.015, 1986.

Summers, A.O., The hard stuff: metals in bioremediation. *Curr. Opin. Biotechnol.* 3, 271-276, 1992.

322 Phytoremediation of Contaminated Soil and Water

Tadesse, B., J.D. Donaldson, and S.M Grimes. Contaminated and polluted land: a general review of decontamination management and control. *J. Chem. Technol. Biotechnol.* 60, 227-240, 1994.

Tuovinen, O.H., Biological fundamentals of mineral leaching processes, in *Microbial Mineral Recovery*, Ehrlich, H.L. and C.L. Brierly, Eds., McGraw-Hill, New York, 1990.

van Bladel, R., H. Halen, and P. Cloos, Calcium-zinc and calcium-cadmium exchange in suspensions of various types of clays. *Clay Minerals* 28, 33-38, 1993.

Vangronsveld, J., F. Van Assche, and H. Clijsters. Reclamation of a bare industrial area contaminated by non-ferrous metals: *in situ* metal immobilization and revegetation. *Environ. Pollut.* 87, 51-59, 1995.

VBBO, Swiss Ordinance Relating to Impacts on the Soil, Verordnung des Schweiz. Bundesrates, Eidg. Drucksachen und Materialzentrale, Bern, Switzerland, SR 814.12, 1998.

Volesky, B., Removal and recovery of heavy metals by biosorbtion, in *Biosorption of Heavy Metals*, B. Volesky, Ed., CRC Press, Boca Raton, FL, 1990.

Wallace, A., R.T. Mueller, and G. Alexander, Effects of high levels of NTA on metal uptake of plants grown in soils. *Agron. J.* 66, 707-708, 1974.

Wenger, K. and S.K. Gupta, Kann eine kontrollierte Mobilisierung die Phytoextraktion von Schwermetallen wesentlich erhöhen? *Bull. BGS 21*, 91-96, 1997a.

Wenger, K., T. Hari, S.K. Gupta, R. Krebs, R. Rammelt, and C.D. Leumann, Possible approaches for *in situ* restoration of soils contaminated by zinc. *ISCO – Proc.*, 1997b, in press.

18 In Situ Metal Immobilization and Phytostabilization of Contaminated Soils

M. Mench, J. Vangronsveld, H. Clijsters,
N. W. Lepp, and R. Edwards

CONTENTS

1-56670-450-2/00/$0 00+$.50
© 2000 by CRC Press LLC

INTRODUCTION

Severe anthropogenic contamination of surface soils by trace elements can occur after several decades of metal input from many different sources, including industries such as smelters and foundries without efficient emission controls, derelict mine sites, urban areas, combustion of urban refuse, waste dumping, dredging of water courses, or organic wastes. Several nonessential elements such as Cd, Pb, Hg, and As, and some essential ones such as Zn and Cu, are generally involved; contamination may also include organic pollutants. Both rural and urban sites can be contaminated by trace elements. Consequently, serious problems may arise for agriculture, domestic horticulture, and adjacent natural ecosystems. Problems of metal phytotoxicity or adverse effects on other compartments of ecosystems do not necessarily imply a high total amount of metal in soil, especially when the input mainly consists of soluble forms (Chlopecka and Adriano, 1996). Adjacent to heavy metal point sources, elevated concentrations of nonferrous metals in the upper soil horizons can be strongly phytotoxic. Natural vegetation completely disappears, and the establishment of a new vegetation may be impossible.

Such bare unvegetated areas occur in many parts of Europe and North America. Apart from observable adverse effects on vegetation cover and other ecosystem components, immediate dangers to adjacent human populations, due to dust ingestion and inhalation and exposure via the food chain, may present additional hazards. In France, policy is directed to rehabilitate the 2000 most polluted sites as soon as possible, and site surveys have shown that most of them contain elevated soil Pb, Zn, and Cr contents (Ministère Environnement, 1994). Metal contamination can occur on a large scale. In northeast Belgium, the surface soil of more than 280 km^2 contains elevated (i.e., above highest background values) levels of Cd, Zn, and Pb due to the activity of four (predominantly pyrometallurgical) Zn smelters over the last 100 years. The industrial legacy of contaminated soils in the U.K. has arisen from base metal mining (Wales, Derbyshire, N. Pennines) and metal refining and smelting, e.g., Zn smelting at Avonmouth (Martin and Bullock, 1994) and Cu refining at Prescot (Dickinson et al., 1996). The widespread application of sewage sludge to agricultural land, coupled with a broad spectrum of industrial emissions, have resulted in significant soil metal pollution in many parts of the country. In certain areas of eastern Europe and the newly independent states, soil metal contamination exists on an enormous scale. Although many contaminated sites have been found to require remedial action, the extent of metal-contaminated soils is not fully documented. It is widely accepted that trace element-contaminated sites have to be monitored as a safety measure and rehabilitated as completely as possible.

In several countries, actual and potential risk are evaluated in order to decide the necessity/urgency of rehabilitation. This means that sites with the highest risks for human health and the natural environment will receive highest priority. Sites with very high total metal concentration but only a limited fraction of mobile (bioavailable) metals and limited environmental risks are considered to have fewer problems.

The remediation of metal-polluted soils requires the assessment of current and future remediation alternatives. Soil remediation is defined here as a set of techniques for reducing the mobile and, in consequence, bioavailable fraction of contaminants in soils with the object of minimizing their transfer into food chains and groundwaters. Different strategies can be adopted. The final choice is a function of the nature and degree of pollution, of the desired end use of the redeveloped area, and technical and financial considerations. Environmental, legal, geographical, and social factors further determine the choice of remediation technique.

Remediation can be achieved either by removal of the heavy metals (cleanup) or by preventing their spread to surrounding soil and groundwater (isolation and/or immobilization). Over the last decade, considerable attention has focused on techniques to decontaminate excavated soils. The most important cleanup techniques currently available for the treatment of metal-polluted soils are excavation (followed by disposal at a controlled disposal site), encapsulation and covering with clean soil, *ex situ* or *in situ* extraction, wet separation (by means of flotation or hydrocyclonic techniques) of excavated soil, thermal treatment (e.g., evaporation of mercury), *in situ* extraction of soil, and electroreclamation. Some other interesting techniques for the removal of excess metals from soils are still in an early stage of development: the bioleaching and bioextraction of metals from soils using soil bacteria and the extraction of metals using hyperaccumulating plants. Soil removal or encapsulation is prohibitively expensive. At the site of a former Zn smelter and sulfuric acid production unit in Belgium (15 ha of Zn, Cd, As, Pb contamination, Dilsen), waste and soil encapsulation cost $6.7 million. Thus more cost-effective technologies that can satisfy compliance requirements, particularly those geared to restore soil quality and protect human health, are highly desirable. Speciation in soil is a key factor for understanding the ecotoxicology of trace elements. Generally, plant uptake and mesofauna exposure parallel the available fractions of elements in soils.

The use of phytoremediation techniques, including the use of metal immobilizing soil additives which can be classified as "soft" or "gentle" approaches for soil remediation, shows some promise. There are two potential methods available for metal-polluted soils, both of which are designed to reduce the size of the bioavailable soil metal pool: (1) phytostabilization, or *in situ* metal immobilization by means of revegetation, either with or without nontoxic metal-binding or fertilizing soil amendments (Czupyrna et al., 1989) and (2) phytoextraction (metal bioextraction by means of hyperaccumulating plants). *In situ* metal immobilization (or metal inactivation) has the potential for slightly polluted soils in order to reduce metal uptake by (crop) plants, reducing metal transfer to higher trophic levels.

For heavily contaminated bare sites, soil application of strong immobilizing agents and subsequent revegetation of the area can be an efficient and cost-effective alternative remediation method, especially for agricultural soils, kitchen gardens, large former industrial sites, and dumping grounds. Effective and durable immobilization of metals reduces leaching and bioavailability. Subsequently, vegetation can develop which stabilizes the soil. Besides the aesthetic profit, vegetation cover provides pollution control and soil stability. Lateral wind erosion is completely prevented and a beneficial effect on metal percolation is evident (Mench et al., 1994a; Sappin-Didier, 1995; Vangronsveld et al., 1995b).

The aim of this chapter is to summarize the state-of-the-art concerning *in situ* immobilization and to highlight special advantages and specific problems related to this technique. Immobilization is not a technology for cleaning up contaminated soil but for stabilizing (inactivating) trace elements that are potentially toxic. This should lead to an attenuation of their impact on site and to adjacent ecosystems.

GENERAL PRINCIPLES OF *IN SITU* IMMOBILIZATION OF TRACE ELEMENTS IN CONTAMINATED SOILS

There are three main objectives for successful *in situ* immobilization: (1) to stabilize the vegetation cover and limit trace element uptake by crops, (2) to change the trace element speciation in the soil and thus minimize the possibility of surface and groundwater contamination, and (3) to reduce the direct exposure of soil organisms and enhance biodiversity. Direct human exposure should also be assessed. Restoration of vegetation cover may inhibit lateral wind erosion, reduce trace element percolation, and enhance biogeochemical cycles.

Sorption, ion exchange, and precipitation can be used to convert soluble and preexisting potentially soluble solid phase forms to more geochemically stable solid phases, reducing the metal pool for root uptake. Previous and projected uses of the soil should be taken into account when considering treatment options. Sorption on a mineral surface may result from various mechanisms (for a review, see Manceau et al., 1992a,b; Charlet and Manceau,1993; Hargé, 1997). Sorbent ions can form either an outer or inner sphere surface complex with the surface reactive groups. When the inner sphere complex involves sorbed polymers, surface nucleation and subsequent precipitation eventually occur. When the sorbed metal ion is found within the sorbent matrix, lattice diffusion and/or coprecipitation may have occurred. These processes determine the probable chemical status of trace elements, their solubility and, as a consequence, their behavior within and impact upon the natural environment.

The amount of trace elements sorbed on a solid phase is primarily dependent on three parameters: the nature of the solid, the pH level, and the concentration ratio between the sorbed element and the ligand (Kabata-Pendias and Pendias, 1992). One must also account for metal type, ionic strength, and competing ions. In most cases, reduction of root exposure to a trace element will depend upon a decrease in its concentration in the soil solution and on the reaction of the most chemically and/or biologically labile solid forms following the soil treatment. The choice of additive can be based on total element concentration, knowledge of the physico-chemical character of the soil, and appraisal of potential site end-use. However, it is useful to determine the speciation of elements by physical techniques, such as Extended X-ray Absorption Fine Structure (EXAFS) and x-ray diffraction, and to combine this information with the behavior of trace elements in plants and their interactions with macro- and micronutrients. Today, the predominant method is to use insoluble chemicals that are spread, and then tilled or mixed in the topsoil. Inorganic materials which create permanent charges and induce less reversible chemical bindings are highly interesting. Many natural or synthetic materials have been screened in batch experiments for their ability to decrease trace element mobility and phytoavailability, e.g., aluminosilicates (zeolites, beringite, clays; Gworek, 1992; Chlopecka and Adriano, 1997; Rebedea and Lepp, 1994; Vangronsveld et al., 1990, 1991, 1993, 1995a,b, 1996a,b; Vangronsveld and Clijsters, 1992; Krebs-Hartmann, 1997), iron and manganese oxides and hydrous oxides (Didier et al., 1992; Mench et al., 1994a,b; Manceau et al., 1997; Sappin-Didier, 1995), phosphates (Mench et al., 1994a,b; Xu and Schwartz, 1994; Sappin-Didier, 1995; Laperche et al., 1996; Chlopecka and Adriano, 1997), and lime (Didier et al., 1992; Mench et al., 1994a; Sappin-Didier, 1995).

The effectiveness of these ameliorants has been assessed in several different ways: by changes in chemical parameters such as exchangeable metal fraction and biological parameters such as plant growth and dry-matter yield, and plant metabolism; by ecotoxicological assays (Boularbah et al., 1996); and by structure and function of microbial populations. Metal mobility in soil is characterized in this chapter by a distribution coefficient (Kd) defined as the ratio of the metal concentration in the solution to that in the solid phase at equilibrium. For all soils presented in Figure 18.1, the Kd values for either Cd or Zn were calculated by dividing the metal concentration in the 0.1 M calcium nitrate-extractable fraction by total metal content. Low values of Kd indicate high metal retention by the solid phase through sorption reactions, hence low potential availability for plant uptake.

When screening soil amendment materials in batch and pot experiments, there is no consensus on standard methods to rank treatment effectiveness. Soil amendments may change element availability by direct surface reaction, pH effects, or by a combination of both. Changes in soil properties (e.g., pH, specific surface, adsorption capacity), the amount of sorbed element by either ameliorant weight or total element content, as well as element concentration in extractable fraction vs. amount of ameliorant added in the soil are possible tools. Studies by Chlopecka and Adriano (1996) and Mench et al. (1997) have shown that the addition of ameliorants can often induce pH changes in the soil, affecting speciation of metals such as Zn, which in turn influences their uptake by plants. For enhanced Zn immobilization, final soil pH should be above 6.5. The effectiveness of ameliorants for reducing metal mobility, generally based on changes in soluble and exchangeable fractions, will also depend on initial element speciation in the unamended soil. Lime, apatite, and zeolite appear less effective when the exchangeable Zn fraction increases in the soil (Chlopecka and Adriano, 1996). In pot experiments, Mn oxides, Fe-bearing amendments such as steel shots, beringite, and hydroxyapatite, consistently outranked all the other additives. Promising results on Cu and Cd immobilization were obtained with a synthetic zeolite (Rebedea and Lepp, 1994), but comparative studies of this material with other amendments have not been made.

In this chapter, the percentage of material added into the soil is based on soil dry weight. Materials used in several comparable studies are summarized in Table 18.1. All metal contents in soils are expressed on dry weight basis. Studies reported were generally carried out in pot experiments, unless stated to the contrary.

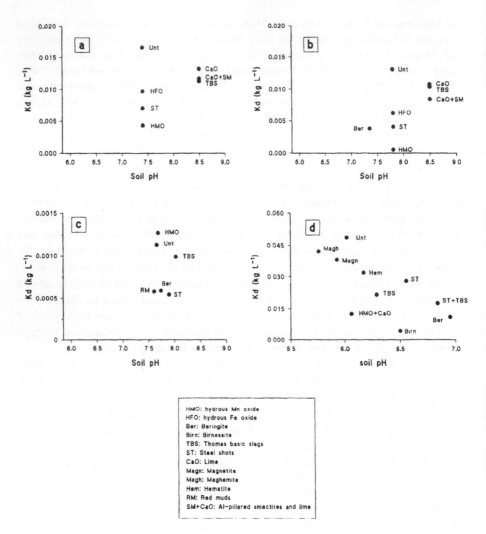

FIGURE 18.1 (a-d): Changes in Kd_{Cd} (0.1 M Ca(NO$_3$)$_2$ — extractable Cd vs. total soil Cd content) in relation to soil pH as a consequence of the incorporation of selected ameliorating agents into four metal-contaminated soils: Soil A, Louis Farges; Soil B, Evin; Soil C, Seclin; and Soil D, Ambares.

SOIL AMENDMENTS AND THEIR EFFECTIVENESS AS IMMOBILIZING AGENTS

Mn and Fe Oxides

Structure of hydrous Fe and Mn oxides

Hydrous ferric and manganese oxides have coherent small-sized scattering domains composed of mixed cubic and hexagonal anionic packing, where each pair of the anionic layer contains, on average, the same number of cations (Manceau et al., 1992b; Charlet and Manceau, 1993). At least five distinct local structures (~0.5 – 1

TABLE 18.1

Summary of the Additional Rates (% Added by Soil Weight) of Amendments Used for Treating the French Metal-Contaminated Soils in Pot Experiments

Soil Ameliorants	Seclin	Ambares	Evin	Louis Fargues	Mortagne du Nord
Lime		0.06	0 05	0.06	
Basic slags	0 25	0 02	0.028	0.02	
Al-pillared smectites			1 (+0 05 lime)	1 (+0.06 lime)	
HMO	0 5	1 (+0.05 lime)	1 (or 0.5)1	1	
HFO			1	1	
Steel shots	1	1 (±0 02 basic slags)	1	1	1 (±5 beringite)
Beringite	5	5	5		5
Maghemite		1			
Magnetite		1			
Hematite		1			
Birnessite		1			
Red muds	5		5		

nm) have been reported for hydrous Fe oxides, i.e., ferric gels with a lepidocrocite-like local structure, so-called "2-line" gels that possess either a goethite-like (αFeOOH) or akaganeite-like (βFeOOH) local structure and that, either aged at neutral pH or heated, converted into a feroxyhite-like (δFeOOH) form, followed by further transformations into hematite. The feroxyhite structure is similar to that of hematite, but shows octahedral vacancies and layer defaults. "2-line ferrihydrite" (HFO) was described as a mosaic of single and double octahedral chains of varying length, ranging from 1 – n octahedra, linked at the corners of the chains (Spadini et al., 1994). The large number of high affinity-free edges found in HFO results from the extreme shortening of these octahedral chains. In contrast, the local structure of hydrous Mn oxides (HMO) does not seem to be related to that of a well-crystallized MnO_2 polymorph (e.g., pyrolusite, ramsdellite, todorokite, chalcophanite). A three-dimensinal framework of randomly distributed edge- and corner-sharing MnO_2 octahedra is the most probable structure.

Trace element sorption on hydrous Fe and Mn oxides

The hydroxyl groups of the hydrous oxides form an ideal template for bridging trace metals because the OH-OH distance matches well with the coordination polyhedra of trace metals (Manceau et al., 1992a,b; Charlet and Manceau, 1993; Spadini et al., 1994; Hargé, 1997). $As_2O_4^{2-}$ and Pb^{2+} form isolated innersphere surface complexes with ferrihydrite (HFO), while Cu^{2+} forms similar complexes on MnO_2. Zn^{2+}, Cd^{2+}, and Pb^{2+} form similar mononuclear complexes on goethite and ferrihydrite surfaces (Manceau et al., 1992a; Spadini et al., 1994; Hargé, 1997). Pb^{2+} binds to

HMO at various surface sites: edge-, double-, corner-, and triple corner-polyhedra linkages being observed (Hargé, 1997). At a similar density of surface coverings, Pb(II) showed greater polymerization on the surface of Mn-dioxide birnessite than on HFO. The birnessite group of minerals are commonly occurring Mn oxides characterized by mixed Mn valency and disordered structures. Zn, Pb, and Cu form innersphere complexes with birnessite ($Na_4Mn_{14}O_{27} \cdot 9H_2O$) or birnessite-like structures (Manceau et al., 1997). In a sludged sandy soil, Zn was mainly bound at lattice vacancy sites of the phyllomanganate chalcophanite ($ZnMn_3O_7 \cdot 3H_2O$), whose structure shows similarities to birnessite (Manceau et al., 1997; Hargè, 1997).

Mn oxides

Birnessite and HMO were used as additives in metal-contaminated soils with differing physical properties and pollution profiles (Didier et al., 1992; Mench et al., 1994a,b; Sappin-Didier, 1995). Two sandy soils, Ambares and Louis Fargues, were obtained from a field trial with long-term sewage sludge application (INRA Couhins experimental farm, Bordeaux, France). The plots were established in 1974 (Ambares) and 1976 (Louis Fargues) and cultivated with maize. Ambares sludge has an elevated Zn (2914 mg kg^{-1}) and Mn (4916 mg kg^{-1}) content; amended soil contains 1080 mg Zn kg^{-1}, compared to 19 mg kg^{-1} in the control soil. The Louis Fargues plots were amended with a Cd/Ni rich sludge, but sludge applications ceased in 1980 due to problems of phytotoxicity. Several other soils were also investigated. Evin soil was collected from the vicinity of a nonferrous metal smelter. This is a limed, silty clay with elevated Zn (1434 mg kg^{-1}), Pb (1112 mg kg^{-1}), and Cd (18 mg kg^{-1}). Seclin soil was collected from an agricultural field that had received dredged sediments from the canalized River Seclin. This was contaminated with Zn (817 mg kg^{-1}), Pb (232 mg kg^{-1}), Cd (3.7 mg kg^{-1}), and Ni (150 mg kg^{-1}). Mortagne du Nord soil is an organic sandy soil from a former Pb/Zn smelter in the north of France; the characterization of this soil is in progress, but it is highly polluted by Pb, Zn, and Cd.

In the sludged soils, addition of Na-birnessite (1%) produced a significant decrease in Kd_{Cd} and Kd_{Zn} values (Figure 18.1d). The high sorption capacity of birnessite results from the replacement of Mn^{IV} by Mn^{III} and the presence of layer vacancies in the structure, creating a deficit of positive charges (Sylvester et al., 1997). When equilibrated in neutral to acidic conditions, Na-birnessite loses its exchange capacity and can absorb large amounts of metals. Metals form three Me-O-Mn bonds at the birnessite surface; this mechanism accounts for the high binding affinity, with low reversibility, reported in metal sorption experiments (Manceau et al., 1997; Sylvester et al., 1997). Shoot Cd and Zn uptake by dwarf beans and ryegrass was investigated using Ambares soil (Tables 18.2 and 18.3). The beans did not germinate in the untreated soil. Both birnessite and HMO combined with lime were effective in reducing Zn and Cd accumulation in aerial plant parts. Surprisingly, the effect of the birnessite addition on Zn availability did not persist beyond the third harvest of ryegrass (four months after initial amendment). Zn sorption may be affected by the roots being potted in a relatively small soil volume (1 L). In addition, roots of some plant species are able to release Mn from Mn oxides in the rhizosphere, and inorganic elements may also be recycled from root decomposition. Among soil treatments, the highest relative increase (four times) in shoot Mn uptake by ryegrass between the first and third harvests was found for the birnessite-treated soil. This may suggest the alteration of birnessite. Subsequent ryegrass cultures in this pot experiment show birnessite to be less effective, but further information must be gained from other contaminated soils, either using greater soil volumes or in field trials over an extended cropping period.

Hydrous Mn oxides can bind metals such as Pb or Cd even in acidic conditions (Manceau et al., 1992a,b). Their surface layers display permanent reactive sites and zero point charge values for HMO ranged from 1.5 to 2.0. Therefore, variable negative charges that may bind cations are also expected in most soils and increase with increasing pH. More Cd was bound to Mn oxides than to Fe oxides between pH 4.5 to 6.5 (Fu et al., 1991). The addition of HMO (1%) to a range of metal-polluted soils reduced levels of Cd in plant tissues, regardless of soil type or plant species (Table 18.2). Moreover, HMO showed the highest efficiency in reducing Cd availability to ryegrass shoots irrespective of soil type or time of harvest. Ryegrass responded more strongly to the HMO soil treatment than did tobacco (Sappin-Didier et al., 1997a). This may be a combined effect of soil properties, soil–root interactions, and plant metabolism. Reduction in Cd uptake partly related to Cd–Mn interactions during root uptake cannot be ruled out, and the effect of Cd on plant metabolism could be reversed by Mn application. For other metals such as Ni and Pb, the reduction of shoot metal uptake following HMO addition was mainly evident with ryegrass (Sappin-Didier et al., 1997a). HMO combined with lime was found to be more effective for immobilizing Pb (Mench et al., 1994a).

Fe oxides

Reactions between iron oxides and trace elements are well documented (Gerth and Brümmer, 1983; Kabata-Pendias and Pendias, 1992; Manceau et al., 1992a,b; Spadini et al., 1994). Electron-microprobe studies confirm that metals in contaminated soils accumulate in iron oxides (Hiller and Brümmer, 1995). Early studies were concerned with contaminated soils treated with either iron sulfate or iron oxides (Czupyrna et al., 1989; Förster et al., 1983; Juste and Solda, 1988; Didier et al., 1992), and initial tests showed iron oxide addition to the soil had some promise for trace metal immobilization (Didier et al., 1992; Sappin-Didier et al., 1997a,b). Soluble, calcium nitrate-exchangeable and EDTA fractions of Cd, Ni, and Zn in two contaminated soils decreased following a single application (1%) of HFO, but to a lesser extent than with HMO (Figure 18.1). However, HFO did not generally reduce shoot metal uptake (Tables 18.2 and 18.3). Only shoot Ni uptake by ryegrass decreased by 50% following the addition of HFO in a sludged soil compared to the untreated soil. Iron oxide treatment of As contaminated garden soils resulted in a 50% reduction of water-extractable As and a comparable decrease of As accumulation in dwarf bean leaves (Mench et al., 1998).

Fe- and Mn-bearing amendments

Release of either iron or manganese by inorganic and organic materials could be another method for application of Fe or Mn oxides to contaminated soils. This has been investigated using steel shots, an industrial material used for shaping metal surfaces that contains mainly iron (97% α-Fe) and native impurities such as Mn,

TABLE 18.2
Changes in Shoot Cd Uptake by Plants as a Result of a Single Application of Amendments in Several French Metal-Contaminated Soils

Soil	Seclin		Ambares	Evin			Louis Fargues	Mortagne du Nord
Reference	3	3	4	2	1	3	2	7
Total Cd	3.7		5.7	18			108	7
Soil pH	7.6		5.8	7.8			7.4	
Plant Species	**RG**	**B**	**B**	**RG**	**B**	**B**	**RG**	**M**
Soil treatments								
Untreated	0.92 a	0.029 d	ng	4.9 bc	1.8 a	0.48 a	16.2 ab	9.4 a
Alkaline materials								
Lime	1.0 a		ng	5.6 b			14.8 b	
Basic slags		0.036 bc	0.28 c	5.0 b		0.21 cd	12.3 c	
Aluminosilicates								
Al-smectites (+lime)				7.0 a			16.9 ab	
Beringite	0.65 b	0.046 b	0.12 d	(6.3)	1.1 b	0.27 bc		6.6 b
Red muds	0.56 b	0.035 c		(5.4)		0.28 b		
Fe and Mn oxides								
HMO	0.79 ab	0.023 d	0.26 c	1.2 e	0.4 c	0.31 b	6.9 d	
Birnessite			0.17 cd					
Steel shots	0.56 b	0.061 a	0.27 c	2.9 d	0.8 bc	0.16 d	9.1 d	6.3 b
Steel shots + basic slags			0.33 b					
HFO				4.3 bc			18.5 a	
Maghemite			0.61 a					
Magnetite			0.80 a					
Hematite			0.39 b					

Note: Plant species: RG (ryegrass, *Lolium perenne* L.); B (dwarf bean, *Phaseolus vulgaris* L.); M (maize, *Zea mays* L.); and ng — no germination. Within a column, mean values followed by the same letter are not statistically different ($p < 0.05$) – Newman-Keuls test.

[1]Mench et al. (1994). [2]Didier et al. (1992) and Sappin-Didier (1995). [3]Gomez et al. (1997). [4]Mench et al. (1997).

TABLE 18.3
Changes in Shoot Zn Uptake by Plants as a Result of a Single Application of Amendments in Several French Metal-Contaminated Soils

Soil	Seclin		Ambares		Evin			Louis Fargues	Mortagne du Nord
Reference	3	3	4	4	1		2	2	2
Total Zn	817		1074		1434			149	700
Soil pH	7.6		5.8		7.8			7.4	7
Plant species	RG	B	RG	B	RG	B	RG	RG	M
Soil treatments									
Untreated	182 a	30 b	246 d	ng	171 b	29	553 a	93 a	611 a
Alkaline materials									
Lime				ng	186 b			83 abc	
Basic slags	189 a	44 a	126 b	90 c	201 a. 189 b		508 b	75 bc. 71 c	
Aluminosilicates									
Al-smectites (+ lime)						39	371 c		502 b
Beringite	147 b	32 b	94 a	62 d			381 c		
Red muds	168 ab	31 b							
Fe and Mn oxides									
HMO	171 ab	30 b	141 c	96 c	105 c	35	483 b	57 d	
Birnessite			130 bc	82 c			148 d		
Steel shots	102 c	31 b	87 a	104 b	94 c	36		54 d	491 b
Steel shots + basic slags			92 a	93 bc					
HFO					195 ab			88 ab	
Maghemite			171 c	139 ab					
Magnetite			162 c	143 a					
Hematite			134 bc	138 ab					

Note: Plant species: RG (ryegrass, *Lolium perenne* L.); B (dwarf bean, *Phaseolus vulgaris* L.); M (maize, *Zea mays* L.); and ng — no germination. Within a column, mean values followed by the same letter are not statistically different (*p* <0.05) — Newman-Keuls test.

[1]Mench et al. (1994), [2]Sappin-Didier (1995), [3]Gomez et al. (1997), [4]Mench et al. (1997).

but very little Cd, Zn, and Ni. These corrode readily, oxidizing into several iron oxides (maghemite, magnetite, lepidocrocite) and Mn oxides in soils (Sappin-Didier, 1995). A bag experiment, with steel shots contained in Durapore filter membranes buried in soils, was used to study patterns of oxidation (Sappin-Didier et al., 1997b). All membranes recovered after 9 months burial showed extensive oxide deposition and elevated Fe and Mn concentrations. Iron and Mn may be released into the soil solution and diffuse, subsequently forming oxides in the soil. These may coat soil particles, creating a large reaction surface for trace element binding from the soil solution. However, the increase in cation exchange capacity in the Ambares soil, with and without 1% steel shot amendment, (8.4 to 9.3 cmol kg^{-1} [cobaltihexammine method]) was very limited (Mench et al., 1999).

Single applications of steel shots (average size 0.35 mm) have been made to contaminated soils and subsequent changes in trace element mobility and plant availability monitored. In a comparative trial, crystallized iron oxides such as maghemite, magnetite, lepidocrocite, and hematite were tested in addition to water-oxidized steel shots and stainless steel shots (Prolabo). The test soils were Ambares, Louis Fargues, Seclin, Evin, and Mortagne du Nord (Mench et al., 1994a,b; 1999; Didier et al., 1992; Gomez et al., 1997). Kd$_{Cd}$ values varied by a factor of 50, ranging from 0.00055 to 0.048 kg^{-1} l (Figure 18.1). In general, maximum values were found for the untreated soils, and amendment with steel shots decreased Cd mobility. Ameliorants such as HMO with or without lime, beringite, and birnessite were more efficient than steel shots. However, Mn oxides are not readily available and present application problems in the field, and the effective application rate of beringite is generally 3 to 5 times higher than that of steel shots.

Changes in Zn mobility have also been reported. In the Ambares soil, even though beringite (5%) and birnessite (1%) delivered by far the most pronounced reduction in Kd$_{Zn}$ (88 and 83%) compared to untreated soil, steel shots (1%) caused a 43% decrease in Kd$_{Zn}$ (Mench et al., 1999). Data dealing with the Louis Fargues soil demonstrated no differences in Kd$_{Zn}$ values between the 1 and 5% application rates of steel shots, whereas the greatest decrease occurred at 10% (Sappin-Didier, 1995). Application rates greater than 5% by weight can lead to problems with soil structure such as aggregate cementation and changes in porosity.

In all soils, steel shots decreased shoot Cd uptake by at least 40% relative to the untreated soil (Table 18.2). There were generally no significant differences in dry-matter yield between the treatments. Therefore, neither dilution effects due to changes in the biomass nor plant evapotranspiration are obvious explanations. In the Ambares soil, steel shots were more effective than Fe oxides for reducing Cd in the primary leaf of dwarf beans; the effect was similar to that induced by basic slags and Mn oxides such as HMO and birnessite. Beringite treatment also reduced foliar Cd concentrations to a greater extent than steel shots, but this occurred following a three- to fivefold greater application rate.

Steel shots (0.35 mm), either native (ST) or previously oxidized by a 15-min pretreatment in water (STO), produced decreased Cd and Zn concentrations in foliage of ryegrass cultivated in Louis Fargues soil, especially for the first and second harvests (Table 18.4). In contrast, Zn and Cd concentrations in ryegrass shoots were not affected by the addition of either lepidocrocite or stainless steel shots. This

TABLE 18.4

Changes in the Chemical Composition and Shoot Yield of Ryegrass Grown in a Pot Experiment with the Louis Fargues Soil Following the Application of Either Various Steel Shots or Lepidocrocite

Element	Cd		Zn		P		Total Dry-Matter Yield g DM pot^{-1}
			(mg kg^{-1})				
Harvest number	1	3	1	3	1	3	3
Soil treatments							
Untreated	17 a	17 a	92 a	90 b	59 a	65 abc	15.8 bc
Lepidocrocite	16 a	18 a	84 a	116 a	54 ab	74 a	16.1 abc
Steel shots (1%, 0.35 mm)	8 bc	10 b	52 b	75 bc	37 c	58 bc	15.8 bc
Steel shots (1%, 0.35 mm)	6 c	10 b	48 b	67 c	35 c	61 bc	12.5 d
Water-oxidized steel shots (1%, 0.35 mm)	10 b	10 b	54 b	70 bc	32 c	58 bc	17.7 ab
Stainless steel shots (1%, 0.3 mm)	15 a	15 a	82 a	84 bc	43 bc	75 a	18.4 ab
Steel shots (1%, 1.70 mm)	15 a	19 a	88 a	92 b	52 ab	67 ab	18.4 ab
Steel shots (1%, 2.36 mm)	14 a	15 a	86 a	80 bc	48 ab	70 ab	19.4 a

Note: Within a column, mean values followed by the same letter are not statistically different ($p < 0.05$) – Newman-Keuls test.

Source: From Sappin-Didier, V. 1995. Ph.D. thesis, Analytical Chemistry and Environment, Bordeaux I University, France.

demonstrates the importance of Fe and Mn release into the soil solution and perhaps the *in situ* crystallization of some Fe and Mn oxides for immobilizing mobile metals in contaminated soils. Water-oxidized steel shots were less effective than ST in reducing shoot Cd uptake in the first ryegrass cut, but no differences were found at subsequent harvests. Microscopic observations showed that the reaction time with water only led to a superficial oxidation of the steel shots. Therefore, it is probable that STO still released Fe and Mn to the soil solution when added to soil. Because the addition of crystallized iron oxides such as lepidocrocite, maghemite, and magnetite has not been shown to be successful in decreasing Cd and Zn mobility to a significant degree, changes induced by steel shots might be attributed to the presence of manganese. EXAFS indicates some Mn from steel shots was transformed to a birnessite-like phyllomanganate compound (Manceau et al., 1997), and birnessite addition into the soil changes Cd and Zn mobility (Mench et al., 1997). Thus, metal mobility and plant availability in steel shots-treated soils may be controlled by Mn oxides as in the HMO- and birnessite-treated soils. Still, steel shots have a polymetallic effect compared to HMO and were notably effective with copper. Therefore, additional mechanisms may also occur.

The particle size of steel shots and their addition rate to soil are significant factors for decreasing plant availability of Cd and Zn in contaminated soils (Sappin-Didier, 1995). Despite a similar chemical composition, steel shots with larger particle size were less effective in reducing shoot Cd and Zn uptake compared to the finest ones (Table 18.4). Indeed, 2% steel shots were more efficient than 1% for decreasing Cd and Zn uptake by ryegrass shoots; this reduction in metal availability must be balanced with increased costs for treating contaminated soils and could be a significant argument for reducing application rates to <1%. However, the calcium nitrate exchangeable fraction of Cd sharply decreased when the addition rate of steel shots increased from 0.01 to 1% (Sappin-Didier, 1995); at rates from 2 to 5% no significant differences in Cd mobility were found. Cd mobility was highly decreased at the 10 and 15% rates but soil texture was greatly affected. In addition, shoot phosphorus uptake by ryegrass was reduced by soil incorporationtion of steel shots (Table 18.4). This produced a decrease in dry-matter yield (20%) with 2% steel shots compared to untreated soil; the 1% rate had no effect.

Shoot Zn uptake in ryegrass can decrease by over 60% following a single application of steel shots (Table 18.3). Similar results were obtained for Cu, Ni, and Pb (Sappin-Didier et al., 1997a). In contrast to beringite and coal fly-ashes, the impact of steel shots on soil pH is less important, and thus it can be used in either alkaline- or lime-contaminated soils without a negative effect on the status of nutrients such as P, Mn, and Fe. Combination of alkaline materials such as basic slags with steel shots increase soil pH and may enhance metal sorption on Fe and Mn oxides initially present in the contaminated soil or newly formed following the addition of steel shots. Indeed, such combinations were more effective in reducing Zn and Cd mobility than when materials were used separately (Figure 18.1d). Combination may also limit possible Mn phytotoxicity in sensitive plant species. However, the combination of steel shots and basic slags did not display a synergistic effect in reducing the plant-available Cd and Zn pools in the Ambares soil (Tables 18.2 and 18.3).

Steel shots and ferrihydrite were also very effective for As immobilization in contaminated garden soils (Vangronsveld et al., 1994; Mench et al., 1998). Water-extractable arsenic decreased by 83 and 95%, respectively, following incorporation of these materials. Subsequent greenhouse and field experiments were performed with steel shots. In both cases, marked reductions in arsenic uptake were observed (Tables 18.5 and 18.6).

TABLE 18.5

Arsenic Contents (mg kg^{-1} fresh wt.) of Lettuce Foliage and Radish Tubers Cultivated in Soil from Various As-Polluted Gardens, Amended with 1% (by Soil Weight) of an As-Immobilizing Soil Additive

	Lettuce		Radish	
Garden	Untreated	Treated	Untreated	Treated
As 1	0 78	0 094	0 343	0.181
As 2	0.443	0 061	0 193	0.094
As 3	0 224	0.077	1 007	0 583
As 4	0 225	0,051	1 851	0.801
As 5	0.313	0 054	1 652	0.56
Control	0.02	N/A	0 15	N/A

Note: N/A — trial not carried out. Untreated soils are unamended garden soils and control is from a comparable garden soil without As contamination. Plant material grown under greenhouse conditions.

Other Fe-bearing materials are available. Pot experiments and a field trial were carried out within the French–German Cooperation Network on soil contaminated by trace elements. Five Fe-bearing materials were added (1% pure Fe in soil) to harbor dredgings from a settling basin (Bremen [Germany]; Müller and Pluquet, 1997); this contained Cd (4.2 to 7.1 mg kg^{-1}) and Zn (453 to 790 mg kg^{-1}). The Fe sources used were red mud from the aluminium industry, sludge from drinking water treatment, bog iron ore, native steel shot, and steel shot waste from descaling of untreated steel plate. Red mud and sludge from drinking water reduced the 1 M NH_4NO_3 extractable fractions of Cd and Zn by over 50%. The other treatments showed less impact on extractable metals, and steel shot waste caused a small increase in extractable Zn. All treatments caused a marked reduction (>30%) in Cd concentration in wheat grain and straw. The most effective treatments for decreasing grain Cd content were red mud, steel shot waste, and sludge from drinking water. All treatments produced lesser decreases in Zn uptake (10 to 15% in wheat grain). Red mud, sludge from drinking water treatment, and steel shot waste showed the best results for reducing Cd uptake by spinach (20 to 50%) and ryegrass (25 to 30%). Again, Zn content was less affected than Cd. In a field trial, Cd concentration in ryegrass was reduced by amendment with sludge from drinking water treatment

TABLE 18.6

**Arsenic Contents (mg kg⁻¹ fresh wt.)
of Vegetables Cultivated *in situ* in an
As-Polluted Garden Soil Amended
with 1% Steel Shots by Soil Weight**

	Untreated	Treated
Radish		
Garden 1	0.079	0.021
Garden 2	0.144	0.05
Control	0.005	N/A
Lettuce		
Garden 1	0.41	0 085
Garden 2	0.208	0.059
Control	0.012	N/A
Carrot		
Garden 1	0 092	0.023
Garden 2	0.072	0.018
Control	0.005	N/A
Potato		
Garden 1	0.019	0 005
Garden 2	0.033	0.009
Control	<0.001	N/A

Note: N/A — trial not carried out. Untreated soils
received no amendment Control data from a comparable garden soil not polluted with As. No
amendments were applied to the control soil.

(20%) and steel shot waste (30%), Zn concentrations fell by about 10%. Soil
treatments in the field trial were less effective than in the pot tests for reducing Cd
and Zn concentrations in both plants and soil extracts. Similar results were obtained
in pot experiments with two other German soils contaminated by either mining
effluents transported by a river or fallout from a former Pb/Zn smelter. Red mud
was also tested in pot experiments using French soils (Tables 18.2 and 18.3). In all
soils studied, both ryegrass and bean showed a decrease in shoot Cd and Zn. After
seven successive harvests, total Cd and Zn uptake by ryegrass was reduced by over
60% and 30% in Evin soil, and by 51% and 18% in the Seclin soil, amended with
1% Fe as red mud (Gomez et al., 1997). Native steel shots were only more effective
than red mud for immobilizing Cd and Zn in the Evin soil. X-ray diffraction analysis
of the German soils showed that steel shots application led to the appearance of
hematite, magnetite, and pure iron in soil samples (Müller and Pluquet, 1997,
personal communication). Steel shot waste contains hematite, magnetite, magnesio-
ferrite, wuestite, pure iron, and zinc. Precipitated sludge from drinking water may
be ferrhydrite. Moreover, changes in outer surface area occurred, from 10 to 15 m²

g^{-1}, following a single application of either sludge from drinking water or native steel shots (Müller and Pluquet, 1997, personnal communication). In batch experiments, red mud and sludge from drinking water were the most effective when the time effect of Cd-adsorption on Fe-bearing materials was determined (Müller and Pluquet, 1997).

Fe-rich (du Pont de Nemours™), a byproduct from the processing of TiO$_2$ pigment, was tested using a silt loam soil spiked with increasing quantities of flue dust, increasing Zn content from 30 to 2400 mg kg^{-1} (Chlopecka and Adriano, 1996). Fe-rich had a pH of 8.5 and a calcium carbonate equivalence of 33.5%. It contains poorly crystalline ferrhydrite, 31.7% Fe, 1.76% Mn, 10.3% Ca, and some metals (20 mg Cd, 1272 mg Cr, 655 mg Pb, 104 mg Ni, and 260 mg Zn per kg). Fe-rich (5%) decreased the concentration of the exchangeable form of Zn at each level of flue dust. The greatest decrease (>80%) occurred with the lowest flue dust dose, but the ameliorative effect was retained up to the highest dose rate. Compared to lime, natural zeolite and hydroxyapatite, Fe-rich was the most effective ameliorant in reducing the availability of Zn to maize, barley, and radish. Only Fe-rich enhanced growth of radish at all flue dust rates. The effectiveness of Fe-rich could have been partly due to its creation of alkaline conditions in amended soil, as well as its Fe–Mn fraction. Concomitant with the largest decrease of exchangeable Zn by Fe-rich were substantial increases in Zn associated with the Fe–Mn oxide and carbonate fractions. This may be indicative of their role as sorbents.

ALUMINOSILICATES

Clays, Al-Pillared Clays

Increasing rates of montmorillonite, Al-montmorillonite and gravel sludge, incorporation resulted in a decrease of NaNO$_3$-extractable Zn and Cd fractions accompanied by a reduction in Zn uptake in red clover (*Trifolium pratense*; Krebs-Hartmann, 1997; Lothenbach et al., 1998). A reduction in Cd uptake was produced only by gravel sludge. Al-montmorillonite and gravel sludge were the most effective binding agents for both metals; there was greater metal desorbtion (in a pH range from 4.0-5.5) from montmorillonite. However, at a lower pH (<4.5), montmorillonite was more effective for immobilization of Cd than Al-montmorillonite. Gravel sludge incorporation increased the acid buffering capacity of the contaminated soils. Macronutrient (Ca, Mg, K, and P) uptake in *Trifolium* was mainly unaffected by the amendments. In field studies, increasing rates of gravel sludge caused an increase in soil pH and a reduction in NaNO$_3$-extractable Zn and Cd fractions. These were accompanied by decreases in Zn, Cu, and Cd uptake in plants. However, in some instances, there appeared to be mobilization of Cu.

Zn and Pb can be adsorbed on a matrix containing clay and Al-hydroxide polymers on the clay surface. Al-pillared smectites were obtained by adding *in situ* aluminium polymers within the clay layers (Bergaya and Barrault, 1990). They were added at 1% in combination with lime to Evin and Louis Fargues soils (Didier et al., 1992). This produced a small decrease in Kd$_{Cd}$ (Figure 18.1a and b), but Al-pillared smectites were less effective than Fe and Mn oxides and steel shots. There was no significant decrease observed in shoot Cd uptake by ryegrass in both contaminated soils, and a decrease in shoot Zn uptake was only found in the Louis Fargues soil (Table 18.3). This effect could be due to increased soil pH from liming. In an arsenic-contaminated garden soil from Reppel (Belgium), the addition of 1% Al-smectite resulted in a 75% decrease in water-extractable arsenic and a 50% reduction of arsenic concentrations in test plants (Vangronsveld et al., unpublished results).

Beringite

Beringite is a modified aluminosilicate that originates from the fluidized bed burning of coal refuse (mine pile material) from a former coal mine (Beringen, northeast Belgium). The combusted material contains approximately 30% coal. The remaining fraction is inorganic and mainly consists of schists. Minerals present in the schists are quartz, illite, kaolinite, chlorite, calcite ($CaCO_3$), dolomite (($Ca,Mg)CO_3$), anhydrite ($CaSO_4$), siderite ($FeCO_3$), and pyrite (FeS_2; De Boodt, 1991). Illite is the dominant clay present. The schists are burned by heating in a fluidized bed oven at about 800∞C, undergoing partial breakdown and recrystallization during the process.

Most of the particles with a median diameter of less than 0.2 mm (clay fraction) are separated in a cyclone; these are the so-called cyclonic ashes, representing about 25% of the total ash fraction. The cyclonic ashes (mainly the modified clay fraction) were shown to possess a very high capacity for immobilizing several trace metals (De Boodt, 1991; Mench et al., 1994c; Vangronsveld et al., 1990, 1991, 1993, 1995a,b, 1996a,b; Vangronsveld and Clijsters, 1992). This is not surprising, since the minerals mentioned above are known to possess high sorption capacities. In terms of chemical composition, the product contains the same elements as the original schists; SiO_2 and $Al2O_3$ represent 52 and 30%, respectively, of the whole product. The high metal immobilizing capacity of beringite is supposed to be based on chemical precipitation, ion exchange, and crystal growth. The combination of these three mechanisms can explain its high metal sorption capacity (De Boodt, 1991). Recent results from laboratory simulations and field trials show that a modified aluminosilicate has a long-lasting effect on bioavailability and leaching of metals from both heavily contaminated industrial soils and garden soils (Vangronsveld et al., 1995a,b, 1996). Experiments are in progress to further elucidate the working mechanism of the product in the field.

Other aluminosilicates which originate from coal mine wastes (Elutrilite, Metir) have been shown to possess limited capacity for metal immobilization (Vangronsveld and Clijsters 1992).

Zeolites

Synthetic and natural zeolites have been investigated with respect to the reduction in uptake of Cu, Pb, Zn, and Cd (Gworek, 1992; Chlopecka and Adriano, 1997; Rebedea and Lepp, 1994). Zeolites are crystalline, hydrated aluminosilicates of alkali and alkaline earth cations that possess infinite three-dimensional crystal structure. Nearly 50 natural species of zeolites have been recognized, and more than 100 species have been synthesized in the laboratory (Chlopecka and Adriano, 1996). The use of zeolites for pollution control primarily depends on their ion exchange capabilities. These result from a substitution of Al^{3+} for Si^{4+} in the silicate tetrahedra, creating fixed negative charged sites throughout the structure. The negative charges are balanced by an equivalent number of mobile cations loosely bonded in the crystal structure and free to exchange with other cations in solution. The general formula of a zeolite is $(nMO \cdot nAl_2O_3 \cdot xSiO_2 \cdot yH_2O)$, where M is a divalent cation, x is greater than or equal to the number of Al atoms present, and y is the water content. Greater substitution of Al for Si gives a higher ion exchange capacity; the higher the charge of the cation, the more strongly it interacts with the zeolite framework. Ion exchange in zeolites can be represented by Equation 1:

$$Na_2Ze(s) + Cd^{2+}(aq) \Leftrightarrow CdZe(s) + 2Na^+(aq) \qquad (1)$$

Zeolites also possess ion sieve properties. Their internal structure consists of a series of interconnecting channels and cages with specific dimensions. These either trap or exclude ions depending on size. For example, the decreasing order of selectivity of zeolite Y for certain transition metals is: $Ni^{2+} > Cu^{2+} > Co^{2+} > Zn^{2+} > Mn^{2+} > Cd^{2+}$ (Coughlan and Caroll, 1976). Thus, it is possible to synthesize zeolites with selective trapping properties.

The metal-binding capacity of three synthetic zeolites, 4A, P, and Y, has been investigated by Rebedea et al. (1997). Each showed specific binding of Cd, Cu, Pb, and Zn, with rapid binding at lower solution concentrations of all metals. In contrast, Beringite has a lower affinity for these metals. Comparative isotherms for Cd and Cu are given in Figures 18.2(a-e). These properties are reflected in pot experiments using zeolite-amended polluted soils. Rebedea et al. (1997) demonstrated a reduction in phytotoxicity for *Zea* in soil polluted with Cd and Cu, with incorporation of 0.5 and 1% of either zeolite 4A or P and 5% incorporation of Zeolite Y. Reductions in metal uptake by other plants (*Helianthus, Lolium, Salix*) were also demonstrated (Rebedea, 1997).

Zeolite-treated soils show a general alkalinization. Reduced metal availability in zeolite-amended soils may be due to changes in soil pH, with the cation binding properties of the zeolites being of lesser importance. Water-displaced metals in two acidic soils polluted by either Cu/Cd deposition (Prescot) or long-term application of sewage sludge (Gateacre) amended with either zeolites or lime were measured following a 90-day incubation period (Rebedea, 1997). Metals displaced by water after 90 days of incubation revealed the significant effect of the zeolite amendments in reducing this fraction. In contrast, 1% lime had little effect on any of the elements investigated (Table 18.7). The observed effects of zeolite amendments reflect the metal-binding properties of these molecules rather than changes in soil pH.

The efficacy of lime and zeolite amendments to Prescot soil were compared in pot trials. *Helianthus* seedlings were planted in 4 kg of Prescot soil amended with either 0.5% lime or 0.5% Zeolite 4A. Plants were grown for 3 months, and then harvested, weighed, and analyzed. No plants grew in the unamended Prescot soil, but plants grew successfully in both of the amendment treatments. The zeolite

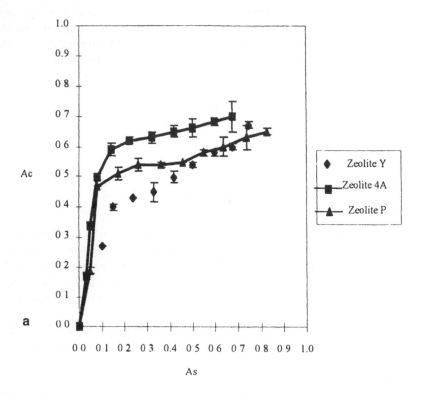

FIGURE 18.2(a-e) Ion-exchange isotherms for Zn, Cd, Cu, and Pb with three synthetic zeolites and beringite. Each isotherm represents the mean of three replicate determinations. Ac represents mole fraction in solid phase and As represents mole fraction in solution.

treatment reduced root and foliar Cd and Cu contents significantly when compared to lime application and also reduced total metal removed per pot (Table 18.8).

Interactions between zeolite amendment and fertilizer treatment were investigated using an infertile mine spoil soil (Trelogan). Zeolite 4A was incorporated at 1%, John Innes Base Fertilizer (5:8:10 NPK) was applied at the rate recommended as a compost amendment (0.16 g kg⁻¹ soil) at either 10 or 20 times this rate. Composition of soil and fertilizer are given in Table 18.9. Amended soils were sown with *Lolium perenne* (cv. Elka). All shoot biomass from each treatment was harvested after 6 weeks growth. From the results (Table 18.9), it can be seen that the combination of a nutrient source with a zeolite amendment is beneficial for plant growth on a heavily polluted soil and that the zeolite does not interfere with the action of the fertilizer by binding nutrient cations.

Laboratory and greenhouse experiments are indicative of a long-term retention of mobile metals by applied synthetic zeolites as opposed to metal immobilization due to pH changes. The long-term soil amendment experiments clearly indicate a plateau of immobilization after 2 to 3 months following application, and pot experiments indicate that this is not reversed by active root growth.

TABLE 18.7

Water-Extractable Metal Concentrations in Soil Solutions from Two Polluted Soils Amended with Either 1% Zeolite or 1% Lime

Soil and treatment	Cd	Cu	Pb	Zn
Prescot				
Unamended	0.8 ± 0.2	53 ± 3 6	3.0 ± 0.3	30 ± 6.6
Initial 3 months	0.8 ± 0.2	51 ± 3 5	3 0 ± 0.3	28 ± 4.7
1% Zeolite P 3 months	0.3 ± 0.1	25 ± 2 3	1.5 ± 0.3	12 ± 2.3
1% Zeolite 4A 3 months	0.4 ± 0.1	26 ± 3.2	1 6 ± 0.2	13 ± 2.2
1% Lime 3 months	0.5 ± 0.1	40 ± 3.0	2.5 ± 0.4	23 ± 3.5
Gateacre				
Unamended	1 7 ± 0.3	1.7 ± 0.3	5.0 ± 0.4	39 ± 4.8
Initial 3 months	1 7 ± 0.3	1.8 ± 0 3	5.0 ± 0.4	39 ± 4.8
1% Zeolite P 3 months	0 7 ± 0 1	0.8 ± 0.2	3 6 ± 0.3	21 ± 3.0
1% Zeolite 4A 3 months	0.6 ± 0 1	0.8 ± 0.2	3.4 ± 0.6	19 ± 1.7
1% lime 3 months	1 0 ± 0 3	1.2 ± 0.2	4.4 ± 0.3	31 ± 3.4

Note: Results from analysis of 50-ml leachate collected after leaching through a 10-g column of air-dried soil values expressed as µg element leached/g dry wt soil (SD (n = 3).

TABLE 18.8

Effects of Amendment with Either 0.5% Zeolite 4A or 0.5% Lime on Yield and Metal Uptake in *Helianthus* Cultivated in Prescot Soil

Treatment	Tissue	Yield	Tissue Metal Concentrations			
			Cd	Cu	Pb	Zn
Prescot	Root	No growth	No data	No data	No data	No data
	Foliage	No growth	No data	No data	No data	No data
+ 0.5% Zeolite 4A	Root	0.21 ± 0.1	0.6 ± 0.2	136 ± 12	50 ± 8.4	76 ± 18
	Foliage	1.0 ± 0.1	0.24 ± 0.8	28 ± 6.0	25 ± 6.2	23 ± 4.3
+ 0.5% lime	Root	0.15 ± 0.2	2.3 ± 1.2	212 ± 28	62.5 ± 17	120 ± 21
	Foliage	1.07 ± 0 1	0 46 ± 0.9	56 ± 9	25 ± 7.5	31 ± 5 6

Note: Plants failed to grow in unamended soil; yields given as g dry wt/pot. (2 plants/pot); and metal concentrations in µg g dry wt. (n = 2).

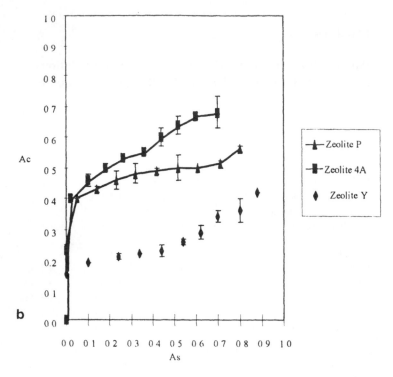

FIGURE 18.2b (Continued)

Other studies complement the above findings. A natural phillipsite was less effective than Apatite for decreasing Cd and Zn uptake by maize but was equally effective for Pb (Chlopecka and Adriano, 1996). This zeolite was also more effective than lime for reducing plant Zn uptake. It is also reported that natural and synthetic zeolites can reduce Cd uptake in lettuce (Gworek, 1992), Cd, Pb, and Zn uptake in various crop plants (Mineyev et al., 1991), and reduce phytotoxicity in a mine spoil soil (Rebedea and Lepp, 1994).

A synthetic cation exchange resin (Lewatit) was unsuccessfully used in a Cu/Zn polluted soil. It did not reduce plant metal uptake and showed a detrimental effect at increasing application rates by reducing available nutrient levels (Geiger et al., 1993).

Coal Fly-Ashes

Information relating to the environmental impacts associated with the disposal and utilization of coal combustion residues have been reviewed by Carlson and Adriano (1993). The composition of fly-ash varies considerably, but it is frequently composed of a ferro-aluminosilicate mineral in which Al, Si, Fe, Ca, K, and Na are the predominant elements. Trace elements derived from the parent coal also occur in this material. The most immediate effects of fly-ash incorporation into soil are increases in both alkalinity and salinity. The former property has led to fly-ash use in neutralizing highly acidic soils. This can be a significant factor in the establishment

TABLE 18.9

Interactions Between Fertilizer and Zeolite Treatments on Yield and Metal Concentrations in Ryegrass Foliage Grown on an Amended Infertile Mine Spoil Soil (Trelogan)

Treatment	Yield (g)	Foliar Metal Concentrations (µg g dry wt)			
		Cd	Cu	Pb	Zn
Trelogan unamended	No Growth				
+ 1% Zeolite 4A	No Growth				
+ 1F	0.25	14 0	48	200	370
		(13-15)	(46-50)	(200-200)	(365-375)
+ 1% Zeolite 4A/1F	0 42	57 5	27	135	220
		(7-8)	(25-29)	(130-140)	(215-225)
+ 10F	0 26	20	50	240	430
		(20-20)	(50-50)	(230-250)	(420-440)
+ 1% Zeolite 4A/10F	0 425	8.0	32	165	250
		(7 5-8 5)	(30-34)	(160-170)	(250-250)
+ 20F	0 365	22	60	350	500
		(20-24)	(60-60)	340-360)	(500-500)
+1% Zeolite 4A/20F	0 5	10	40	235	300
		(10-10)	(40-40)	(225-245)	(300-300)

Note: 1F, 10F, 20F — normal, 10 × and 20 × rates of application; fertilizer – John Innes Compost base (5:8:10 NPK); grass failed to grow in either unamended Trelogan soil or Trelogan soil amended with 1% Zeolite 4A; and yield and metal concentration values are means of duplicate trays. Range of metal concentrations given in parentheses.

of vegetation. However, fly-ash application may also result in excessive soluble salt concentrations, excessive B, adverse effects on soil properties (e.g., cementation), and may generate increased concentrations of potentially toxic trace elements.

Phosphate Minerals

Phosphate minerals may sorb and/or coprecipitate trace metals (Ma et al., 1993, 1994). Sorption of metals (Cd, Cu, Ni, Pb, and Zn) on hydroxyapatite surfaces indicate that surface complexation and coprecipitation are the most important mechanisms to explain Zn and Cd immobilization (Xu and Schwartz, 1994). Availability of Cd and Zn to maize was reduced subsequent to amendment with apatite in several contaminated soils (Chlopecka and Adriano, 1997). Apatite was found to be more effective than lime in decreasing Zn content of barley leaves, especially at high soil Zn concentrations (Chlopecka and Adriano, 1996). Reactions of apatite with aqueous Pb, resin-exchangeable Pb, Pb-contaminated soil materials, and selected Pb minerals have also been reported (Laperche et al., 1996).

Thomas basic slags (TBS) are different from other phosphate minerals. These are made from steel metallurgy wastes (Mench et al., 1994a). In acid soils, TBS

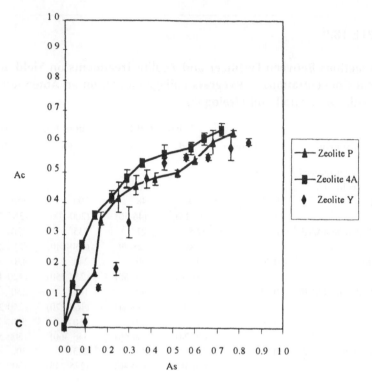

FIGURE 18.2c (Continued)

affect Kd_{Cd} (Figure 18.1). Decreases in shoot Cd and Zn uptake by ryegrass and restoration of dwarf bean growth were evident in TBS-amended soils (Tables 18.2 and 18.3). There was less reduction in plant Cd availability than Zn. Amendment of a Pb-contaminated soil with TBS reduced the transfer factor for Pb in ryegrass by an order of magnitude (Mench et al., 1994a,b). The processes of metal fixation by TBS require clarification.

Alkaline Materials

Lime is the most widely used inorganic ameliorant in agriculture. Dolomitic lime changes the soil pH, induces metal hydrolysis reactions and/or coprecipitation with carbonates, and thus acts as a precipitating agent for metals in the soil. However, the rise in pH may also mobilize toxic anions such as arsenates and chromates, or organic pollutants by increasing soluble organic matter fractions. Lime shows a significant decrease in effectiveness with time, especially at high levels of soil metal contamination (Chlopecka and Adriano, 1996). Lime has been compared to steel shots, iron, and manganese oxides under a similar soil pH regime (Sappin-Didier 1995). Lime was less effective than the metal amendments for reducing Cd mobility (Figure 18.1) and plant Cd and Zn uptake (Tables 18.2 and 18.3); these findings have been confirmed by Müller and Pluquet (1997), Chlopecka and Adriano (1996),

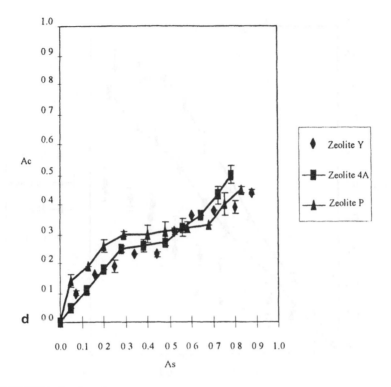

FIGURE 18.2d (Continued)

and Gomez et al., (1997). Lime was also shown to be less effective than synthetic zeolites in reducing soil metal mobility at a similar soil pH (Rebedea, 1997).

BENEFITS AND LIMITS OF *IN SITU* REMEDIATION

The so-called "hard" (or "high-impact") technologies currently available for treating soils contaminated by trace elements are often expensive, destructive, and can generate byproducts. *In situ* immobilization of metals by strong immobilizing agents of heavily contaminated bare sites and subsequent revegetation of the area can be an efficient, economically realistic, and cost-effective alternative remediation method, especially for agricultural soils, kitchen gardens, former industrial sites, and dredged sediment dumps. Strong immobilization of metals reduces leaching and bioavailability. Subsequently, vegetation can develop which stabilizes the soil. Besides the aesthetic profit, such a vegetation cover provides pollution control and stability to the soil. Lateral wind erosion is completely prevented, and a beneficial effect on metal percolation is observed (Vangronsveld et al., 1991, 1993, 1995b). Compared to "hard" remediation techniques, organic matter, microorganisms, and soil texture are not destroyed or separated from the soil. Generally, any byproducts are created by the *in situ* immobilization technique, which can be classified as a "soft" ("low impact") rehabilitation technique. Lastly, this can be a standby process

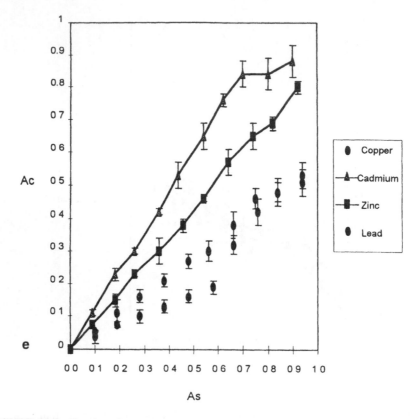

FIGURE 18.2e (Continued)

to reduce the impact of trace element-contaminated soil prior to the use of the most appropriate technologies for cleanup.

APPLICATION RATE, REACTION TIME, AND APPLICATION METHOD

The application rates of the additives, which generally range from 0.5 to 5% by soil weight, and subsequent mixing with different compounds are very important. Combinations of two or more additives were tested in several pot experiments, which were successful when steel shots or Mn oxides were combined with lime and basic slags. Another promising combination is beringite and steel shots. However, experiments are generally needed to optimize chemical additives and either single or multiple metal interactions. Reaction time between the additive and the substrata is also questionable. Metals sorption subsequent to steel shots addition is rapidly effective (less than 50 days), but kinetic changes continue over several months (Sappin-Didier, 1995). Beringite also has a very rapid effect, but changes observed in results of selective and sequential extractions at different time intervals after soil treatment suggest that kinetic changes in metal speciation occur over several years (see Sustainability section below). Soil moisture content, organic matter, and metal content may be limiting factors, and metal immobilization is not always successful as a consequence. On the other hand, nutrient deficiency, which may disturb plant metabolism, must be avoided. In some cases, metal concentrations are reduced but detrimental increases in nutrient deficiencies or salinity due to the treatment may have adverse effects on plant growth. In most pot experiments, fertilizers were supplied to plants. This point is questionable in field conditions, especially at industrial sites; the mixing of the additive into the soil is a delicate operation. A good homogenization is required in order to enhance the reaction. It is sometimes better to split the additive input and to make subsequent mixings. Usually, only the surface soil can be treated, i.e., 0 to 0.40 m depth. This can be sufficient to implement the growth of annual plant species such as grass; however, it may be inadequate for establishment of woody plant species which develop deeper root systems with time. In this case, the soil may need to be excavated and treated on-site before replacement.

EVALUATION OF EFFICACITY

The overall effect of ameliorants and the sustainability (durability) of metal immobilization in contaminated soils must be evaluated using various living organisms from different trophic levels; existing ecotoxicity tests can be used, but new ones may need to be developed. Some microbial assays can detect specific categories of toxicants such as trace metals, and a solid-phase version of MetPLATE™ has been developed to assess metal toxicity (Bitton et al., 1996). Solid-phase MetPLATE was used as a tool to study the effectiveness of steel shots (1%), compared to that of basic slags (0.28%) and HMO (1%), for reducing metal toxicity in the Evin soil (Boularbah et al., 1996). This soil was highly toxic to bacteria (72% inhibition using MetPLATE). All three ameliorants were effective in reducing metal toxicity. However, HMO was the most efficient treatment for reducing the toxicity of Evin soil (44% inhibition) compared with basic slags (55% inhibition) and steel shots (62% inhibition). Another microbial assay using bacterial biosensors (Corbisier et al., 1994, 1996) was shown to be a reliable test for the efficacy of various soil ameliorants (Corbisier, unpublished results).

SUSTAINABILITY

Extrapolation to field conditions should be made with caution. Monitoring the fate of trace elements, as well as the additive itself, over long periods of time is required, particularly with respect to leaching by acid rain. Varying conditions in the soil such as low pH or a varying Eh may tend to solubilize metals. Plots have been established at derelict sites, in kitchen gardens, and on agricultural fields (Vangronsveld et al., 1995b, 1996a, unpublished results; Müller and Pluquet, 1997; Rebedea 1997). Results are available for beringite, steel shots, and synthetic zeolites for, respectively, 5 years, 2 years, and 18 months following a single application.

In 1990, 3 ha of a highly metal-polluted acid sandy soil at the site of a former pyrometallurgical zinc smelter was treated with a combination of beringite and compost (Vangronsveld et al., 1995b). After soil treatment and sowing of a mixture of metal-tolerant *Agrostis capillaris* and *Festuca rubra*, a healthy vegetation cover developed. Five years later, an evaluation was made based on soil physicochemical parameters, potential phytotoxicity, floristic and fungal diversity, and mycorrhizal infection of the plant community (Vangronsveld et al., 1996a). Phytotoxicity was shown to be maintained at the low level observed immediately after soil treatment. The water-extractable metal fraction of the treated soil was up to 70 times lower compared to the nontreated soil. The vegetation was still healthy and regenerating by vegetative means and by seed. Diversity of higher plant species and saprophytic fungi was extremely low in the untreated area due to the high soil toxicity and the absence of metal tolerant ecotypes of plants and fungi. On the treated soil, in contrast, there was greater species richness of higher plants; several perennial forbs which are not noted as metal tolerant had colonized on the revegetated area. Most of these species belong to mycotrophic families, so the presence of a mycorrhizal network in the soil promotes their establishment. The ubiquity of the mycorrhizal fungi in the roots showed that a functioning ecosystem was establishing. In non-treated soil, the mycorrhizal infection rates of the roots were consistently lower during the growing season.

Durability of metal immobilization by beringite was also tested in a simulation experiment (Vangronsveld et al., 1996a,b). Column tests were performed using Zn- (730 mg kg^{-1}), Cd- (8 mg kg^{-1}), and Pb- (300 mg kg^{-1}) contaminated sandy garden soil originating from Lommel (Belgium). Comparisons were made of soil treated with 5% beringite and original nontreated soil. The effects of natural rainfall (600 mm/year) were simulated and accelerated. Mildly acidic rain water was used as a percolating fluid. The collected percolate was quantified and metal concentrations were determined. The plant available fraction of metals was evaluated using both chemical extraction methods and biological methods (phytotoxicity test, metal accumulation in bean, spinach, and lettuce). Over a 30-year simulation period, Zn and Cd percolation from the treated soils were reduced by 84 and 91%, respectively. During the period under investigation, both water and ammonium acetate-EDTA extractable metal fractions decreased significantly, especially during the last 10 years of the simulation period. Compared to the values obtained 1 year after the soil treatment, water and ammonium acetate-EDTA extractable metal fractions decreased by 70 and 30%, respectively. On the treated soils, bioavailability of the metal decreased to or was maintained at a low level (data of zinc and cadmium concentrations in lettuce are given as an example in Table 18.10). On the untreated soil, the bioavailability of zinc slightly decreased, whereas for cadmium a very drastic increase was observed at the end of the 30-year period.

A long-term field trial of two synthetic zeolites, 4A and P, was established in an area of copper-contaminated grassland at the BICC Rod and Wire (formerly BICC Rod Rollers) Copper Rod plant, Prescot, Merseyside in June 1995. This site receives continual atmospheric copper deposition (as copper oxides) from the furnace chimney. Full details of the site history and current trends of Cu deposition are given in Lepp et al. (1997). An area of depauperate grassland adjacent to the main factory building was subdivided into nine 2 m^2 plots, arranged in a square and separated by buffer zones of 0.5 m diameter. Then 3 plots received zeolite 4A at 1%, 3 zeolite P at 1%, and a further 3 were left as unamended controls. The plots were thoroughly

TABLE 18.10

Zinc and Cadmium Contents (mg kg^{-1}) of Lettuce Grown on Soils from the Percolation Experiment

| | Metal Content Leaves | | | | | |
| | Zinc | | | Cadmium | | |
Soil	(Zn/DW)	(%DW)	(Zn/FW)	(Cd/DW)	(%DW)	(Cd/FW)
Control	66 0	7.31	4.82	1 05	7.31	0.08
Untreated	146.5	6.17	9 00	3.75	6.17	0.23
K1 (1 year)	80.0	8.74	6.99	2.10	8.74	0.18
K2 (3 years)	88.5	7.37	6.49	1.65	7.37	0.12
K3 (6 years)	89.5	7.49	6.67	1.95	7.49	0.15
K4 (10 years)	90.0	8.71	7.84	1.85	8.71	0 16
K5 (15 years)	88.0	9 63	8.47	1.86	9 63	0.18
K6 (20 years)	45.5	4 35	1.98	2.01	4 35	0.09
K7 (30 years)	55.5	4.47	2.48	1.99	4.47	0.09
Untreated (30 years)	95.0	6.13	5.82	9.05	6.13	0 55

Note: Figures in parentheses refer to the number of years simulated acid precipitation received.

scarified with a mechanical lawn slitter; large quantities of dead, undecomposed thatch was removed, and the zeolites were watered into the appropriate plots as a slurry, using a watering can. This allowed for precise application, with no spread to adjacent plots. Soil samples (0 to 10cm depth) were collected randomly across the site prior to preparation to provide details of initial total and extractable Cu concentrations at the start of the experiment. The plots were resampled 12 months after the original zeolite applications (June 1996) and have been retained for further analysis in the future.

Soil analyses conducted 12 months after the original zeolite application clearly demonstrated that both the amendments had reduced water-extractable Cu fractions by up to 75% of the initial value. At the same time, continuing emissions from the factory had increased this fraction in the control plots (Table 18.11). This is a clear indication of the durability of zeolite effects, even under conditions of ongoing metal input to soil. Due to this ongoing deposition, foliar metal analysis could not be used as an indicator of changing soil metal fractions. The water-extractable Cu concentrations are too high for the rapid spread of seedlings from adjacent less polluted sites (Lepp et al., 1997). As a result, the extensive natural colonization reported by Vangronsveld et al. (1996a,b) on the beringite-amended zinc smelter waste was not observed. Qualitative evaluation of root growth in *Agrostis* from each of the three treatments indicated a much greater proportion of new root growth in the two zeolite plots in comparison to the control plots (Rebedea, 1997). Further field trials, on sites with multiple metal contamination where plant material can be established on prepared ground, are clearly desirable.

TABLE 18.11

Water-Extractable Metal Concentrations in Soil from Prescot Field Trial Plots 12 Months After Zeolite Application

Treatment	Water-Extractable Metals (μg g dry wt. soil)			
	Cu	Cd	Pb	Zn
Initial content	10 ± 0 67	0 8 ± 0.54	2.7 ± 0.09	2.9 ± 0.93
Untreated (12 months)	14.5 ± 0.74	0 75 ± 0.45	2 7 ± 0.04	2.8 ± 0.82
1% Zeolite P (12 months)	3.7 ± 0.72	0.42 ± 0 26	2.1 ± 0.02	1 4 ± 0 48
1% Zeolite 4A (12 months)	4.1 ± 0 93	0.46 ± 0.29	2 3 ± 0 03	1 4 ± 0 57

Note: Values represent means ±SD of 15 replicates for each treatment.

COMMERCIAL AVAILABILITY AND CHEMICAL COMPOSITION

Numerous ameliorants exhibit clear potential for reducing the transfer of metals from soil solution, their eventual entry into the food chain via plant uptake, their impact on pedofauna, and percolation to groundwaters. Nevertheless, only few efficient products are currently available in quantity. Iron-bearing materials (e.g., steel shots, sludges from drinking water treatment plants, red muds) and beringite applications are technically feasible. There are large-scale manufacturers of synthetic zeolites in Europe and the U.S. where these materials find widespread industrial and commercial use. Large zeolite and apatite deposits can be mined; however, it is questionable whether red mud can be applied in practice because of its high content of Cr, Al, and Na, especially in the soluble form of NaOH (Müller and Pluquet, 1997). The content of As in sludge from drinking water treatment should be checked whereas the application of steel shots often led to an increase in Ni and Mn plant uptake. Secondly, for Fe-bearing materials there is insufficient data from field trials. In practice, the treatment with such material should exceed the 1% pure Fe tested and affect a depth of more than 10 cm (Müller and Pluquet, 1997).

DIVERSITY OF TECHNIQUE

From the literature survey, it appears to be difficult to elaborate a standard soil treatment protocol. Based on this review of pot experiments and field trials, remediation of metal-contaminated soils needs to be tailored to a particular soil-plant system, especially when considering subsequent land uses. The effect of the soil amendment on plant yield and uptake is generally influenced by plant species (Chlopecka and Adriano, 1996). The data available today suggest that none of the immobilizing agents narrowed the differences in Cd accumulation between the plant species. This demonstrates that both plant and soil factors control the soil–plant transfer, and the efficiency of ameliorant must be considered as a combination of soil properties, including element speciation and the metabolism and ecophysiology of plants. Amelioration of tolerant species, therefore, is as important as optimization of trace element immobilization in contaminated soils. In addition, it was noticed in some studies that soil treatment by inorganic amendments can change trace element partitioning in the plant. For example, after cultivation on a Zn-contaminated silt loam soil, Zn content in maize roots was higher than in old leaves after soil liming, whereas the opposite occurred when either zeolite or apatite was added to this soil (Chlopecka and Adriano, 1996). Direct (ionic competition) or indirect effects such as iron and carbonate precipitates on root epidermial cells might be involved.

It is frequently claimed that the greatest reduction in soil-exchangeable metal leads to the largest reduction in metal uptake by plant species. Consequently, the question arises whether single or selective sequential extractions are reliable predictors of the effectiveness of ameliorants. Numerous case studies have demonstrated that amendments can decrease the soluble and exchangeable metal fractions, but that changes in metal uptake by plants were not significant (Sappin-Didier et al., 1997; Müller and Pluquet, 1997; Chlopecka and Adriano, 1996). Therefore, single (selective) extractions can only be used as a first approach. For the evaluation of metal-contaminated soils and the effectiveness of soil amendments, it is essential to investigate the availability of metals to plants and other living organisms.

COMPLIANCE OF *IN SITU* IMMOBILIZATION WITH REGULATIONS FOR SOIL AND PLANTS

In several countries, guidelines have been set for total trace element content in the soil. Indeed, *in situ* immobilization is not a method for reducing total trace element content in the substrate. Thus, in such countries there can be a conflict between environmental regulation and the techniques of immobilization and phytostabilization. This point must be adapted according to forthcoming results of field trials. Because of the large extent of some types of historical contamination, more realistic and pragmatic approaches need to be considered. In some regulations (e.g., Flanders, Belgium) specifically for historical contamination, the idea of risk evaluation and risk reduction has been introduced. *In situ* immobilization is perfectly in accordance with this approach. Alternative noninvasive methods such as phytoextraction may be proposed, but they require further development prior to becoming a widely applicable commercial reality.

Recommended guideline values have also been published for permissible trace element concentrations in edible plant parts and foodstuffs. One pathway, besides root uptake and atmospheric deposition, is contamination by soil particles re-entrained during rainfall events. This could be significant for total trace element content of forages and vegetables. Forages can often contain 3% of soil particles. Once again, *in situ* immobilization is not effective in limiting soil particle ingestion as topical coatings on plant products. However, as mobile fractions in soil are decreased by soil treatment, the bioavailability of trace elements in the animal digestive systems might be reduced also. Further research is needed to elucidate this point.

RESEARCH NEEDS

Up to the present time for most pot experiments and field trials carried out with contaminated soils and soil amendments, main results have related to dry-matter yield of plant parts, trace element uptake, and distribution in soil fractions. Information has also been gained on the effectiveness of soil amendments to reduce the indirect transfer to animals and humans. However, changes in the trace element availability of soil particles ingested directly by living organisms must be investigated. Few published studies deal with *in situ* immobilization and both the pedo- and mesofauna. Some studies are in progress. In the case of plants, little work has been carried out on trees (Rebedea, 1997), which are important for restoring bare areas, limiting wind erosion and percolation, as well as in general urban landscaping and as potential non-food crops to turn reclaimed sites into productive use.

Studies on the speciation of trace elements should focus on bioavailable organic species that may be more toxic than the inorganic forms. In addition, the mechanisms of *in situ* immobilization of trace elements should be investigated at the molecular level as an important aspect of our understanding of the sustainability of soil remediation techniques.

Methods of application, together with physical characteristics of potential amendments, also require further study. Those amendments with the greatest long-term potential should be developed for use in conventional mechanized delivery systems, or new delivery methods should be developed.

Finally, we need to manage and monitor remediated sites with emphasis on changes in exposure pathways. To be credible, this generic technique must prove that it is not necessary to remove and replace soils to alleviate risk from soil metal pollution.

ACKNOWLEDGMENTS

The INRA Agronomy Unit is grateful to the Association Interprofessionnelle du Plomb (Paris, France), Association Française Interprofessionnelle du Cadmium (Paris, France) for financial support. LUC is grateful to OVAM (Openbare Afvalst-offenmaatschappij van het Vlaamse Gewest, Mechelen, Belgium) and Union Minière (Brussels, Belgium) for support for several projects. LJMU acknowledges the support and cooperation of Crosfield Chemicals, Warrington, U.K., especially Dr. Tony Lovell, and BICC Rod and Wire, Prescot, U.K., especially Arthur Baker. The INRA Agronomy Unit and LUC wish to thank the Commission of the European Community for the support under the program Environment & Climate, PHYTOREHAB project, Contract No. ENV4-CT95-0083. The authors wish to extend their thanks to the coordinators of the 4th International Conference on the Biogeochemistry of Trace Elements, Berkeley, CA .

REFERENCES

Bergaya. F. and Barrault, J. 1990. Mixed Al-Fe pillared laponites: preparation, characterization and their catalytic properties in syngas conversion. In *Pillared Layered Structures. Current Trends and Applications.* L. V. Mitchell, Ed. Elsevier Applied Science, London, 167-184.

Bitton, G., Garland, E., Kong, I.C., Morel, J.L., and Koopman, B. 1996. A direct solid phase assay specific for heavy metal toxicity. I. Methodology. *J. Soil Contamin.*, 5, 386-394.

Boisson, J., Sappin-Didier, V., Solda, P., Bussière, S., Vangronsveld, J., and Mench, M. 1997. Influence of beringite and steel shots on the mobility and bioavailability of Cd and Ni: a field study. *Proc. 4th Int. Conf. Biogeochem. Trace Elements,* Iskandar, I., Hardy, S.E., Chang, A.C., and Pierzynski, G., Eds., Berkeley, CA, 309-310.

Boularbah, A., Morel, J.L., Bitton, G., and Mench, M. 1996. A direct solid-phase assay specific for heavy-metal toxicity. II. Assessment of heavy-metal immobilization in soils and bioavailability to plants. *J. Soil Contamin.* 5, 395-404.

Carlson, C.L. and Adriano, D.C. 1993. Environmental impacts of coal combustion residues. *J. Environ. Qual.,* 22, 227-247.

Charlet, L. and Manceau, A. 1993. Structure, formation and reactivity of hydrous oxide particles: insights from x-ray absorption spectroscopy. In *Environmental Particles. Environmental Analytical and Physical Chemistry Series Vol. 2.* Buffle, J. and van Leeuwen, H.P., Eds. Lewis Publishers, Boca Raton, Fl., 117-164.

Chlopecka, A. and Adriano, D.C. 1996. Mimicked *in situ* stabilization of metals in a cropped soil: bioavailability and chemical form of zinc. *Environ. Sci. Technol.,* 30, 3294-3303.

Chlopecka, A. and Adriano, D.C. 1997. Inactivation of metals in polluted soils using natural zeolite and apatite. *Proc. 4th Int. Conf. Biogeochem. Trace Elements,* Iskandar, I., Hardy, S.E., Chang, A.C., and Pierzynski, G., Eds., Berkeley, CA, 415-416.

Corbisier, P., Thiry, E., Masolijn, A., and Diels, L. 1994. Construction and development of metal ion biosensors. In *Bioluminescence and Chemoluminescence. Fundamental and Applied Aspects.* Campbell, A.K, Kricka, L.J., and Stanley, P.E., Eds. John Wiley & Sons, Chichester, U.K., 150-155.

Corbisier, P., Thiry, E., and Diels, L. 1996. Bacterial biosensors for the toxicity assessment of solid wastes. *Environ. Toxicol. Water Qual.,* 11, 171-177.

Coughlan B. and Caroll, W.M. 1976. Water in ion exchanged L, A, X and Y zeolites. *J. Chem. Soc. Faraday Trans.,* 72, 2016-2030.

Czupyrna, G., MacLean, A.I., Levy, R.D., and Gold, H. 1989. *In situ* immobilization of heavy-metal-contaminated soils. Noyes Data Corporation, Park Ridge, NJ.

De Boodt, M.F. 1991. Application of the sorption theory to eliminate heavy metals from waste waters and contaminated soils. In *Interactions at the Soil Colloid — Soil Solution Interface.* Bolt, G.H., De Boodt, M.F., Hayes, M.H.B., and McBride, M.B., eds. NATO ASI Series, Series E: Applied Sciences, Vol. 190, 293-320.

Dickinson, N.M., Watmough, S.A., and Turner, A.P. 1996. Ecological impact of 100 years of metal processing at Prescot, North West England. *Environ. Rev.,* 4, 8-24.

Didier, V., Mench, M., Gomez, A., Manceau, A., Tinet, D., and Juste, C. 1992. Rehabilitation of cadmium-contaminated soils: efficiency of some inorganic amendments for reducing Cd-bioavailability. *C. R. Acad. Sci. Paris,* 316 (Série III): 83-88.

Förster, C., Kuntze, H., and Pluquet, E. 1983. Influence of iron in soils on the cadmium-uptake of plants. In *Processing and Use of Sewage Sludge.* L'Hermite, P. and H.D. Ott, Eds. Reidel Publishing, Dordrecht, The Netherlands, 426-430.

Fu, G., Allen, H.E., and Cowan, C.E. 1991. Adsorption of cadmium and copper by manganese oxide. *Soil Sci.,* 152, 72-81.

Geiger, G., Federer, P., and Sticher, H. 1993. Reclamation of heavy metal contaminated soils: field studies and germination experiments. *J. Environ. Qual.*, 22, 201-207.

Gerth, J. and Brümmer, G. 1983. Adsorption und Festlegung von Nickel, Zink und Cadmium durch Goethit (-FeOOH); Fresenius. *Z. Anal. Chem.*, 316, 616-620.

Gomez, A., Vives, A., Didier-Sappin, V., Prunet, T., Soulet, P., and Chignon, R. 1997. Immobilization *in situ* des métaux lourds par ajout de composés minéraux en sols pollués. Final Report 92056, Ministère de l'Environnement, Paris.

Gworek, B. 1992. Lead inactivation in soils by zeolites. *Plant Soil*, 143: 71-74.

Hargé, J.C. 1997. Spéciation comparée du zinc, du plomb et du manganése dans des sols contaminés. Ph.D. thesis, Université Joseph Fourier, Grenoble, France, 178.

Hiller, D.A. and Brümmer, G.W. 1995. Mikrosondenuntersuchungen an unterschiedlich stark mit Schwermetallen belasteten Böden. 1. Methodische Grundlagen und Elementanalysen an pedogenen Oxiden. *Z. Pflanzenernähr. Bodenk.*, 158, 147-156.

Juste, C. and Solda, P. 1988. Influence de l'addition de différentes matières fertilisantes sur la biodisponibilité du cadmium, du manganèse, du nickel, et du zinc contenus dans un sol sableux amendés par des boues de station d'épuration. *Agronomie*, 8, 897-904.

Kabata-Pendias, A. and Pendias, H. 1992. *Trace Elements in Soil and Plants*. CRC Press, Boca Raton, FL, 365.

Krebs-Hartmann, R. 1997. *In situ* immobilisation of heavy metals in polluted agricultural soil — an approach to gentle soil remediation. Ph.D. thesis, Swiss Federal Institute of Technology, Zurich, Switzerland.

Laperche, V., Traina, S.J., Gaddam, P., and Logan, T.J. 1996. Chemical and mineralogical characterization of Pb in a contaminated soil: reactions with synthetic apatite. *Environ. Sci. Technol.*, 30: 3321-3326.

Le Hecho, I. 1995. Décontamination de sols de sites industriels pollués en métaux et arsenic par extraction chimique ou électrocinétique. Ph.D. thesis, Université Pau-Pays de l'Adour, Pau, France.

Lepp, N.W., Hartley, J., Toti, M., and Dickinson, N.M. 1997. Patterns of soil copper contamination and temporal changes in vegetation in the vicinity of a copper rod rolling factory. *Environ. Pollut.*, 95, 363-369.

Logan, T.J. 1992. Reclamation of chemically degraded soils. *Adv. Soil Sci.*, 17, 13-35.

Lothenbach, B., Krebs, R., Furrer, G., Gupta, S.K., and Schulin, R. 1988. Immobilisation of cadmium and zinc in soil by Al-montmorillonite and gravel sludge. *Eur. J. Soil Sci.* 49, 141-148.

Ma, Q.Y., Traina, S.J., and Logan, T.J. 1994. Effects of aqueous Al, Cd, Cu, Fe(II), Ni, and Zn on Pb immobilization by hydroxyapatite. *Environ. Sci. Technol.*, 28, 1219-1228.

Ma, Q.Y., Traina, S.J., Logan, T.J., and Ryan, J.A. 1993. *In situ* lead immobilization by apatite. *Environ. Sci. Technol.*, 27, 1803-1810.

Manceau, A., Gorshkov, A.I., and Drits, V.A. 1992a. Structural chemistry of Mn, Fe, Co, and Ni in Mn hydrous oxides: II. information from EXAFS spectroscopy, electron and x-ray diffraction. *Amer. Mineral*, 77, 1144-1157.

Manceau, A., Charlet, L., Boisset, M.C., Didier, B., and Spadini, L. 1992b. Sorption and speciation of heavy metals on hydrous Fe and Mn oxide: from microscopic to macroscopic. *Appl. Clay Sci.*, 7, 201-223.

Manceau, A., Hargé, J.C., Bartoli, C., Sylvester, E., Hazemann, J.L., Mench, M., and Baize, D. 1997. Sorption mechanism of Zn and Pb on birnessite: application to their speciation in contaminated soils. *Proc. 4th Int. Conf. Biogeochem. Trace Elements*, Iskandar, I., Hardy, S.E., Chang, A.C., and Pierzynski, G., Eds., Berkeley, CA, 403-404.

Martin, M.H. and Bullock, R.J. 1994. The impact and fate of heavy metals in an oak woodland ecosystem. In *Toxic Metals in Soil-Plant Systems*. Ross, S.M., Ed. John Wiley & Sons, Chichester, U.K., 327-365.

Mench, M., Didier, V., Löffler, M., Gomez, A., and Masson, P. 1994a. A mimicked *in situ* remediation study of metal contaminated soils with emphasis on cadmium and lead. *J. Environ. Qual.*, 23: 58-63.

Mench, M., Didier, V., Löffler, M., Gomez, A., and Masson, P. 1994b. A mimicked *in situ* remediation study of metal-contaminated soils. In *Workshop 92 Soil Remediation*, Avril, C. and Impens, R., Eds. Faculté des Sciences Agronomiques, Gembloux (B), 48-56.

Mench, M., Vangronsveld, J., Didier, V., and Clijsters, H. 1994c. Evaluation of metal mobility, plant availability and immobilization by chemical agents in a limed-silty soil. *Environ. Pollut.*, 86: 279-286.

Mench, M., Vangronsveld, J., Lepp, N.W., and Edwards, R. 1998. Physico-chemical aspects and efficiency of trace elements immobilization by soil amendments. In *Metal-Contaminated Soils: In Situ Inactivation and Phytorestoration*, Vangronsveld, J. and S.D. Cunningham, Eds., Springer-Verlag and R.G. Landes Company, 151-182.

Mench, M., Manceau, A., Vangronsveld, J., Clijsters, H., and Mocquot, B. 1999. Capacity of soil amendments in lowering the plant availability of sludge-borne zinc. *Agronomie*, In press.

Ministère de l'Environnement. 1994. Recensement des sites et sols pollués. Service de l'Environnement Industriel, Ministère Environnement, Paris, 277.

Mineyev, V.G., Kochetavkin, A.V., and Nguyen, Van Bo. 1990. Use of natural zeolites to prevent heavy metal pollution of soils and plants. *Sov. Soil Sci.*, 22, 72-79.

Müller, I. and Pluquet, E. 1997. Immobilization of heavy metals in mud dredged from a seaport. *Int. Conf. Contam. Sediments*, 09/7-11/1997, Rotterdam.

Pierzinski, G.M., and Schwab, A.P. 1993. Bioavailability of zinc, cadmium, and lead in a metal contaminated alluvial soil. *J. Environ. Qual.*, 22, 247-254.

Rebedea, I. 1997. An investigation into the use of synthetic zeolites for *in situ* land reclamation, Ph.D. thesis, Liverpool John Moores University, 244.

Rebedea, I. and Lepp, N.L. 1994. The use of synthetic zeolites to reduce plant metal uptake and phytotoxicity in two polluted soils. In *Biogeochemistry of Trace Elements*, D.C. Adriano et al., Eds. *Sci. Tech. Lett.*, Northwood, 81-87.

Rebedea, I., Edwards, R., Lepp, N.W., and Lovell, A.J. 1997. Potential applications of synthetic zeolites for *in situ* land reclamation. 121.PDF In *Contaminated Soils, 3rd Int. Conf. Biogeochem. Trace Elements*, Paris May 15-19, 1995. Prost, R., Ed., Les Colloques n°85, INRA Eds., Versailles, France.

Sappin-Didier, V., Mench, M., Gomez, A., and Lambrot, C. 1997a. Use of inorganic amendments for reducing metal bioavailability to ryegrass and tobacco in contaminated soils. In *Remediation of Soil Contaminated with Metals*. Iskandar, I.K. and D.C. Adriano, Eds. Advances in Environmental Science, Science Reviews, Northwood, U.K., 85-98.

Sappin-Didier, V., Tremel, A., and Mench, M. 1997b. A bag method for studying the influence of steel shots on cadmium mobility in a field experiment. *Proc. 4th Int. Conf. Biogeochemistry Trace Elements*, Berkeley, CA, 569-570.

Sappin-Didier, V. 1995. Utilisation de composés inorganiques pour diminuer les flux de métaux dans deux agrosystèmes pollués: étude des mécanismes impliqués par l'emploi d'un composé du fer. Ph.D. thesis, Analytical Chemistry and Environment, Bordeaux 1 University, France.

Spadini, L., Manceau, A., Schindler, P.W., and Charlet, L. 1994. Structure and stability of Cd^{2+} surface complexes on ferric oxides. 1. Results from EXAFS Spectroscopy. *J. Colloid Interface Sc.*, 168, 73-86.

Sylvester, E., Manceau, A., and Dits, V.A. 1997. The structure of monoclinic Na-birnessite and hexagonal birnessite. Part 2. Results from chemical studies and EXAFS spectroscopy. *Amer. Mineral.*, 82, 962-978.

U.S. Environmental Protection Agency, 1991. Rod annual report FY 1990. Report 540/8-91/067, USEPA, Washington, D.C.

Vangronsveld, J. and Clijsters, H. 1992. A biological test system for the evaluation of metal phytotoxicity and immobilization by additives in metal-contaminated soils. In *Metal Compounds in Environment and Life, 4 (Interrelation Between Chemistry and Biology)*, Merian, E. and Haerdi, W., Eds. *Sci. Technol. Lett.*, Northwood, U.K., 117-125.

Vangronsveld, J., Carleer, R., and Clijsters, H. 1994. Transfer of metals and metalloids from soil to man through vegetables cultivated in polluted gardens: risk assessment and methods for immobilization of these elements in soils. In *Environmental Contamination*, 142-145, Varnavas, S.P., Ed., CEP Consultants, Edinburgh.

Vangronsveld, J., Colpaert, J., and Van Tichelen, K. 1996a. Reclamation of a bare industrial area contaminated by non-ferrous metals: physico-chemical and biological evaluation of the durability of soil treatment and revegetation. *Environ. Pollut.*, 94, 131-140.

Vangronsveld, J., Ruttens, A., and Clijsters, H. 1996b. Study of the efficiency and durability of metal immobilization in soils. Report of a study performed for Union Minière, Brussels, Belgium.

Vangronsveld, J., Van Assche, F., and Clijsters, H. 1990. Immobilization of heavy metals in polluted soils by application of a modified alumino-silicate: biological evaluation. *Environmental Contamination*, (Barcelo, J., Ed.), 283-285. CEP Consultants, Edinburgh.

Vangronsveld, J., Van Assche, F., and Clijsters, H. 1995a. Reclamation of a bare industrial area contaminated by non-ferrous metals: *in situ* metal immobilization and revegetation. *Environ. Pollut.*, 87, 51-59.

Vangronsveld, J., Sterckx, J., Van Assche, F., and Clijsters, H. 1995b. Rehabilitation studies on an old non-ferrous waste dumping ground: effects of revegetation and metal immobilization by beringite. *J. Geochem. Expl.*, 52, 221-229.

Vangronsveld, J., Cammaer, C., Van Assche, F., and Clijsters, H. 1991. Reclamation of a "desert-like" site in the North East of Belgium: evolution of the metal pollution and experiments *in situ*. *Heavy Metals in the Environment*, Vol. 1 (Farmer, J.G., Ed.), 58-61. CEP Consultants, Edinburgh.

Vangronsveld, J., Sterckx, J., Van Assche, F., and Clijsters, H. 1993. Rehabilitation studies on an old non-ferrous waste dumping ground: effects of metal immobilization and revegetation. In *Heavy Metals in the Environment*, 563-566. CEP Consultant Edinburgh.

Xu, Y. and Schwartz, F.W. 1994. Sorption of Zn^{2+} and Cd^{2+} on hydroxyapatite surfaces. *Environ. Sci. Technol.*, 28: 1472-1480.

19 Phytoextraction and Phytostabilization: Technical, Economic, and Regulatory Considerations of the Soil–Lead Issue

Scott D. Cunningham and William. R. Berti

CONTENTS

INTRODUCTION

Lead (Pb) is a naturally occurring element and, as a result of anthropogenic activities, a ubiquitous environmental contaminant. Elevated Pb levels in the soil cause concern for human health and the environment. Soils with elevated levels of Pb were contaminated primarily through mining and smelting activities, but also through the widespread use of Pb-based paints, the manufacturing and testing of explosives, manufacturing and combustion of antiknock agents in gasoline, and occasionally with Pb bullets or shot.[1,2] Another source of Pb is the demolition of industrial buildings containing Pb-based paint, Pb pipes, and Pb linings used as antispark coatings.

Lead is an extremely difficult soil contaminant to remediate because it is a "soft" Lewis acid that forms strong bonds to both organic and inorganic ligands in soil. For the most part, Pb-contaminated soils are remediated through civil engineering techniques that require the excavation and landfilling of the contaminated soil.[3] Soils that present a leaching hazard in the landfill are either placed in a specially constructed hazardous waste landfill, or treated with stabilizing agents, such as cement, prior to disposal in an industrial landfill. Many innovative site decontamination techniques have been tried, including electroreclamation and several variations of soil washing, but, with the exception of soil washing, most of these techniques appear ineffective for the remediation of Pb-contaminated soils in a cost-effective manner. Soil washing as a particle separation technique may effectively remediate soils in which the Pb is associated with small soil particles (e.g., the clay fraction in soils). This technique uses particle segregation technologies (e.g., hydrocyclones, fluidized-bed separation) to separate the fine materials from coarser materials such as the sand and silt fractions of soil.[3] The separated Pb-enriched fraction is disposed of in a landfill, resulting in a significantly smaller volume of landfilled material when compared to the excavated whole soil. Often, the remaining coarser fractions of soil are relatively clean and replaced at the site. Thus, the savings from soil washing are primarily in reduced landfill costs. This technique is most applicable in soils that have a relatively small portion of fine materials. All remediation techniques currently available for Pb-contaminated soil are expensive and disruptive to the site. Additionally, they require further steps to return a healthy ecosystem to the remediated area. Increasingly, there is a need for the development of sound technologies that are low-cost and noninvasive, yet are equally protective of human health and the environment.

Not all Pb in the soil is chemically and biologically equivalent.[4] The premise of much of the newer remediation techniques is to understand, exploit, or alter the different chemical forms of Pb that exist in soil to reduce health and safety hazards. Lead contaminants originated in a number of different forms including Pb metal, inorganic Pb salts, and organic Pb compounds such as the antiknock agents tetraethyl and tetramethyl Pb. These Pb materials entered a variety of different environments, including soils and sediments, each with its own characteristic chemistry that may further alter the chemistry of the Pb contaminants once deposited. To a large degree, the chemical interactions that occur when Pb contaminants are introduced to a soil determine the eventual fate of the Pb in the soil. Lead in soil may be present in many different forms, in different oxidation states, and associated with different complexation states on soil surfaces. Divalent Pb is often complexed with organic matter, adsorbed onto cation exchange sites on the soil surfaces, or precipitated as relatively insoluble salts.[5] Divalent Pb may also be occluded with iron (Fe) and manganese (Mn) oxide coatings on soil particles, or associated with free Fe or Mn oxides. In some cases, Pb^0 metal and some very small amounts of organic lead (e.g., alkyl Pb antiknock agents) can still be found in the environment. The various forms present in soils have different solubilities and bioavailabilities, and each presents a unique environmental risk. An understanding of the species of Pb present in a soil, and of how these forms of Pb interact with biological organisms that are sensitive to Pb toxicity (e.g., a child, a site worker, or an earthworm), are crucial to the selection or development of remediation techniques.

The purpose of this chapter is to explore two emerging remediation technologies based on "agronomic" or "low-tech" solutions. These are known as phytoextraction and phytostabilization. Phytoextraction, also called biomining, is a site decontamination technique in which Pb is gradually removed from the site by plant uptake and harvesting (Figure 19.1). In contrast, phytostabilization is a site stabilization technique that does not remove Pb from the soil, but may reduce environmental and health risks by an alternative strategy. This technique, also called in-place inactivation or natural land restoration, uses soil amendments to sequester Pb in such a manner that it does not interact biologically with target organisms. Both of these techniques are relatively simple and low-cost; however, each resolves the risk to the environment and human health through contrasting approaches.

PHYTOEXTRACTION

In phytoextraction, soil Pb is taken into plant roots, translocated into the top of the plant, and removed by plant harvesting (Figure 19.1). Several technical parameters affect how efficiently this process functions. First, soil Pb must be in a form that is available to the plant root. Often at a site, the total Pb level is quite high, but the fraction of Pb availability for root uptake is exceedingly low. Second, the plant must be able to transfer Pb in the roots to the xylem stream to be carried to harvestable plant tissues. In most plants, this is difficult due to the chemical and physical environment in their tissues. At any stage of the uptake process, a number of physicochemical processes in the soil and plant can influence the transport of Pb from the soil to the aerial plant tissues. Some of these parameters are only marginally understood, and some are still undefined. Finally, the Pb-containing harvested plant material must be processed. Presently, the technical and logistic points regarding post-harvest treatment are mostly conceptual. The plant may be viewed as a solar driven extracting system that moves Pb from a silica-, iron-, or aluminum-based matrix (the soil), into a primarily carbon-, hydrogen-, or oxygen-matrix (the plant). This transfer of Pb to a wholly different matrix alters the engineering technologies that can be used to separate the Pb from the matrix. Several different pathways have been proposed to separate the organic plant material from the Pb, including the removal of the organic material by microbial degradation, incineration, low temperature ashing, or direct smelting from which the Pb could be recovered. Along with these technical concerns, there are still a number of regulatory, economic, and logistical questions that must be answered adequately before phytoextraction is accepted as a cost-effective and viable remediation technology. At the present, no Pb-contaminated site has been fully remediated using phytoextraction.

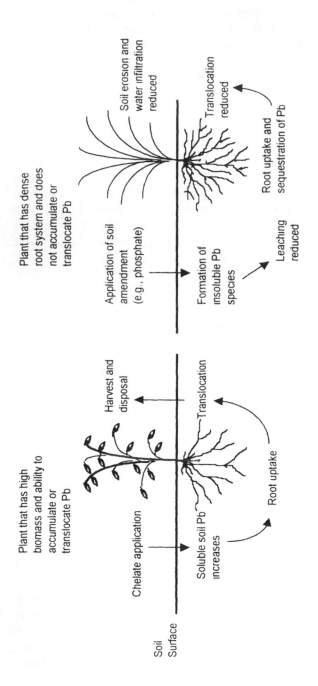

FIGURE 19.1 Comparison of the processes involved in phytoextraction and phytostabilization for Pb-contaminated soils.

TECHNICAL CONSIDERATIONS: LEAD UPTAKE IN PLANTS

In the current concept of phytoextraction, the removal process will only be practical, relatively timely, and effective if the plant has a high biomass and can accumulate at least 1% Pb in its above-ground tissues.[5] At one time, achieving such high Pb concentrations in plant tissue was thought a remote possibility at best. However, literature produced over the past 20 years showed that some unique plants can accumulate large amounts of many heavy metals in plant tissue, Pb among them.[6-10] These plants, called hyperaccumulators, are a widely divergent group of species that occur naturally throughout the world. The most dramatic examples of hyperaccumulation occurs with hyperaccumulators of metals other than Pb. *Sebertia accuminata* L., which grows in New Caledonia on a weathered outcropping rich in nickel, can accumulate in its sap over 22% nickel by dry weight without exhibiting toxic effects.[8] Another example is a Belgian *Thlaspi caerulescens* L. that can accumulate more than 3% zinc by dry weight of plant tissue without suffering any harm.[9,10]

The discovery of these and other metal-hyperaccumulating plants promoted the search for hyperaccumulators of Pb with similar tissue concentrations. The literature defines a Pb-hyperaccumulating plant as one that can accumulate at least 1000 mg Pb/kg.[6] These concentrations of Pb are substantial when compared to nonhyperaccumulators that rarely accumulate more than 100 mg Pb/kg regardless of the soil Pb concentration. However, even this level of hyperaccumulation does not approach the desirable concentration of 1% for phytoextraction, as do the hyperaccumulators of copper, nickel, and zinc. Despite extensive searches for Pb hyperaccumulators by these authors and others, no plant found in nature consistently accumulates more than 1000 to 2000 mg Pb/kg. We have been unable to confirm some literature reports of Pb hyperaccumulating germplasm under controlled laboratory experiments. Furthermore, investigations to determine the potential of known hyperaccumulators in commercial level phytoextraction have generally yielded disappointing results. Despite their fascinating biochemistry, many hyperaccumulators have extremely low biomass and slow growth rates. Additionally, because these plants are mostly uncultivated, their agronomic, breeding, and disease and pest control requirements are not optimized. Thus far, these constraints have made it difficult to translate the hyperaccumulating phenotype of these plants directly into a remediation strategy.

TECHNICAL CONSIDERATIONS: OPTIMIZING PHYTOEXTRACTION

Several approaches have been investigated to overcome the obstacles described above. Some researchers have examined the fundamental soil and plant processes influencing plant uptake of metals and tried to optimize these processes by agronomics, and more recently by genetic engineering. Results from plant screening tests, including hydroponic studies with solutions of varying Pb concentrations, have shown that plants differ widely between their ability to take up Pb from the solution and accumulate it in the roots.[11-15] Some plants, such as corn, may accumulate at most a few hundred mg Pb/kg in the roots. Other plants grown under identical controlled conditions, such as *Thlaspi rotundafolium* L., may accumulate over 3% in their roots. Shoot concentrations for most plants are disappointingly low. Relatively little Pb is transferred from the root to shoot. A shoot-to-root Pb concentration ratio of over one would greatly increase the effectiveness of phytoextraction. Lead is only sparingly soluble in solution, and even at the most contaminated sites, Pb in the soil solution is often less than 4 mg/l. At these low concentrations, many plants can effectively remove soil solution Pb into root tissues; however, very little Pb is translocated to aerial plant tissues that can be harvested. The chemistry of plants provides a logical reason for poor Pb translocation in most plants. When Pb enters the plant root, it immediately comes in contact with high phosphate concentrations, relatively high pH, and high carbonate-bicarbonate concentrations in the intracellular spaces. Under these conditions, Pb precipitates out of solution as phosphates or carbonates that can be seen in electron micrographs of roots from plants grown hydroponically in Pb solutions. These plant roots show inclusion bodies of these forms of Pb in the tissue, resolving the question of the cause of limited Pb translocation in plants.[16]

Phytoextraction appears to be complicated by three factors: (1) the low solubility of Pb in soils causing Pb to be unavailable for plant uptake, (2) the poor Pb translocation in plants to harvestable plant portions, and (3) the toxicity of Pb to the plant tissue. A technique should be developed to increase soil Pb bioavailability in plant roots and increase internal plant translocation from root-to-shoot by sequestering Pb in such a way that it is not precipitated either in the soil or in the root tissue. Recent research suggests that certain organic chelates may directly address all three of the factors mentioned above and mobilize Pb into aerial plant tissues.[5,14,15,17-19] Many chelates have a great affinity for Pb and form strong bonds with the element,[20] increasing soil solution concentrations. More surprisingly, chelates prevent Pb from precipitating as an insoluble salt or adsorbing in the plant tissue. At one site, HEDTA not only increased Pb in soil solution from around 4 mg Pb/l to over 4000 to 5000 mg Pb/l, but dramatically enhanced its translocation from roots to shoots.[18] However, it appears that many plants suffer damage from the elevated Pb concentrations in their tissues produced with chelate application. Strategic use of chelates may address all three current limitations of Pb phytoextraction. Chelates increase Pb bioavailability to plant roots, dramatically alter Pb translocation in the plant, and may make plant Pb tolerance irrelevant. Chelates may be applied after plants have produced sufficient biomass. After chelate application, the plant accumulates Pb and may be severely damaged, yet the plant remains harvestable. This approach to chelate application may have additional benefits in that the plant does not contain large amounts of Pb throughout most of its life cycle. However, this may cause some concern for herbivorous organisms that might feed on the plant.

PRACTICAL CONSIDERATIONS

Although the use of chelates in Pb phytoextraction still requires a considerable amount of refinement, these chemicals appear to provide a plausible commercial level process that may be adaptable to a range of site conditions. However, phytoextraction, even with the use of chelates, is not adaptable to every Pb-contaminated site. Land use, water table characteristics, depth of contamination, and contaminant concentration must be considered. Additionally, sites such as those containing Pb bullets and other large Pb objects almost certainly require a different remediation approach. Research to date indicates that the effectiveness of phytoextraction depends on agronomic practices, plant selection, chelate selection, soil pH, ionic balance of the soil solution, climatic factors, and the use of fertilizers. Mycorrhizal infection of the plant roots may also influence phytoextraction.[21,22] Under low Pb levels, mycorrhizae increase Pb uptake into the plant, but they actually decrease Pb uptake under higher Pb levels.[23] Even by altering some of these parameters to optimize Pb removal, phytoextraction at relatively contaminated sites may require a decade or longer to remediate a site. Given the typical yields of plants grown on contaminated soils and the Pb concentrations of the harvestable plant tissues in most plants, it has been estimated that 300 to 1000 mg Pb/kg soil may be removed in 7 to 10 years. From this estimation, it appears that there is an upper Pb concentration limit for this technique, and that phytoextraction is perhaps most applicable at low to moderately contaminated sites where restoration of a "clean" soil may have additional benefits.

PHYTOSTABILIZATION

In-place inactivation stabilizes soil Pb both chemically and physically through the use of soil amendments and a vegetative cover (Figure 19.1). Soil amendments alter the existing Pb chemistry in the soil and reduce the biological availability of Pb by inducing the formation of very insoluble Pb species. Plants with dense canopies and rooting systems, such as grasses, physically stabilize the soil against rain impact and erosion or leaching, as well as restrict off-site migration of the contaminants. This technique is based on a sound fundamental knowledge of soil Pb chemistry, agricultural practices, experience with the reclamation and revegetation of mining sites, and industrial plant ecology.

TECHNICAL CONSIDERATIONS

Amendment Selection

Currently, the most promising amendments for phytostabilization of Pb appear to be phosphate materials,[24-27] materials containing hydrous iron oxides,[27,28] steel shot,[28] inorganic clay minerals,[28-30] and organic material.[26,27] These amendments immobilize soil Pb by different approaches (Table 19.1). Phosphate amendments, such as phosphoric acid, calcium phosphates, and other fertilizers, may enhance the formation of essentially insoluble forms of Pb such as chloropyromorphite.[24] These materials only produce immediate changes in Pb chemistry and availability when they are applied at rates much higher than those used in agriculture. Hydrous Fe oxide materials have a great capacity to adsorb Pb and reduce the metal's bioavailability. One material, Iron Rich, a byproduct of TiO_2 production, contains 50 to 60% hydrous Fe oxide by weight. Currently, several research groups are investigating this amendment for its use in phytostabilization.[26,27] Organic materials such as composts, manures, or the coproducts from the burning of fossil fuels are readily available to help reduce the bioavailability of Pb in the environment.[26,27] Additionally, some of

TABLE 19.1

Soil Amendments Proposed for Phytostabilization and Their Suggested Mode of Pb Inactivation

Amendment Type		Suggested Mode of Inactivation
Phosphate materials	H_3PO_4,[27] apatite, calcium orthophosphates,[26] Na_2HPO_4,[24] KH_2PO_4,[26] and other phosphate fertilizers	Formation of pyromorphites (e.g., chloropyromorphite)
Hydrous iron oxides	Iron Rich,[26] corroded steel shots[28]	Sorption of Pb on oxide surface exchange sites, or formation of Pb-Fe compounds
Inorganic clay minerals	Synthetic zeolites,[29] aluminosilicate byproducts from burning of coal refuse[30]	Sorption of Pb on mineral surface exchange sites, or incorporation into the mineral structure
Organic materials	Manures, composts, sludges, and other biosolids[29]	Sorption of Pb on exchange sites, or incorporation into the organic material

these materials improve soil fertility by supplying nutrients and increasing soil moisture-holding capacity. Generally, soil amendments are selected that have little or even "negative" economic value. Amendments with negative value are waste products or byproducts that someone currently pays to have disposed. These latter materials include biosolids from wastewater treatment plants, some manures, and byproducts from industrial processes.

Determination of Risk Reduction

After amendments have been applied and a chemical equilibrium is established, the relative intrinsic hazard of the contaminant must be measured. Three types of measurement techniques are used to measure risk. First, chemical tests may be used to show that the solubility, availability, or leachability of Pb has been reduced. These tests include chelate buffers, simulated leaching tests, and regulatory tests such as the Toxicity Characteristic Leaching Procedure (TCLP)[31] or the Synthetic Precipitation Leaching Procedure (SPLP).[32] Another chemical test that has also been used is sequential extraction.[33] Methods for sequential extraction vary from laboratory to laboratory, but all of these methods sequentially extract the soil with a set of solutions of increasing harshness. In most methods, the extraction process begins with water or a dilute salt solution, and the last step often is a strong acid digest. The results of sequential fractionation yield a profile of Pb in the soil that suggests the relative solubility, bioavailability, or leachability of the soil Pb. Furthermore, it suggests the Pb chemistry of the soil and the relative risk a soil may pose under different conditions (e.g., the acidic solution in a child's stomach, or the mildly reducing condition of a waterlogged soil). Sequential fractionation is particularly useful when comparing the profiles of the Pb before and after amendment of the same soil, and is less useful when comparing across soils.[26,33] Another chemical test is becoming more widely used to indicate Pb bioavailability to a human. The Physiologically Based Extraction Test (PBET)[34,35] is an *in vitro* test that simulates the fasting conditions in the gastrointestinal tract of a young child. The Pb bioavailability measurements obtained from this test appear to correlate well with animal feeding studies, currently the regulatory standard for determining bioavailability. The PBET has not yet gained wide regulatory acceptance.

The second approach to measure intrinsic hazard involves determination of the actual Pb species present in the soil following amendment. If a soil can be shown to contain Pb in highly insoluble forms that are known to have low bioavailability, then risk reduction at the site can be inferred. Techniques such as x-ray diffraction or x-ray absorption spectroscopy are being used to describe Pb chemistry on soil fractions and on whole soils. In relatively noncomplex systems such as isolated clay minerals, these techniques are used to determine the oxidation state, chemical bonds, and nearest neighbors of a particular element. These techniques are still being developed for wide use on more complex systems such as whole soils, but may be particularly valuable as fingerprinting tools to characterize soil Pb species and to compare across soils.[36,37] Currently, scanning electron microscopy (SEM) using a microprobe is used more widely to determine, in a qualitative sense, the physical and chemical nature of Pb in the soil.[38,39]

The third type of risk measurement involves some estimation of soil Pb bioavailability using bioassays. These may include plant assays, earthworm assays, Microtox analysis, or animal feeding studies. These assays determine Pb availability to specific target organisms and may have limited applicability to Pb bioavailability to other organisms. Currently, animal feeding studies on pigs or rats are most widely accepted by the regulatory community for indicating Pb bioavailability to humans. However, the cost and time required for animal feeding studies prohibits their use at many sites. *In vitro* surrogate tests for bioavailability, such as the PBET, are considerably less expensive and time consuming. This test could be performed in many laboratories on many soils without extensive research efforts, and should become a valuable alternative to animal feeding studies.

The establishment of phytostabilization as an acceptable remediation alternative is gaining favor as the regulatory acceptability of site-specific adjustments in Pb bioavailability and land use restrictions (e.g., an "industrial" vs. "residential" classification) are being considered. In many countries, the need for remediation is based on the level of hazard a site presents to human health.[40] For Pb-contaminated sites in the U.S., the hazard is often determined for young children using a computer model, the Integrated Exposure and Uptake Biokinetic model (IEUBK).[41,42] This model considers several parameters such as a child's nutritional status, exposure frequency and rate, and Pb bioavailability, to predict the percentage of children in an exposed population that may have a blood Pb level above 10 μg/dl. Children with blood Pb levels above this value are considered to be at risk of suffering Pb-related health problems. The IEUBK model uses a default value of 30% for Pb bioavailability; however, this may greatly overestimate Pb bioavailability for a given site, such as a site remediated by phytostabilization. A change in the default value for

Pb bioavailability in the IEUBK model may change the amount of children with blood Pb levels above 10 μg/dl (Figure 19.2). In the U.S., it is generally acceptable that no more than 5% of the children have blood Pb levels above the critical level. For a soil with 2000 mg Pb/kg and a Pb bioavailability of 30%, the IEUBK predicts that over 50% of the children will be above the critical blood Pb level. However, when the Pb bioavailability value is reduced to 10%, the percent of children who are at risk drops to 11%. While this hypothetical scenario still exceeds the U.S. acceptable limit for the percent of children above the critical blood Pb level, it clearly indicates that assumptions on Pb bioavailability may yield incorrect predictions concerning children at risk at a given site. Lower Pb bioavailability may also lead to a higher acceptable level of total Pb in the environment with equivalent protection of human health. If regulatory authorities continue to allow the increased use of site-specific Pb bioavailability values to be substituted for the default value, then inactivation techniques that reduce bioavailability can be a cost-effective technique to reduce risk.

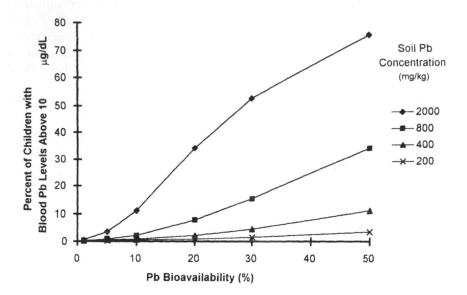

FIGURE 19.2 Predictions of the percentage of children with blood Pb levels above 10 μg/dl with changing soil Pb concentrations and Pb bioavailability. These predictions were generated with the IEUBK model and assume unchanging or default values for all parameters other than soil Pb concentration and Pb bioavailability.

PRACTICAL CONSIDERATIONS

Phytostabilization is becoming an attractive possibility for the remediation of many different sites. However, critical questions remain that must be resolved before this technique will be widely accepted by site managers, regulatory agencies, and the public. The first of these questions concerns what simple test most appropriately measures Pb bioavailability while adequately reflecting risk. It is impractical to consider animal feeding studies as a basis for establishing appropriate regulatory guidelines because of the cost and time required to conduct these studies. The second question regards the longevity of inactivation treatments and the possible need for re-treatment or restrictions on future land use of the site following phytostabilization. The third question concerns the practicality of optimizing treatment effectiveness. Currently, the choice of amendments and rates is empirically derived, adding some uncertainty to the process.[33] Future efforts should be made to develop a sound theoretical and practical basis for treatment selection. The fourth question concerns the concentration limit above which phytostabilization is not practical or effective. Currently, most research concentrates on soils in the 1000 to 3000 mg Pb/kg range. The practical upper soil Pb concentration limit for phytostabilization is unknown. Until these questions are adequately answered, regulatory and public acceptance of phytostabilization will be unattainable.

REGULATORY CONSIDERATIONS

Both phytoextraction and phytostabilization are relatively new techniques, and to date, no site has actually been remediated by either strategy to the full satisfaction of the regulatory community. Generally, both site managers and regulatory authorities would prefer to have a site remediated by removing the contaminant from the site rather than by a stabilization technique. When the contaminant is removed there remains little question as to its relative hazard. Decontaminated sites are more easily managed and have fewer land use restrictions. Many regulatory authorities are showing interest in phytoextraction for sites where this technique may be feasible. However, due to the present technical obstacles in the development of phytoextraction as a viable technique, phytostabilization is perhaps a more widely attainable remediation strategy at the current time. It is technically more difficult to explain to a concerned audience the risk reduction provided by phytostabilization. Phytostabilization requires a more sophisticated understanding of the chemical, biological, and physical processes that occur in soil. A receptive regulatory and public audience must therefore be provided with extensive proof of the effectiveness of phytostabilization. Researchers investigating phytoextraction and phytostabilization are actively working with state and federal regulators at many sites to push these techniques forward and demonstrate them on a full-scale basis. Hopefully, with sufficient field validation and some changes in current regulatory concepts regarding Pb bioavailability, these techniques will soon be available as alternatives to current practices.

ECONOMICS

Economics are a necessary consideration in the choice of remediation technique for any site. A generalized economic analysis comparing phytoremediation techniques to four commonly practiced engineering techniques is provided in Table 19.2. The site decontamination techniques considered in this analysis include solidification and stabilization off-site, soil washing by particle separation, and phytoextraction.

TABLE 19.2

Estimated Economic Analysis of Remediation Alternatives for a 1-ha Pb-Contaminated Site

Alternative	Clearing	Excavation to 30 cm	Soil Disposal	Variable Costs (US$)		Net Present Cost
				First Year	10 y Recurring[a]	
Site Decontamination						
Solidification and stabilization off-site[b,c,e]	8800	43,000	1,300,000	270,000	0	1,600,000
Soil washing[c,e]	8800	43,000	290,000	450,000	0	790,000
Phytoextraction	8800	0	0	0	270,000	279,000
Site Stabilization						
Asphalt capping[d,e] (parking lot)	8800	0	0	150,000	0	160,000
Soil capping[d,e]	8800	0	0	125,000	6700	130,000
Phytostabilization	8800	0	0	44,000	6700	60,000

[a] Costs include a 12% annual discount rate and 3% annual inflation rate.

[b] Estimates for solidification and stabilization off-site assumes a landfilling cost of $226 per metric ton of soil; costs are for a regular landfill.

[c] Estimate for soil washing assumes a landfilling cost of $250 per metric ton of soil; costs are for a hazardous waste landfill.

[d] Racer/ENVEST™ Delta Technologies Group Inc., Denver, CO, 1996.[44]

[e] Gary Quinton, DuPont Corporate Remediation Group, personal communication.

Site stabilization techniques include asphalt capping, soil capping, and phytostabilization. To compare these techniques equally, a number of general assumptions must be made. The site is defined as a 1-ha area contaminated primarily with Pb to a depth of 30 cm, and covered with some rubble on the surface. The total volume of soil to be remediated is 3000 cubic meters, weighing 4×10^6 kg. For all alternatives except phytoextraction, total soil Pb is assumed to be >0.2% (w/w). Phytoextraction estimates are instead calculated for a soil containing 0.14% Pb (w/w). The Pb-contaminated soil exceeds the regulatory TCLP limit of 5 mg Pb/l and would require stabilization prior to disposal in a hazardous waste landfill. All techniques require initial clearing of the site at a cost of $8800. Economic estimates reflect a discount rate of 12% per year and an annual inflation rate of 3%. The discount rate accounts for the interest paid on funds borrowed for remediation. These are general estimates, and actual costs may vary.

This comparison is intended to compare the costs for the remediation technologies alone, apart from costs associated with site assessment or monitoring. For all of the technologies, however, costs for site assessment should be equal, with no effect on the comparison. Costs for site monitoring are difficult to generalize, since they may vary considerably with the site, the technology, and local, state, or federal regulations. Some technologies, such as solidification and stabilization off-site, require no site monitoring following treatment. Soil washing and capping strategies may or may not require monitoring. Most likely, phytostabilization and phytoextraction, at least for the duration of the remediation process, will require some monitoring. Currently, it is difficult to predict how much monitoring would be required, and at what cost, since these technologies have not yet been used to remediate an actual site.

SITE DECONTAMINATION ALTERNATIVES

Of the site decontamination techniques, the most expensive alternative by far is excavation and landfilling of the contaminated soil. The costs associated with this process are associated with excavation ($43,000), disposal ($1,300,000), and stabilization ($270,000). The soil is stabilized with cement and other chemicals prior to placement in a regular landfill at $226 per metric ton of soil. Remediation in this manner is completed within the first year, with no annual operating and maintenance requirements, and incurs a net cost exceeding $1.5 million per hectare. This cost does not include any additional site restoration measures such as backfilling with uncontaminated off-site soil or revegetation.

Soil washing to remove the fine material is the second most expensive decontamination technique. This technology separates the relatively uncontaminated larger particle size fractions (e.g., silt and sand) from the contaminant-bearing finer particles (e.g., clay). The finer particles are disposed of in a landfill, while the coarser fractions are returned to the site. For this alternative, 3000 cubic meters of soil are excavated and treated at a soil washing facility that processes 22 metric tons of soil per hour. The fine material is assumed to comprise 20% of the contaminated soil, and is disposed of without further treatment in a hazardous waste landfill at $250 per metric ton of soil. This technique includes the same excavation costs as solidification and stabilization, but has a higher variable cost of $450,000 because of the additional equipment needed to process the soil. However, the cost of landfilling is considerably lower than the first alternative because of the smaller volume of material to be disposed. The net present cost of this alternative is approximately $790,000 per hectare.

The least expensive decontamination alternative is phytoextraction. For this scenario, the soil contains 0.14% Pb (w/w) and must be remediated to a final Pb concentration of 0.04% Pb (w/w) for residential land use. It is assumed that phytoextraction can remove 0.1% Pb (w/w), for a total of 4000 kg Pb to be removed from the site in 10 years. The technique involves three harvests per year with a total annual harvested biomass of 40 ton/ha. Plants are assumed to accumulate 1% Pb in their biomass. For this scenario, Na_4EDTA is applied with irrigation to assist Pb removal at a ratio of 1 mol Pb to 1 mol Na_4EDTA. The total amount of Na_4EDTA required is 7400 kg, at a cost[43] of $43,000. The harvested biomass is disposed of in a hazardous waste landfill following on-site moisture reduction. Total landfilling costs are slightly below $70,000. Annual operating and maintenance also includes plowing, seeding, and spraying. The net present cost of phytoextraction is roughly $279,000. If an alternative scenario is used with plants producing 20 t/ha annually, the time required for remediation increases to 20 years, and the net present cost rises to $416,000. Not only is the cost of phytoextraction less than that of the engineering techniques for decontamination, but the cost is spread out over 10 (or more) years, requiring less money up front. Phytoextraction is the least expensive of the three techniques, and it is the only technique that actually removes the Pb from the soil as opposed to removing the contaminated soil or soil portions from the site.

SITE STABILIZATION ALTERNATIVES

Of the site stabilization techniques considered in this comparison, the most expensive alternative is asphalt capping for site use as a parking lot with 240 spaces. In this technique, the contaminated soil is covered by a base layer of asphalt approximately 20 cm thick, followed by a subgrade 25 cm thick, and finally by a surface layer of asphalt 4 cm thick. This alternative also includes installation of water drainage and parking curbs. Asphalt capping effectively seals the Pb-contaminated soil to prevent any environmental contact. It provides a cost of approximately $160,000 per hectare, and is considerably less expensive than the decontamination strategies.

A second, less expensive capping alternative uses a 60-cm thick cap of uncontaminated soil from off-site to cover the area of concern. A simple vegetative cover is established on the soil cap surface to prevent erosion and restrict water infiltration. No geotextiles, membranes, or other restrictive barriers are included in the estimate. Soil capping requires annual reseeding and remulching of 10% of the site, and four mowings per year. The total cost per hectare for this technique falls below $140,000.

The last stabilization technique considered is phytostabilization. Phytostabilization involves some initial site preparation to optimize growing conditions. This includes three plowings and the application of 11 t/ha lime ($865), 0.18 t/ha ammonium nitrate ($40), and 0.17 t/ha potassium muriate ($15). Amendments to inactivate soil Pb include 90 t/ha triple super phosphate ($23,000) and 396 t/ha Iron Rich ($15,000). Annual operating and maintenance includes reseeding and remulching of 10% of the site, as well as four mowings per year. The net present cost of phytostabilization is $53,000, and it is the least expensive option for stabilization.

Phytoextraction and phytostabilization are the least expensive alternatives in each of their respective categories, however, phytoextraction is the more expensive of the two. The choice of which of these alternative techniques should be implemented at a site is not solely a matter of economics, for they have different constraints and applications, and are sensitive to different site parameters such as soil Pb concentration, soil chemistry, contamination depth, or the time frame required for remediation. If an immediate reduction in risk were required, phytostabilization would be chosen because of the length of time required for plants to remove Pb in phytoextraction. However, at sites where decontamination is desired and feasible, phytoextraction is the more appropriate technique, despite the higher cost.

CONCLUSIONS

Phytoextraction and phytostabilization are emerging technologies that integrate soil chemistry, engineering, plant biology, and agricultural practices for the remediation of Pb-contaminated soils. Phytoextraction is a promising technology that could meet an increasing need for a low input decontamination technique. The development of this technique to a full-scale basis is anxiously awaited. However, questions still remain regarding the optimization of Pb uptake by chelate addition, cropping systems, and cultivation techniques. In laboratory studies and on small field plots, phytostabilization has been shown to reduce Pb bioavailability and the risk to human health. At the current stage of development, this technique is ready for field validation, and some full-scale demonstrations are in progress. Hopefully, these demonstrations will yield positive results that will encourage regulatory and public acceptance and help establish this technique as a viable remediation alternative. Currently, these technologies, particularly phytoextraction, still require some technical and logistical refinement. Their establishment as viable remediation alternatives depends on the cooperative efforts of many scientists, regulators, and remediation technologists. These technologies are in part a result of a recent approach to remediation that seeks the most environmentally benign techniques to clean up the contamination caused by human activities while protecting human health.

ACKNOWLEDGMENTS

The authors would like to thank Gary Quinton for his invaluable assistance in the economic analysis of the various remediation alternatives, and Ellen Cooper for her excellent help in preparing and editing the manuscript.

REFERENCES

1. Adriano, D.C., *Trace Elements in the Terrestrial Environment*. Springer-Verlag, New York, NY, 1986.
2. Alloway, B.J., The Origin of Heavy Metals in Soils, in *Heavy Metals in Soils*. 2nd ed., Alloway, B.J., Blackie Academic and Professional, New York, NY, 1995, 38-57.
3. Smith, L.A., J.L. Means, A. Chen, B. Alleman, C.C. Chapman, J.S. Tixier, Jr., S.E. Brauning, A.R. Gavaskar, and M.D. Royer, *Remedial Options For Metals-Contaminated Sites*. Lewis Publishers, New York, NY, 1995.
4. Nriagu, J.O., *Properties and the Biogeochemical Cycle of Lead*, Part A and Part B, Elsevier/North-Holland, New York, NY, 1978.
5. Cunningham, S.D., W.R. Berti, and J.W. Huang, Phytoremediation of contaminated soils. *Biotechnology*, 13: 393-397, 1995.
6. Baker, A.J.M. and R.R. Brooks, Terrestrial higher plants which hyperaccumulate metal elements — a review of their distribution, ecology, and phytochemistry. *Biorecovery*, 1: 81-126, 1989.
7. Reeves, R.D. and R.R. Brooks, Hyperaccumulation of lead and zinc by two metallophytes from mining areas of central Europe. *Environ. Pollut.*, 31: 277-285, 1983.
8. Jaffre, T., R.R. Brooks, J. Lee, and R.D. Reeves, *Sebertia accuminata*: a nickel-accumulating plant from New Caledonia. *Science*, 193: 579-580, 1976.
9. Brown, S.L., R.L. Chaney, J. S. Angle, and A.J.M. Baker, Phytoremediation potential of *Thlaspi caerulescens* and bladder campion for zinc- and cadmium-contaminated soil. *J. Environ. Qual.*, 23: 1151-1157, 1994.
10. Baker, A.J.M., R.D. Reeves, and A.S.M. Hajar, Heavy metal accumulation and tolerance in British population of the metallophyte *Thlaspi caerulescens* J. & C. Presl (Brassicaceae). *New Phytol.*, 127: 61-68, 1994.
11. Kumar, N., V. Dushenkov, H. Motto, and I. Raskin, Phytoremediation: the use of plants to remove heavy metals from soils. *Environ. Sci. Technol.*, 29: 1232-1238, 1995.
12. Salt, D.E., M.J. Blaylock, N.P.B.A. Kumar, V. Dushenkov, B.D. Ensley, I. Chet, and I. Raskin, Phytoremediation: a novel strategy for the removal of toxic metals from the environment using plants. *Biotechnology*, 13: 468-474, 1995.
13. Berti, W.R. and S.D. Cunningham, Remediating soil with green plants, in *Trace Substances, Environment and Health*. Cothern, C.R., Science Reviews, Northwood, 1994, 43-51.
14. Huang, J.W. and S.D. Cunningham, Lead phytoextraction: species variation in lead uptake and translocation. *New Phytol.*, 134: 75-84, 1996.
15. Huang, J.W., S.D. Cunningham, and S.J. Germani, Plant based soil remediation: preliminary study of Pb phytoextraction from Pb-contaminated soils, in *Proc. 3rd Int. Conf. Biogeochem. Trace Elements*. May 15, 1995, Prost, R., Ed., Collogue 85, INRA Edition, Paris, 1997.
16. Qureshi, J.A., K. Hardwick, and H.A. Collin, Intracellular localization of lead in a lead tolerant and sensitive clone of *Anthoxanthum odoratum*. *J. Plant Physiol.*, 122: 357-364, 1986.
17. Blaylock, M.J., D.E. Salt, S. Dushenkov, O. Zakharova, C. Gussman, Y. Kapulnik, B.D. Ensley, and I. Raskin, Enhanced accumulation of Pb in indian mustard by soil-applied chelating agents. *Environ. Sci. Technol.*, 31: 860-865, 1997.
18. Huang, J.W., J. Chen, W.R. Berti, and S.D. Cunningham, Phytoremediation of lead-contaminated soils: role of synthetic chelates in phytoextraction. *Environ. Sci. Technol.*, 31: 800-805, 1997.

19. Huang, J.W., J. Chen, and S.D. Cunningham, Phytoextraction of Pb from contaminated soils, in *Phytoremediation of Soil and Water Contaminants*. Kruger, E.L., T.A. Anderson, and J.R. Coats, American Chemical Society, Washington, D.C., 1997, 2-19.

20. Norvell, W.A., Reactions of metal chelates in soil and nutrient solutions, in *Micronutrients in Agriculture*. 2nd ed., Mortvedt, J.J., F.R. Cox, L.M. Shuman, and R.M. Welch, Soil Science Society of America, Madison, WI, 1991, 145-186.

21. Gildon, A. and P.B. Tinker, A heavy metal-tolerant strain of a mycorrhizal fungus. *Trans. Br. Mycol. Soc.*, 77: 20-21, 1981.

22. Killham, K. and M.K. Firestone, Vesicular arbuscular mycorrhizal mediation of grass response to acidic and heavy metal depositions. *Plant and Soil*, 72: 39-48, 1983.

23. Heggo, A. and J.S. Angle, Effects of vesicular-arbuscular mycorrhyzal fungi on heavy metal uptake by soybeans. *Soil Biol. Biochem.*, 22: 865-869, 1990.

24. Rabinowitz, M.B., Modifying soil lead bioavailability by phosphate addition. *Bull. Environ. Contam.*, 51: 438-444, 1993.

25. Ruby, M.V., A. Davis, and A. Nicholson, *In situ* formation of lead phosphates in soils as a method to immobilize lead. *Environ. Sci. Technol.*, 28: 646-654, 1994.

26. Berti, W.R. and S.D. Cunningham, In-place inactivation of Pb in Pb-contaminated soils. *Environ. Sci. Technol.*, 31: 1359-1364, 1997.

27. Chaney, R.L. and S.L. Brown, Memorandum: Further information on the Joplin site being considered for field plots for the IINERT study of *in situ* inactivation of soil Pb, 1996.

28. Mench, M.J., V.L. Didier, M. Loffler, A. Gomez, and P. Masson, A mimicked *in-situ* remediation study of metal-contaminated soils, with emphasis on cadmium and lead. *J. Environ. Qual.*, 23: 58-63, 1994.

29. Gworek, B., Lead inactivation in soils by zeolites. *Plant and Soil*, 143: 71-74, 1992.

30. Vangronsveld, J. and H. Clijsters, A biological test system for the evaluation of metal phytotoxicity and immobilization by additives in metal contaminated soil, in *Metal Compounds in Environment and Life, 4 (Interrelation between Chemistry and Biology)*. Merian, E., and W.M. Haerdi, Eds., *Science Reviews*, Wilmington, DE 1992, 117-125.

31. Federal Register, U.S. Environmental Protection Agency. Part V. 40 CFR Part 261 et al. 55(126): 26986-26998, 1990.

32. U.S. Environmental Protection Agency, Office of Solid Waste and Emergency Response, *Test Methods for Evaluating Solid Waste. Physical/Chemical Methods*. 3rd ed. U.S. EPA, Washington, D.C, 1987.

33. Berti, W.R., S.D. Cunningham, and L. W. Jacobs, Sequential chemical extraction of trace elements: development and use in remediating contaminated soils, in *Contaminated Soils*, 3rd Int. Conf. Biogeochem. Trace Elem., May 15, 1995, Prost, R., Ed., Collogue 85, INRA Editions, Paris, 1997, 121-132.

34. Ruby, M.V., A. Davis, T.E. Link, R. Schoof, R.L. Chaney, G.B. Freeman, and P. Bergstrom, Development of an *in vitro* screening test to evaluate the *in vivo* bioaccessibility of ingested mine-waste lead. *Environ. Sci. Technol.*, 27: 2870-2877, 1993.

35. Ruby, M.V., A. Davis, R. Schoof, S. Eberle, and C.M. Sellstone, Estimation of lead and arsenic bioavailability using a physiologically based extraction test. *Environ. Sci. Technol.*, 30: 422-430, 1996.

36. Cotter-Howells, J.D., P.E. Champness, J.M. Charnock, and R.A.D. Pattrick, Identification of pyromorphite in mine-waste contaminated soils by ATEM and EXAFS. *Eur. J. Soil Sci.*, 45: 393-402, 1994.

37. Manceau, A., M. Boisset, G. Sarret, J. Hazemann, M. Mench, P. Cambier, and R. Prost, Direct determination of lead speciation in contaminated soils by EXAFS spectroscopy. *Environ. Sci. Technol.*, 30: 1540-1552, 1996.

38. Davis, A., M.V. Ruby, and P.D. Bergstrom, Bioavailability of arsenic and lead in soils from the Butte, Montana, Mining District. *Environ Sci. Technol.*, 26: 461-468, 1992.

39. Davis, A., J.W. Drexler, M.V. Ruby, and A. Nicholson, Micromineralogy of mine wastes in relation to lead bioavailability, Butte, Montana. *Environ. Sci. Technol.*, 27: 1415-1425, 1993.

40. Chaney, R.L. and J.A. Ryan, *Risk Based Standards for Arsenic, Lead and Cadmium in Urban Soils,* Dechema, Frankfurt, Germany, 1994.

41. U.S. Environmental Protection Agency, Guidance Manual for the Integrated Exposure Uptake Biokinetic Model for Lead in Children. EPA/540/R-93/081, PB 93-963510. U.S. Environmental Protection Agency, Office of Emergency and Remedial Response, Research Triangle Park, NC, 1994.

42. U.S. Environmental Protection Agency, Technical Support Document: Parameters and Equations Used in the Integrated Exposure Uptake Biokinetic Model for Lead in Children (v0.99d). U.S. Environmental Protection Agency, Office of Emergency and Remedial Response, Research Triangle Park, NC, 1994.

43. *Chemical Economic Handbook,* SRI International, 1996.

44. Racer/ENVEST™, Delta Technologies Group Inc., Denver, CO, 1996.

Index

A

C